W9-BTS-135

To Roxi –
with love
and thanks!

Jean Stapleton
"Trying Times"
4/89

The Technique
of the
Professional
Make-up Artist

The Technique
of the
Professional
Make-up Artist
for Film, Television, and Stage

Vincent J-R Kehoe

FOCAL PRESS

Boston • London

LARRY EDMUNDS
CINEMA BOOKSHOP, INC.
HOLLYWOOD, CA 90028
(213) 463-3273

Focal Press is an imprint of Butterworth Publishers.

Figures 3.4, 4.1, 7.1, 7.11, and 8.5 were drawn by Gail Burroughs.

Copyright © 1985 by Vincent J-R Kehoe.
All rights reserved.

No part of this publication may be reproduced, stored in a retrieval system, or transmitted, in any form or by any means, electronic, mechanical, photocopying, recording, or otherwise, without the prior written permission of the publisher.

Library of Congress Cataloging in Publication Data

Kehoe, Vincent J-R.
  The technique of the professional make-up artist for film, television, and stage.

   Bibliography: p.
   Includes index.
   1. Film make-up.   2. Make-up, Theatrical.   I. Title.
PN1995.9.M25K43 1985   792'.027   84-28785
ISBN 0-240-51244-8

Butterworth Publishers
80 Montvale Avenue
Stoneham, MA 02180

Focal Press
Borough Green
Sevenoaks, Kent TN15 8PH
England

10 9 8 7 6 5 4 3

Printed in the United States of America

This book is gratefully dedicated in friendship and deep respect to my colleagues and great innovators in make-up artistry:

DICK SMITH,

who has always, freely and unselfishly, shared his vast knowledge and most intimate research with us all so as to better, and to advance, the quality of our profession,

EDDIE SENZ,

who has aided me in countless ways to achieve any success I have enjoyed, and

THE WESTMORE FAMILY,

whose great contributions to our profession have inspired us all.

CONTRIBUTING AUTHORS

Dr. George Gianis, "Cosmetic Lenses and Eyes"
Dr. Robert O. Ruder, "Cosmetics and Surgery"
Dr. Jean M. Doherty, "Cosmetic Dentistry"
Dick Smith, "Special Effects—Dick Smith"
Christopher Tucker, "British and European Make-up Materials"

Contents

Preface   xi

PART I   BACKGROUND AND BASICS   1

1. An Approach to Professional Make-up
    Artistry   3

2. Screen and Stage Production Methods   5
    Motion Pictures   5
    Television   7
    Theater   7
    Make-up Assignments   8
    *The Physical Position of the Make-up Artist on the
    Set*   9

3. Mediums and Color Relationships   10
    Color   10
       Paint Colors   11
       Light Colors   11
       Stage Lighting Gelatins   11
       Color and Light Relationships   12
    Black-and-White Film and Television   12
    Color Television   12
    Amount and Placement of Light   13
    Color Temperature   15
    Color Values of Sets, Costumes, and Make-up   16
       Wardrobe   16
    Film Types   17
       Production Filming and Procedures   17
       Film Processing   18
    Audience Viewpoint   19
       Diffusion   20

4. The Make-up Department   22
    Lighting the Make-up Room   22
    Room Furniture   23
    On-Set Furniture   24
    Location Make-up Room   24
    Make-up Records   24

5. Make-up Products and the Kit   26
    The Make-up Kit   27
    Historical Progression of Modern Make-up   27
    Foundation Systems   28
       Color Matching   29
       Color Perception   29
       Foundation Shades   30
       Foundation Thinner   32
    Shading and Countershading   32

Beardcovers   32
Cheekcolors   32
Lipcolors   33
Eyecolors   33
Pencils   33
Mascaras   34
False Lashes   34
Powder   34
Cleansers   34
   Special Cleansers   34
Skin Care Products   35
Tools and Equipment   35
   Sponges   35
   Puffs   35
   Brushes   35
   Special Tools   36
Character Make-up Materials   37
Dress and Appearance of Make-up Artists   40
Kit Maintenance   40
Obsolete Terms and Items   40

Conclusion to Part I: A Lesson for Learning   43

Part II   BEAUTY MAKE-UP   45

6. Faces and Their Structures   47
    *Facial Shape Theory for Black-and-White
    Mediums*   47
    Facial Anatomy   47
    Skin and Hair Types   48
    Color Shaping and Contouring Concepts   49
    Color Principles   49
    Color Charts   50

7. Straight Make-up   51
    Women   51
       Facial Coloration   51
       Blends and Lines   52
       Facial Contouring   52
       Facial Definition   52
       Sequence of Make-up Application   52
       Make-up Styles   53
       Make-up Application   54
       High Fashion Make-up   68
       Women with Dark Skins   68
    Men   70
       Hairline Adjustments   72

Children   73
Body Make-up   74
Special Notes on Stage Make-up   74
   Straight Make-up   75
   Women   75
   Men   76
   Character Make-up   76

8.   **Personal Revitalization Techniques**   77
Physical and Cosmetic Skin Care   77
Skin Treatments   79
Medical Multitherapy Approaches   79
Cosmetic Lenses and Eyes by George V.
   Gianis, O.D.   80
Hair Removal and Hair Transplants   81
Cosmetic Surgery   81
   General Plastic Surgeons   81
   Head and Neck Surgeons   81
   Eye Surgeons   81
   Skin Surgeons   81
Cosmetics and Surgery by Robert O. Ruder,
   M.D.   81
   Nasal Surgery   82
   Eyebrow and Eyelid Surgery   82
   Facelift Procedures   82
   The Skin   84
   Other Procedures   85
Cosmetic Dentistry by Jean Marie
   Doherty, D.M.D., D.Sc.   86
   Jaw Relationships   86
   Orthodontics   86
   Crowns or Caps   86
   Bridges   87
   Bonding   87
   Periodontics   87

Conclusion to Part II: The *Right* Make-up   88

Interlude   88

Part III   CHARACTER MAKE-UP   89
Character Make-up   91
Facial Distortion   91
Multiple Characters   93
Make-up Special Effects   94

9.   **Age with Make-up**   97
Middle Age   97
Older Age   98
Make-up Application   100
   Foundations   101
   Age Spots and Coloration   101
   Highlights and Shadows   101
   Wrinkles   102
   Eyes   102
   Nasolabial Folds   102
   Chins and Jawlines   102

Foreheads   102
The Head   102
The Hands   107
Old Age Stipple   107
Prostheses   110
Progressive Age   110
Reversal Techniques   110

10.   **Racial and National Aspects**   119
Basic Head and Facial Shapes   119
Hair   119
Make-up to Change Racial Type   120
   Skin Coloration   120
   Eye Shapes   120
   Nasal Changes   121
   Jawlines and Cheekbones   121
   Lips   121
   Hair Shape and Color   121
National Types   122
   Europe   122
   North America   122
   Central and South America   125
   Southwestern Asia   125
   Southern Asia   125
   Southeastern Asia   126
   Eastern Asia   126
   Central and Northern Asia   126
   Oceania   127
   Africa   127

11.   **Period and Historical Characters**   129
Prehistoric Man   129
Ancients   130
   Egyptians   131
   Mesopotamians   131
   Greeks   132
   Romans   133
   Hebrews   133
Dark Ages to Present Time   133
   The Dark Ages   133
   Renaissance   133
   Seventeenth Century   133
   Eighteenth Century   134
   Nineteenth Century   135
   Twentieth Century   135
   Twenty-first Century and Beyond   138
Special Historical Characters   138

12.   **Special and Popular Characters**   141
Horror Characters   141
Skulls, Devils, and Witches   144
Animal Men-Creatures   146
Fantasy Types   151
Dolls, Toys, and Tin Soldiers   152
Statuary   152
Surrealistic Make-ups   155
Pirates   157

Clowns   157
Old-time or Early Film Types   158

Conclusion to Part III: Dedication   159

PART IV   LABORATORY TECHNIQUES   161

13. **Casting and Molding**   165
   Impression Materials   165
      Rigid Impression Materials   165
      Flexible Impression Materials   167
   Duplicating Materials   167
      Cold Molding Compounds   167
   Separators   168
      Sealers   169
   Cast of the Face   169
   Sculpture Materials   172
   Molds   178
      Flat Plate Molds   178
      Slush and Paint-in Molds   179
      The Two-piece Mold   183

14. **Latex and Plastic Appliances**   187
   Natural Rubber Materials   187
      Latex   187
      Latex Appliances   189
      Foamed Latex   192
      Basic Procedures for Handling Latex
         Appliances   198
   Plastic Materials   198
      Foamed Urethanes   198
      Solid Polyurethane Elastomers   202
      Cap Material   203
      Molding Plastics   204
      Tooth Plastics   207
      Special Construction Plastics   209
      Gelatin Materials   209
   Adhering Appliances   210
      Matting and Thixotropic Agents   210
      General Applications   211
      Latex and Plastic Bald Caps   212

15. **Special Effects**   217
   Bidding on a Production   217
   Rick Baker   218
   Carl Fullerton   222
   Spasms by Dick Smith   224
   Christopher Tucker   229
   Stan Winston   231
   Basic Make-up Effects   232
      Waxes   232
      Glycerin   234
      Burns   234
      Bruises   235
      Blood Effects   235
      Scars   235
      Tattoos   236
   Volatile Solvents   237
   Thickening Agents   239

**PART V   HAIR GOODS AND
THE HAIR KIT**   241

16. **Types of Hairs and Their Uses**   244
   Wool Crepe Hair   244
   Real Crepe Hair   245
   Yak Hair   245
   Angora Goat Hair   245
   Horsehair   245
   Human Hair   245
   Artificial Hair   247
      Acrylic Hair Shades   247
   Tools of the Trade   248
   The Hair Kit   249
      Basic Items   249
      Optional Items   249

17. **Application Techniques**   250
   Facial Hair   250
      Laid Hair   250
      Prepared Hair Goods   253
      Natural Hair   256
      Stubble Beards   256
      Blocking Brows   256
   Body Hair   256
   Wigs, Falls, and Other Hairpieces   257
      Measuring the Head   257
      Applying a Lace-fronted Wig   257
      The Half-wig or Fall   257
      Women's Wigs and Hairpieces   258
   Hair Coloring   258
      Darkening the Hair   258
      Lightening the Hair   258
      Graying the Hair   258
   Hair Sprays and Other Dressings   259
   Styling Hair and Hair Goods   259
   Basic Women's Hairpieces   259
   Attaching a Hairpiece   259
   Removing a Lace Hairpiece or Wig   259
   Rental Wigs and Hairpieces   000

A New Beginning . . .   265

APPENDICES

   A: Professional Make-up Products   267
   B: Suppliers and Professional
         Addresses   273
   C: Lighting and Filters   275
   D: Research File System   276
   E: Professional Make-up Kits   277
   F: Make-up Artist Examinations   279
   G: British and European
         Make-up Materials   281

SELECTED BIBLIOGRAPHY   286

INDEX   287

Preface

The Technique of Film and Television Make-up, first published in 1958 and fully revised to include color mediums in 1969, had over the years become the classic required edition for the study of professional make-up artist procedures. However, with the advances in technology in all phases of the entertainment industry and media of this decade, as well as the new approaches in the cosmetico-medical field, an entirely new direction and scope of make-up techniques has evolved.

Make-up for women's street wear had always been influenced by new screen cosmetic colors and methods, often necessitated or guided by changes in film and television systems' receptivity to colors. But today, the introduction of such a large variety of shades of color for the eyes, lips, and cheeks for street make-up has reversed this trend and has affected the fashions for film and television use. With the addition of color enhancement techniques in television, and current reproduction quality of film negative and printing, many people who often do not wear make-up (such as sports personalities or street interviewers) can still appear reasonably presentable in many cases. True, those on screen *with* make-up properly applied still can appear to better advantage, but in some manner the strict necessity of make-up has lessened. With this in mind, it becomes strongly evident that good make-up techniques must take advantage of both fashion influences and so-called natural appearance for straight make-up in all mediums.

Males, especially those engaged in politics—where they must appear in person on the street, followed possibly by a television or filmed interview in the studio, and then maybe pose for some still photographs in black and white or color, in succession, on the same day — must employ a very naturalistic make-up if they wish to take the best advantage of their appearance potential. Foundation bases and other products must be applied to appear as a skin color, not a coat of make-up.

Even stage make-up has greatly changed in recent years because of the redesign of most theaters to eliminate footlights, the better application of frontal and side lighting on proscenium stages, and the close-up quality of theater-in-the-round design. A far more realistic display to the completed make-up for both straight and character make-up (with the phrase *theatrical make-up* becoming obsolete) is necessary. Today correct theater make-up (with the exception of the large opera-type theater) is almost identical in application, design, and result with screen make-up.

Many individual products are mentioned herein by specific trade names and numbers due to the fact that *professional* products, such as are made directly for studio and make-up artist use, should be employed for professional make-up application rather than *commercial* (drug or department store cosmetics) products. Many commercial cosmetics do not provide sufficient coverage for studio use, and even commercials touting them have models with professional make-up applied. Therefore, all recommendations of products or shades have been based on my personal experience and preference, so inadvertent omissions of other trade name products should not be construed as reflecting detrimentally on such products. Equivalency charts will be found in the appendices to describe the differences in names and numbers of the professional and theatrical lines of make-up products.

Many of the products employed in the book are not available in the general theatrical make-up lines and are special effects materials or specific products designed by the Research Council of Make-up Artists. Although there are many outmoded methods and some products that are still utilized by even college-level instructors in make-up courses, the more advanced means and methods are described in detail in this book.

Character make-up techniques have also taken on new dimensions with the invention and use of new plastic materials and *make-up special effects,* which employ the talents of a make-up artist and a make-up technician, in conjunction, to create new visuals. Special chapters on cosmetic surgery and revitalization techniques also show the new approaches in geriatric therapy promoted by cosmetico-medical studies.

As well as my personal research and information, both the photographs and the views of other make-up artists and their techniques offer other advances in make-up. I thank those who have so generously contributed their time, effort, and imagination to this book.

ACKNOWLEDGMENTS

Make-up artists who contributed to this book:

Rick Baker   Bob Philippe
Gary Boham   Craig Reardon
Tom Burman   Tom Savini
Vincent Callaghan   Bob Schiffer
John Chambers   Charles Schramm
Carl Fullerton   Dick Smith
Werner Keppler   Terry Smith
Bob Laden   Bill Tuttle
Leo Lotito   Bud and Frank
Gus Norin   Westmore
Dana Nye   Stan Winston
Bob O'Bradovich

Wigmakers to whom I am indebted:

Ira Senz Inc.
René of Paris
Leon Buchheit

Thanks to:
   All the many chemical suppliers who so generously furnished samples and technical information for this work;

Dr. George Gianis for his section on contact lenses,
Christopher Tucker for the information in Appendix G,
The many motion picture companies whose press departments furnished photos,
Dr. Irwin Lubowe for his book on skin care and Joel Gerson for his book on salon aesthetics for my research,
The many photographers who furnished photos,
The models who posed for photos and the performers in others,
The Eastman Kodak Company for their research by Earl Kage on make-up for films,
Allan Buitekant for his drawings,
The staff of the Editorial, Production, and Marketing departments of Butterworths for their work in producing and promoting this book,
And, finally, my loving and patient wife who not only modelled for the photos on beauty make-up, but also put up with my long stays at the typewriter, make-up lab, and constant research projects for this book.

Part

I

Background and Basics

One should assiduously study and practice the
requirements of a profession before attempting
to make a living by it.

CHAPTER 1

An Approach to Professional Make-up Artistry

TO BECOME A KNOWLEDGEABLE, COMPETENT, AND COMplete professional make-up artist requires, first, a sincere desire to learn everything there is to know—and more—about make-up procedures, products, ingredients, lines, comparative work, and how it is or was done. Next, it is necessary to practice, practice, practice and redo trial and test make-ups until there is some degree of satisfaction and accomplishment with your work, collecting photos of your work and other make-ups and faces to recreate, reading every book and pamphlet on every approach and style, and preparing a full kit for studio and field use. Finally, one must be prepared for competition on an aesthetic level as well as a comparative one with the artists already in the field and realize that persistence as well as talent is necessary to establish a position in the world of make-up.

Make-up artistry is a profession that one must enter into neither lightly nor with any diffidence whatsoever. It is a highly competitive profession, and one must always be prepared to give way to better talent—like it or not! Great make-up artists maintain their positions of excellence by constant and unending research and study as well as accomplishments in their work. Talent must be the criterion and work the incentive—without any other considerations of nationality, sex, or union membership.

Eventually, there is always the question of whether one should join a craft union or not. To work in the networks or many studios in the New York or Los Angeles areas, one must belong to the IATSE (International Alliance of Theatrical Stage Employees) Local 706 in California and Local 798 in New York. However, some of the film and television studios there and about the country employ crews that belong to NABET (National Association of Broadcast Engineers and Technicians) Local 15 in New York and Local 531 in California. In addition, there are actually hundreds of other studios in New York, California, or in between that are nonunion and do not require their employees to belong to, pay dues to, or be examined by a union. Most of the local television stations in U.S. cities outside of New York and Los Angeles, even though they are network affiliates, employ both union and nonunion employees. Although the unions do have varying examination requirements, just being a union member does not guarantee talent. Some are just powder-puff mechanics who have been taken into the unions under the term *organizational* (which simply means a large union absorbs a smaller one or takes the non-union-people in with little or no exam—or talent in some cases). Unfortunately, some union members are taken in due to relationships—brothers, sons, cousins, for example—but again, blood relationship does not guarantee talent. Therefore, should one belong to a union or not? If one wishes to work in the television or film studios in the major production centers, the answer must be yes. However, if an artist wishes to work in other smaller cities in the United States, why pay dues to a union that cannot aid him or her unless he or she wants it for the possible prestige it might carry with some employers (and this is often doubtful!).

In addition to studying make-up strictly as a profession, some actors wish to know more about their craft to add to their knowledge of the theater (where most actors have to do their own make-up), and a wide range of amateurs interested in little theater work, home videotape, or Super 8 film productions are willing to devote considerable time and effort to the study and practice of make-up. This latter group sometimes accomplishes superior work and, more often than not, forms the nucleus of budding professional make-up artists, bringing to the profession that certain avidity that stems from sincere interest and thirst for knowledge. Such amateur productions provide excellent basic training for aspiring make-up artists, but before packing their bags and rushing off to the major production centers, they should gain as much amateur or professional out-of-town experience as possible before attempting competition with established and knowledgeable peers.

Another variety of study, or classification, is that of the so-called *beauty salon make-up artist,* who is often referred to in the fashion magazines as a "face designer," "make-up genius," "cosmetic expert," and the like, who simply do high fashion or beauty corrective make-up for magazine stills or beauty salons. Such is a very lucrative, albeit limited, field relative to the variety of make-up directions and knowledge that must be accomplished to be classified as a full studio make-up artist. For such personnel, character studies would add to their knowledge of facial structure and correction and would aid in designing the proper make-up for each person's face. However, more often than not, these artists, as a group, are not interested or qualified in this advanced and necessary phase of make-up. Beauty salon make-up, however, has one hitch. To do any make-up application in a licensed salon in the United States, and in many foreign countries as well, one must have a *hairdresser/cosmetologist* license, requiring the operator to take about 1,500 hours of beauty school training, mainly in hairdressing and with very little make-up practice required or performed. Some states do not even require any practical make-up work in their examination for a license, with only some true-false questions sufficing. As a result, if one really wants to do make-up application in a beauty salon, one must take additional training *after* one has received one's state licenses to be basically proficient in the work required—somewhat like learning to ride a bicycle and then being *given* a license to drive a car!

There are some cosmetic studios, as well, that do no hairdressing at all and simply specialize in skin care and make-up application in addition to sales. Here, no license is required, but like those who sell for manufacturers in department stores, some states require a demonstrator's permit as a *cosmetician,* and although no exam is given, a fee is charged yearly by the administering state agency.

One other growing field is that of the *aesthetician,* who is a person trained in skin care and make-up by a state-licensed school but who do no hairdressing. Again, these schools train one not to be a studio make-up artist but to perform and sell skin care products, methods, and service in a specialized salon that does not do hairdressing. Well established in Europe, this trend is growing in the United States. However, many skin care machines and nature product fads have infiltrated this field with items of questionable efficacy and results, so such salons do not provide a true basic training area for those who wish to work in the film or television make-up sphere.

Outside the United States, even though many countries have goverment-operated television studios, the film industry is mainly privately owned, and unions exist in most. Work permits are normally required, so Americans should not attempt to gain employment in foreign countries without first checking the requirements.

From a truly professional level of study, the best way to approach the profession is to take seminars with an experienced artist (who will undoubtedly be a union member also) in the field and then to try to become an apprentice to a working make-up artist. Just watching a professional at work is an experience for those with a keen eye for observations, and much can be learned in this manner. I recall my first visit backstage to watch a theatrical make-up artist at work, mixing greasepaints in the palm of his hand and opening a huge suitcase of powder cans (and seeing the haze rise as the cover was opened) to select the *right* shades for the American Indian character he was working on. When I asked why he didn't just use the greasepaint stick marked "American Indian," he replied, "Sonny, don't ever judge a make-up base by the description the manufacturer puts on it, but rather study how skin color actually appears as in life. If they don't make it, then mix your own." This, of course, means that many make-up artists constantly experiment with products to achieve what they consider to be the best materials for them.

Finally, one of the questions that is repeatedly asked is, "Should I go to college to be a make-up artist?" The answer is, as a primary effort, no, but as a secondary adjunct, certainly. No college or university gives a degree for make-up artistry, and most of the stage make-up courses, or even the visual arts courses encompassing film, television, or tape subjects, given in academic institutions are insufficient in scope or taught to performers rather than to persons interested in professional make-up artistry. However, academic courses in anthropology, chemistry, drawing, painting, sculpture, and foreign languages contribute in an educational as well as contributory sense to the professional make-up artist's overall education.

In addition, schools and colleges would better serve their students in a far more competitive and professional sense if they hired a professional studio make-up artist who is working in the field to teach make-up courses. Too often, the educational or scholastic level is the main criterion employed by institutions of learning in judging instructor levels rather than talent and proficiency in the field of endeavor that is being taught. No scholastic credit can equal the ingrained knowledge one gains in a professional competitive sense that the make-up artist must have to earn a living in the field.

CHAPTER 2

Screen and Stage Production Methods

ALTHOUGH THERE ARE NO MAIN DIFFERENCES TODAY BE-tween the actual quality, density, products, or procedures of professional make-up for film, television, or the stage in regard to the degree of excellence required, amount of application or basic techniques, there are production responsibility divergences for the make-up artist on and off the set.

Any properly executed straight or even character make-up applied for film purposes would picturize well on the television screen and also appear properly on stage for most modern proscenium or theater-in-the-round stages. The exceptions are the huge amphitheater or opera type of stages, but even then, from the distant seats, the effect of most facial make-up is lost, and the performer carries the part in shape, motion, wardrobe, and performance. Overemphasis of facial foundation colors, character lines, eyecolors, or eyelines only serves to produce a make-up that appears overdone and extraneous to the closer (more expensive and critics') seats. An outmoded concept of stage make-up was to apply sufficient emphasis of make-up products so that the actor appeared correctly made up from the center of the audience. Now, depending upon the size of the house, the make-up may look fine in some theaters or overdone in others when viewed by the audience half-closest to the stage. For a better criterion, one should make it the rule that whatever you can see in a properly lit make-up mirror is what you will see picturized on the screen or stage, and *this* will be correct emphasis. Don't hedge, but critically and exactly judge each make-up applied.

I remember one of the first critical notes on make-up made by a New York reviewer in *Variety* after viewing the make-up I created for the opera, *The Medium,* for Marie Powers at CBS-TV, when he said, "The Met, by the way, could take a lesson in plausible facial make-up here," as a final sentence to the review, and that was over 30 years ago. Lesson: Don't apply heavy, outmoded make-up for any size stage or any other medium for that matter, unless that is what you are really trying to create.

The *color quality* of a foundation may vary slightly in color intensity, depending on the medium and the lighting. For example, a make-up done for still photos or for some television stations looks just about the same as a good street make-up for a woman, and a *production* make-up for motion pictures (or television when the show is on a stage with an audience in the studio viewing it) is only slightly more saturated with pink-beige tonal value.

The foundation base shades employed for television may be affected electronically by the camera system employed on a specific studio or set. Sometimes the system is plus-red, while others may be minus-red in sensitivity. This simply means that certain television studio systems seem to add redness or warmth to the foundation color, while others appear to lessen the red so colors appear to be colder in effect. However, much of the balance of facial tones is achieved by the technicians in the control room of each studio as well as a monitoring of them by the master control before the signals are sent out to the home receivers. It is amazing how many gradations of skin tone values can be produced that are beyond the ken of the make-up artist for both film and television by printing, filtering, electronics, and such. The make-up artist can normally only set a basic standard that seems to be compatible with the medium for good skin tones by employing tests as determinants, with all other variables fixed. The Eastman Kodak Company has generally advocated that tests made on any skin tone, wardrobe, or background color should be made with only one variable at a time so that results can be predictable and the interaction of elements controlled by known factors. Thus, don't change a dress color *and* a facial skin tone foundation for the same test, but test each separately before a combination is made. Production responsiblities and methodologies vary for film, television, and stage productions, so the make-up artist should know to whom he is answerable when judgments, decisions, or changes are to be made that will alter the production values.

MOTION PICTURES

The hierarchy may vary somewhat in some studios and production companies, but basically it is as follows:

Charge of Production   Runs the studio—has the last word;

Executive Producer   Is in charge of a number of productions;

Producer   Is business head of a particular production;

Associate Producer   Handles some special phase of the production or shares in decisions due to a particular association with the production;

Assistant to the Producer   Is glorified secretary;

Casting Director   Selects performers to be interviewed and passed upon by the director and/or producer.

These personnel constitute mainly the business end of a production, and they are assisted by the film editor who will cut and edit the film, a musical director for the score, writers for the script, designers, and other department heads working under the director's approval.

The *aesthetic* group of a production consists of:

Director   Is in charge of the members of the cast, the crew, and the action of the film;

Stars   Players whose names are usually listed above the title of the film. Today they may *own* the production company;

Featured Players   Important players, not stars;

Bit Players   Performers with small parts of speaking lines;

Extras   Performers with no speaking lines but may have special business to perform;

Stand-ins and Doubles   A person hired to stand in the same areas on the set as it is being lit for a scene so that the stars do not have to be on set for this strictly technical appearance. Stand-ins do not require make-up generally and are selected to be approximately the same height and build of the stars.

A double is a performer who is dressed and made-up to appear as another performer in a scene that might be too dangerous (then it is called a *stunt* person) or in a scene at such a distance that the double would not really be recognized as such (long shots, running sequences, car chases, and so on). Doubles, who must look like the performer for whom they double, must often be given proper make-up time on the schedule.

The production crew consists of:

Production Manager   Has overall charge of carrying out the producer's orders and everyday running of the production of a film;

Assistant Director   (often called the first assistant and sometimes is also the production manager)   Assists the director with the flow of the film's physical action of the players and the crew. He is the one who says,"Quiet down" and, on the director's nod, "OK, roll 'em." Really is an assistant *to* the director rather than aiding him in the aesthetic direction of the film;

Second and Third Assistants   The starting point for one who wishes to learn the production end of the business. In film making, they are often referred to as *gophers* (because they "go for" this or that) by the first assistant. Their job on set is usually to see that the performers, extras, and crew are in their proper places and positions on time to begin shooting.

The camera department is made up of the following people:

Cinematographer or Director of Photography   Is the head cameraman responsible for everything that appears on film. He is accountable only to the director in most cases;

Operating Cameraman   Is the person who sits behind, carries, or stands behind the camera, operating its movements and aiming;

Assistant Cameraman   Keeps the focus on large cameras and loads the camera magazines.

In the theater, most of the following people are known as *stagehands,* but for film or television they are known as the *stage* or *set crew:*

Gaffer   Head electrician who sets up the lighting pattern for the cinematographer;

Best Boy or Second Man   Gaffer's assistant;

Juicers or Electricians   Other personnel who adjust or move the lights;

Key Grip   Head grip. Grips move the scenery or set pieces, camera dolly, and so on. Other grips are under the key grip;

Property Men   Handle small items on the stage and help dress the set. On smaller productions they sometimes handle the special effects, greenery, and other jobs performed by a department in the larger studios;

Sound Department   Handles and operates all the sound equipment such as recorders, microphones, booms, and so forth.

Only special effects personnel sometimes work closely with the make-up artist on the set to create certain effects such as bullet wounds and blood.

The make-up, hair, and wardrobe departments are in a special category of performers' services, both on and off set, and can be a group of three for most commercials and for small industrial or minor theatrical films. However, in a large studio there are the following:

Make-up Director or Department Head   May be a studio department head who acts in an executive capacity to set assignments for other artists and designs the make-up for each production and whose name appears on the credits of a film, or is the prime worker on a film;

Production Make-up Artist   In charge of application of make-up on all performers for a specific film. Any additional persons to aid them and work under their direction are called journeymen or the "second persons" of the make-up crew;

Special Effects Make-up Artist   A new category established in some unions to describe highly specialized work in character make-up, employing make-up specialties, mechanicals, and other materials. Often called in just to do a certain effect that may be beyond the sphere of the regular production make-up person. These effects often entail much outside preparation and laboratory work, which is not just standard make-up application.

Head Hairstylist   Director or head of the hairstyling on a film or for a studio. May work on the same executive and operating level as the head make-up artist. Like the make-up director, may not actually do set work, but like that person, attends the production meetings or conferences, designs styles, and assigns jobs. They must be expert in every phase of hairstyling.

Hairstylist   Does the female hairstyling, wigs, and so on on set and prepares the hair goods for such. Male hairstyling or character hairstyling may be a cooperative venture between the make-up artist and the hairstylist. In true production terms, one does one job—either hair or make-up, but both departments cooperate completely for the good of the production.

Although there is a minor trend of make-up artist/hairstylist, with one person performing both jobs on the set in non-union or small productions, this splitting of the responsibilities may be detrimental to production excellence unless both the hairstyle and the make-up are extremely simple and on one person only. Both time to do the job and the experience necessary are better vested in two people, make-up and hair, than one.

TELEVISION

In television, production of either a live or a taped show has a slightly different group of people and chain of responsibility. The business group can often be identical; however, the show may be produced by the network or an outside producer who sells the idea or show complete to the network, plus influence is often felt through the advertising agency representing the client who will sponsor the show.

Rather than having a film editor, if the show is taped, it may have a tape editor, although much of the cutting is done in the control room operation. The television director (who directs the aesthetic action), the technical director (who has charge of the personnel in the control room—audio, video, and so forth), lighting director (charge of the lights and such personnel on stage), and many other technical people sit in the control room where all the pictures of each camera on set are viewed on television screens called *monitors*.

Each camera on set is operated by a cameraman and is placed on a mobile base that is moved by the cameraman. They have cable men who follow the cameras and keep the heavy cables from fouling or tangling each other. These cables carry the signals from the cameras to the control room monitors where they are viewed by the control room personnel.

The lighting director sets up the lights with the aid of the juicers or lighting men for the various playing areas. Only if a certain effect requires a movable light (such as a fill light called a *scoop*) are lighting personnel on the floor of the studio during the taping or live production. Neither an assistant cameraman nor the equivalent of a cinematographer is required for television.

Make-up, hair, and wardrobe staff can be identical and operate in the same manner as for film, but there has always been a tendency (a very bad habit, indeed) of some directors or network executives to think that make-up, hair, and wardrobe take less time to do for television than for film. It does not, so quality suffers. Responsibility for this situation sometimes rests with the make-up artist and the hairstylist who do not assert that they *require* either more time or personnel to do the job properly.

Make-up people must cooperate and coordinate closely with the lighting director and video personnnel to achieve the best possible picturization of the make-up. As such, the make-up and hair people should never make any judgments by viewing the on-set monitor (a television set on the floor of the set to check actor's positions) but should check always on the *line* monitor (each camera's picture is shown on a different control room monitor, and the selected picture to be shown is on the final or line monitor). There, any changes can be discussed with the director. Unlike film, television can be viewed directly and all elements seen before they are put on the air or taped.

In either case, film or television, production meetings are often held in the planning stages to set most of the make-ups and hairstyles and iron out many of the preproduction questions and problems. Appointments for special tests, trying on wigs or other hair goods, or the procurement of special materials for make-up or hair are set well in advance so that all can be ready to go on the production date.

THEATER

The make-up artist for the theater generally takes his or her assignments from the stage director who, in the main, controls all aspects of a production. Amount of time does not vary for doing the make-up, but more often than not, the make-up artist does only a specific character or personality backstage, as most of the minor or even featured players are expected to do their own make-up for each performance. However, sometimes a producer will hire a make-up artist to design the make-ups for the entire cast (to be approved during dress rehearsal by the director). Then the make-up artist is asked to teach the performers to do their own. It is unusual for a stage production to have a make-up artist as part of the daily production crew (which otherwise is similar to the other mediums), but for some specific productions or at the request of the star, it may be necessary. As such, many times the performer must pay for the services of the make-up artist, either from his or her own pocket (or as a clause in the contract if the performer is important enough).

Make-up is more likely done in the dressing rooms of the theater since no specific room is set up as a make-up room. Lighting most often must be augmented or changed, and a make-up chair must be added for comfort. This all depends upon how much the make-up artist can ask for—and get, according to the importance of performer to be made up.

MAKE-UP ASSIGNMENTS

A determination of how many make-up and hair personnel are required for a production needs to be made. Let us say, for example, that the script calls for two stars, one male and one female; four featured players, two male and two female; six bit players, four male and two female; and ten various extras, doubles, and so on. All may not be working on the same day at the same time, but let us consider a possible work day with both stars, two featured players, three bit players, and four extras who will work. A *minimum* crew would be at least three make-up artists, two hairstylists, and two wardrobe people (plus a female body make-up artist, a person who just does body make-up for women as required by some unions) if all performers are required to be on set, ready to shoot at 9 A.M. for a television series or a film. Performers' calls are the responsibility of the first assistant director in most cases after consultation with the head of the make-up department for the production, and are posted or given by 4 P.M. the previous day.

6:00 A.M. Calls for three make-up artists and two hairstylists; Cast: Four extras for hair and make-up.
6:45   Cast: Three bit players.
7:30   Cast: Two featured players (One make-up and one hair):

Female star
7:00 A.M.   to one hairstylist
7:30   To one make-up artist
8:15   Back to hairstylist, wardrobe, and body make-up if required

9:00   On set with make-up, hair, and wardrobe personnel accompanying her

Male star
8:00 A.M.   to one make-up artist
9:00   On set in wardrobe accompanied by make-up artist

On set, the stars are watched by their assigned make-up artist and hair personnel, while all the others in the cast are watched by the other make-up artist and hairstylist. This is, or course, for simple, everyday make-up, hair, and wardrobe. If a period piece is being filmed or televised, additional time and make-up and hair people may be needed if the make-up and hair are more complicated or extensive than usual.

It is bad judgment on the part of management to attempt to require the make-up, hair, and wardrobe people to rush through their work due to too many performers being placed on their schedules because, if the work takes more time, then the entire production crew may be held up and have to wait. This delay often costs more than an extra make-up or hair person.

When make-up artists and hairstylists arrive on the set with the performers, they should report to the assistant director and, if necessary, check their work with the director or cinematographer. Personality conflicts should be avoided between the performers and the make-up and hair personnel, and problems of any nature should be brought to the attention of the assistant director to avoid any delays. Experience has shown that the more seasoned the performers, the more they know personally what they want for hair and make-up (but also the less problem they are in general), and one should defer to their opinions if valid for the production.

There is no difference in the make-up or hair for films or tape made for television series, commercials, or theater release. However, television productions done live—that is, shown on the air as it happens—or even taped for just one showing (such as interviews, newscasts, and sportscasts) may display current styles and fashions in clothes, hair, and make-up, including some high fashion trends. In most production filming or taping, a more middle-of-the-road trend is normally taken for fashion so the production will not become dated for future showing. A film made for theater release will often take six months to a year before such release, so no possibly ephemeral or overt current modes are selected for this reason.

Filmed commercials have other elements in the business end that are inherent "to the beast," so to say. These are the clients (representative of the sponsoring firm), their advertising agency people (who may be producing the commercial), and sometimes, a styles director for the agency that wishes to oversee the make-up, hair, and wardrobe. Coordination and cooperation are necessary for the proper operation with these extra elements, and compromises often must be made aesthetically. Sometimes problems with the female models may occur when the make-up and hair personnel wish to apply their trade as they see it but are told that what they are doing is not the model's *image*—a style created for a specific model by her agency to make her distinctive from others. Such controversy generally requires a decision of the producer or director of the commercial about how the model will appear. Filmed commercials are not the most rewarding experiences in a creative sense for most make-up and hair artists.

Still photography studios have a simpler crew arrangement, and the make-up and hair people are responsible most often only to the photographer. However, the same model agency—client's representative—advertising agency problems may be part of the day's work here too.

THE PHYSICAL POSITION OF THE MAKE-UP ARTIST ON THE SET

Most people studying to be make-up artists ask the question, "Where do I stand or observe the performers when I'm on the set?" This position varies somewhat from medium to medium, set to set, director and/or cinematographer.

For the theater, work on the performers is usually done in a dressing room backstage, and the results can only be properly viewed by going out front into the audience viewing areas. Most of this viewing is done during the full dress rehearsal when all elements of the production are seen together—often for the first time. It is then that adjustments on shades, colors, and definitions are made both on the director's decisions and/or the recommendations of the make-up artist. Some changes may be made with the coordination of the set or lighting designer or the costume designer as well as the hairdresser. However, once the people are on set doing a performance, they cannot be reached for touch-up or change.

If it is a particularly hot stage, make-up may be seriously affected and can run or smudge, hair goods can become unstuck, and all sorts of other untoward things can occur. If such is the case, the make-up artist must often re-adjust or retouch in the wings of the stage or the closest dressing room. If changes must take place in character make-up during the performance, the make-up artist and hairstylist may remain throughout the performance. However, most times, the make-up artist does the work prior to the performance, then packs up and leaves, making certain that some powder or lipcolor are left for the performers to retouch themselves between entrances or acts.

For television, the make-up artist and hairstylist have the advantage of seeing their work immediately by viewing the performer during the dress rehearsal on camera in the control room. Trying to make a judgment on the viewing monitors on the set is not

the best way because they are seldom color corrected. Check with the technical director as to which is the best monitor in the control room to view your subject for decisions. The monitor that has the complete show on it is most often the best for making such determinations. Corrections can then be made during the dress, on set, or if complicated, in the make-up room. Unless changes must be made during a live performance, the make-up artist should find an unobtrusive spot to stand in the control room to view the production and make notes during the dress for correction for the live show.

If quick changes must be made, a set-up on stage in an unused corner is best, with the make-up kit on a table. One must take care that any lights on the make-up table face away from the set. Incidentally, the terms *stage* or *set* are often employed in television and films as well as the theater to denote the playing areas. When television first started, all the dramatic shows, musical productions, and so forth were done live, and many hilarious (and not so laughable to the make-up artist!) mistakes happened. Today, these productions are done on tape so time for redoing, retouching, changes, and so forth is available, and the make-up artist does not have to worry about being caught on camera or having insufficient time for a quick change.

For videotape or motion pictures where the scenes are shot with cuts made in between, the make-up artist should confer with the cameraman and the director as to whether they want the make-up people to step in between each shot and immediately retouch or wait until they are called to do so. Some directors will rehearse a scene until they are satisfied with the performances and then say, "Let's make one." At which time the assistant director will call "Make-up," and the artist steps in to do the work. When a star performer is being made up or retouched, the arrangement may be different, and the star may demand constant attention. As such, the normal place for the make-up artist (and maybe too the hairdresser) is immediately behind the camera, but not too close to interfere with its operation or movement. However, once the make-up artist learns the ways of the cameraman and/or director, he or she must work within the framework of their pattern of operation for a smooth running production.

It is the make-up artist's responsibility to keep an eye on every performer, on and off camera, all the time, and to make adjustments so that everyone is always ready for camera. Watching each shot on some productions is often quite important so that the make-up artist can tell the director if anything is wrong with the make-up while the scene is being filmed or taped. Where the director is concerned with watching the performance and the cameraman is trying to watch everything else, they may not see an error in make-up until it shows up on the film or tape.

CHAPTER 3

Mediums and Color Relationships

THE MEDIUMS OF FILM, TELEVISION, AND THE THEATER ARE lit, recorded, and/or viewed differently from each other in the main. Film and television are viewed on a screen—*directly* in the case of television and *reflected* for motion pictures. The medium of television is one of electronics sent out from a transmitter over the air in waves to a receiver where the transmitted picture is reconstituted for viewing on a screen. Film is a recorded medium on a sensitized strip of material that, when light is shone through it, projects pictures on a screen. Magazines and other photographic mediums show still pictures printed on a material, like paper, for direct viewing. The theater is different in the sense that all actions of the performers or the scene are viewed immediately and directly, in person, with no electronics or recording medium between the action and the viewer.

Lighting is necessary for the viewing of every medium and is the major requirement in producing a picture. Once lights are placed on a scene, the human eye begins to distinguish colors, movements, and differences in persons, articles, or actions. Only the mediums of television or film (or photos printed from film) can express colors in a *gray scale* (Figure 3–1) or *black-and-white* manner. As such, film sensitized to record pictures or electronics designed to reproduce a scene in black and white can translate colors (red to violet) into tones of gray for viewing.

COLOR

To describe color properly, there are some basic terms that must be defined:

Color   A psychophysical property of light;
Brightness   The relative intensity or luminance of a color;
Hue   An attribute of color that allows separations in groups by terms such as red, green, blue, and so on;
Saturation   The degree that a color deviates from a neutral gray of the same brightness or the distinctness or vividness of hue; often referred to as the *purity* of a color;
Shades   A gradation of color;
Chromaticity   Hue and relative saturation.

Light has wavelengths (similar to radio waves) measured in millimicrons, and their *frequency* is the number of cycles per second. The visible light wavelengths recognized by the human eye are between 400 and 700 millimicrons, and in the rainbow of colors of the spectrum, *red* light has the longest wavelength (between 600 and 700 millimicrons) and the shortest frequency. *Blue* light has the shortest wavelength (between 400 and 500 millimicrons) and the highest frequency, while between them is *green,* which is between 500 and 600 millimicrons.

In normal human vision, color is perceived either from light directly or from light reflected by pigmented objects. Each object reflects its own color and absorbs any other from the light it is reflecting. A *primary* color is one that cannot be produced by mixing two other colors. *Secondary* colors are those that are produced by equal mixtures of two primary colors.

Paint Colors

We can recognize *paint colors* (Figure 3–2) (such as foundation colors) of the major shades of red, orange, yellow, green, blue, and violet, plus black and white and combinations of such to produce in-between shades. With a *color circle,* or *color wheel,* such combinations are readily distinguished and are the basis for producing the shades employed for facial make-up.

The *primary* paint colors are red, blue, and yellow, and when equally mixed, they produce gray. The *secondary* paint colors—orange, green, and violet—are made by combining two adjacent primary colors. Opposites on the color circle are *complements,* while in-between colors may be produced by a predominance of one color over another. For example, blue and green

FIGURE 3.1   *Kodak Gray Scale and color patches, showing their translation into black-and-white mediums. (Courtesy Eastman Kodak Company.)*

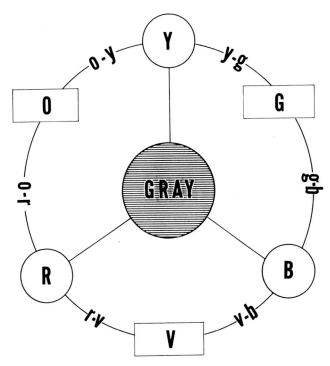

FIGURE 3.2 *Color circle of paint colors.*

mixed with a predominance of green will give a bluish green or, with a predominance of blue, a greenish blue. Varying amounts will produce descriptive colors such as *turquoise, aqua, opal,* and so forth. A *tint* is made by adding white to a color, while a *shade* is made by adding black to one (in the particular parlance of mixing artist's colors).

Light Colors

Light colors combined to form other light colors do not respond to the eye the same as paint colors. They are, however, greatly dependant upon each other in the creation of various effects; that is, a red light will wash out (to white, theoretically) a red paint color, while a green light will make red paint appear to be almost black.

Primary *light* colors are red, green, and blue. Combined in equal intensity, they produce *white* light. This principle was sometimes employed in theaters with borders and footlights with a red, a blue, and a green gelatin placed over incandescent light bulbs. The electrician, controlling the light board with dimmers, could produce a fully red, or blue, or green effect or combinations thereof. When equal voltage was placed on each one at the same time, the stage was then flooded with white light.

Color filters, or *gelatins,* will absorb all the colors in the light spectrum except the one that is the color of the filter; that is, a red gelatin placed over a light will pass only the red of the spectrum through the filter and no other colors, and we get a red light. Below the visible color spectrum of red is invisible *infrared,* while above the visible violet there is invisible *ultraviolet.*

The best reference and study of the principles of paint and light primaries and their references to color film production can be found in *The Elements of Color for Professional Motion Pictures,* published by the Society of Motion Picture and Television Engineers in 1957.

Stage Lighting Gelatins

There are many shades and tints of gelatins. For example:

Frosts	Blue-greens
Pinks	Greens
Magentas	Yellows
Purples	Reds
Lavenders	Chocolates
Blues	

The most commonly employed in the theater are:

#3 Flesh	#17 Surprise Pink or
#62 Bastard	Special Lavender
Amber	#25 Daylight Blue
#27 Steel Blue	#55 Straw

Table 3.1 *Effects of Lighting on Make-up*

	LIGHTS				
MAKE-UP COLOR	RED	YELLOW	GREEN	BLUE	VIOLET
Red	Fades out	Stays red	Darkens greatly	Darkens	Lightens to pale red
Orange	Lightens	Fades slightly	Darkens	Darkens greatly	Lightens
Yellow	Turns white	Turns white or fades out	Darkens	Turns orchid	Turns pink
Green	Darkens greatly	Darkens to dark gray	Fades to pale green	Lightens	Turns to pale blue
Blue	Darkens to dark gray	Darkens to dark gray	Turns to dark green	Turns to pale blue	Darkens
Violet	Darkens to black	Darkens almost to black	Darkens almost to black	Turns orchid	Turns very pale

Note: Black, gray, and brown make-up remain the same except for very light, almost imperceptible, changes in tonal value.

Generally, theater stage lighting is a combination of cold and warm neutrals (such as #17 and #62, respectively) for general illumination, plus time-of-day-effect gelatins (#27 Steel Blue for night and #55 Straw for late afternoon sun effects) that do not, relatively speaking, affect the color of make-up shades appreciably in any adverse manner. However, some stronger lighting gelatins will produce color differences in make-up shades (Table 3.1).

Color and Light Relationships

By focusing three spotlights with selected red, blue, and green primary color gelatins on a white screen, it can be seen that where all three colors intersect or overlap in the center, white light is produced. Where the red and blue lights overlap, magenta is produced; where the blue and green overlap, cyan is produced; and where the red and green overlap, yellow will be produced (Figure 3–3).

Red, green, and blue are called *additive primaries,* while cyan, magenta, and yellow are known as *subtractive primaries.* If light with cyan, magenta, and yellow filters were projected, the full intersection of these would produce black. Where the yellow and magenta overlap, red is produced; where the magenta and cyan overlap, blue is produced; and mixing cyan and yellow produces green.

Color television is based on an additive color system, while color film is based on a subtractive one. The visualization of color in both these mediums, plus in life or on stage, provides a psychological factor of entertainment in the human mind. Color always demands interest and the eye is sensitive to color training to a great degree. One example of this is the red light for stop and the green light for go in our highway traffic system.

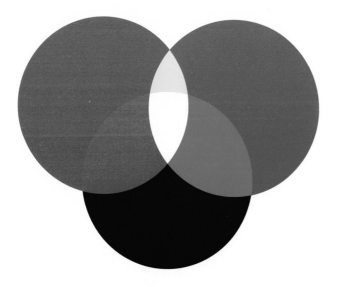

FIGURE 3.3 *Color and light primaries, showing their interaction.*

BLACK-AND-WHITE FILM AND TELEVISION

Monochromatic, or black-and-white, film or television make-up requires that the make-up artist visualize the effect in gradients of the *gray scale* rather than color to ascertain the correctness of the tonal value of the skin. Shadows will picturize a deep gray while highlights appear a lighter gray than the skin tone selected as a norm by the cooperative efforts and determinations of the make-up artists, lighting directors, and video engineers in television or of the director of photography, processing laboratory, and film type in motion pictures. Each craft can control to a certain degree the gray scale skin tone of the performer on the screen, and each, by the operation of his or her craft, complements the others in producing a skin tone effect that will appear natural in the gray balance for the engineer in master control to send out a picture for television or for the projectionist to show the film in a theater (although the latter has little control over the quality of the print shown in a theater).

Picture signals are *painted* on a monochromatic television receiver tube by the modulated intensity of a spot or beam of luminescence that sweeps across from left to right on the face of the viewing tube. This spot travels back and forth in a descending pattern from the top to the bottom of the tube in about a one-sixtieth of a second. As this cycle occurs repeatedly, it gives the effect of motion on the screen.

COLOR TELEVISION

In color television, the additive properties of colored lights are employed to produce tonal values. A plate in the receiver tube, known as the *phosphor plate,* is a glass screen covered with over a million small dots (for the average 21-inch-size television tube). One-third of these dots glow green, one-third glow red, and the remaining one-third glow blue under the influence of electronic beams. These beams are shot from three electron guns placed in the neck of the tube, one for each of these primary light colors.

The dots on the plate are placed in clusters of three, one of each color, over the entire surface of the plate and are so small that they cannot normally be distinguished from each other at the regular viewing distances. When the three guns are emitting proportionately equal strength signals, white light is produced. If these emissions are decreased in intensity gradually to near zero but kept in equal proportions, light gray to black can be produced. If the blue gun beam is strong while the others are low or near zero, the color on the screen will be of a blue tone. By adding or subtracting various amounts of each color as emitted from the electron guns, any shade or tone in the visible spectrum can be reproduced on the face of the tube.

Two factors affect the amount and tonal value of the color on a color television receiver, and they are known as the *luminance,* or light signal, and the *chrominance,* or color signal. They combine to form the video signal that is transmitted for reception on color receivers. As to compatibility, or the reception of a color telecast on a monochromatic receiver, only the luminance portion will be supplied to the single electron gun in the monochrome tube to produce a black-and-white version of the signal, while the visibility of the chrominance portion of the signal will be materially reduced.

Ambient, or surrounding, light in the viewing room (or control room in a television studio) has an effect on the picture being viewed on either television or motion pictures. In cinema theaters or for home movies, the amount of ambient light must be kept low to make possible pictures of good contrast on the screen. Control of ambient light is very important in viewing motion pictures because the reflecting screen upon which the picture is projected does not distinguish to any great degree between image and ambient light.

In television, the image is transmitted rather than reflected to the viewer, so the value of the ratio of image to ambient light is controllable. Because the amount of ambient light in the home when viewing television is higher than for motion picture viewing, certain neutral density filters and electronic compensators are built into television receivers.

AMOUNT AND PLACEMENT OF LIGHT

The amount of light on a subject can be measured by a meter and is rated in *foot-candles*. A reflected light meter, when pointed at a subject, records the light reflected from that subject, while an incident light exposure meter measures the illumination falling on the subject from a light source. The reflected light meter generally finds its greatest use for exterior shooting while the incident light meter is the one more often employed by most camerapeople or lighting directors for interior sets where the ratio of the intensity of one light to another provides the correct balance for proper exposure. In other words, it is not only the amount of light on a set but also the direction from which it falls that illuminates the subject with artistic style. *Painting* with light was one of the great accomplishments of the most famous cinematographers in the days of black-and-white film production, and now the superfast color films allow this same artistry. However, in the early days of black-and-white television, the sets and performers were rather flatly lit to avoid contrast (a property that was inherent to the system, —more so than film at the time), and the pictorial results were far from controllable or pleasing.

Today, basic incandescent lighting for any medium consists of an arrangement of a *key,* or main, light; a *fill,* or secondary, light of less intensity and direction to allow shadow sculpturing of the subject; and varying amounts of *back light* to separate the subject from the background. Back lighting can be placed to shine on the background itself or employed to halo the subject (by being so placed to shine on the back of the subject rather than on the set).

Special lighting techniques are created by the placement of lights—for example, *eye* lighting, which is illumination to produce a specular reflection from the eyes (or the teeth) without adding a significant increase in light to the subject, or *cross* lighting, which is illumination on the front of the subject from two directions at substantially equal and opposite angles with the optical axis of the camera. Such cross lighting, from one side or the other, emphasizes any wrinkles and is often employed to add age level to a face. Any illumination of the background or set other than provided for principal subject or areas is called *set* lighting (Figure 3.4).

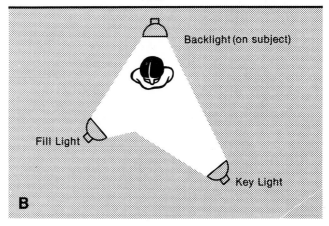

Normal lighting Set-up

Same with backlighting on subject

FIGURE 3.4 (A) *Standard production lighting scheme. (B) A variation with back lighting on the subject.*

For motion picture interiors, this type of lighting is most often used; while for exterior filming, the sun is normally employed as a key light, and either reflectors (coated with dull aluminum foil in most cases) or large arc lamps are utilized for fill lighting. However, even though a cinematographerr may change the lighting pattern for each scene or setup, television lighting is more or less fixed on a set with a general lighting scheme similar to stage lighting. As such, *playing areas* are lit, and a general illumination takes place.

The ratio of the intensity of key to fill light may be less in live or taped television (videotape is a means of permanently recording the television signal on a recording tape for future showing that employs the same television cameras, lighting, and such for a live telecast but that allows the editing and cutting of scenes during production taping or postproduction editing) than for film lighting due to the necessity of having a flatter lighting placement to obtain a proper video level in the picture. Ratios (for interiors) of key to fill intensity of 2:1 to 3:1 for live or taped television are common, while film studio ratios of 3:1 to 4:1 are normal for most scenes. Faster film, of course, allows much higher ratios and even night filming with little

FIGURE 3.6   *How all overhead lighting affected early television make-up. Photo taken of a television monitor in the studio of actress Janet Blair shows unnatural deep shadows in the eyes, cheeks, and under the chin, which did little to enhance her near-perfect face.*

light. In general, television and stage lighting have a great dependence on overhead fill and side key lighting, with floor fill lights sometimes employed to lessen facial shadows (Figures 3.5, 3.6, and 3.7).

Film can be divided into types used for still and motion picture photography. Still camera film can be exposed by incandescent lighting or by electronic flash, which is a very intense and rapid (1000th to 2000th of a second in duration) flash of light actuated by a high voltage source but with a setup often similar to such incandescent fixtures. The freezing action of electronic flash, or strobe lighting, at this high speed of light provides very sharp pictures on the film and is most often employed in commercial photography.

An enormous amount of light is encountered when shooting exterior scenes as compared to interior scenes, with daylight being the sum total of sunlight and skylight. On a clear day, with a blue sky, the illumination from the sun is approximately 9,000 to 9,500 foot-candles if measured on a plane perpendicular to the rays of the sun. In addition to this, approximately 1,500 foot-candles are provided by the blue skylight, making a total of 10,500 to 11,000 foot-candles of illumination. The lighting ratio is about 7:1 on a subject being photographed in this light. When the sky is overcast, obscuring the sun, all the light falling on the subject is from the sky. Even though the illumination may be as low as 200 foot-candles when the sky has dense clouds, the light is almost uniform over the subject. Therefore, as the sunlight decreases in intensity because of the hazy sky conditions, the strength of the fill light source increases. The highest contrast is obtained with a clear sky and a bright sun (unless the subject is shaded by surrounding objects).

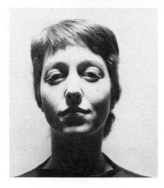
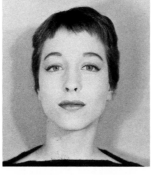

FIGURE 3.5   *Effects of lighting on make-up.* TOP LEFT: *Side lighting may wash out features on one side and place the other side in shadow. It accentuates facial age by emphasizing the wrinkles.* TOP RIGHT: *Lighting from below produces a weird effect of unnatural shadows.* BOTTOM LEFT: *Lighting strictly from overhead often produces unflattering shadows in the eye, nose, and mouth areas.* BOTTOM RIGHT: *A balanced lighting setup with a key and fill light gives a good tonal value to the features for a flattering effect. Note background shadows. No background light was used to wash out shadows in this photo.*

COLOR TEMPERATURE

For visual purposes, the *color quality* of a light source is expressed in degrees Kelvin *(K)* of *color temperature.* The higher the color temperature of a light, the bluer it is, and the lower, the redder it is. Color temperature refers only to the visual appearance of light and does not describe its photographic or television effect. The Eastman Kodak Company has a series of filters for color temperature correction, one of which is for *light balancing* of incandescent to daylight filming with the same film, and the other, called *color-compensating* filters, provides only slight correction changes of red, blue, green, cyan, magenta, or yellow light (see Appendix C for a full listing).

The actual color temperature of a lamp depends upon its construction, the voltage at which it is operated, and the age of the lamp. Although lamp ratings lower after long use, the correct color temperature can be kept by varying the voltage applied to the lamp. The color temperature of a lamp will vary with the voltage, changing about 10° K for each change of one volt.

Skin tones are affected and become unnatural if color film is not exposed under correct conditions of color temperature. Also, some television studios, to lengthen the life of their lamps, burn them at 3,100° K by lowering the voltage. This practice is questionable because it does affect the skin tone on the red side. Electronics and film cannot always *see* colors in the same manner as the human eye, and while the eye can compensate for such color differences, film and electronics cannot without some adjustment.

The sensitivity to the color spectrum of black-and-white film or the electronics of black-and-white television cameras varies with the film or television system, and as such, certain colors will register on the gray scale with varying amounts of gray and contrast depending upon such sensitivity. Color film and color television, as well, do not picturize colors the same as they might appear to the human eye; that is, to the eye, a green material or color will appear approximately the same in daylight as in artificial light as our vision compensates for changes in the quality of light to some degree. However, color film and television, having no such automatic compensation, will reproduce colors as the eye sees them only when in the illumination that is a correct color temperature for the which the film or television system is balanced.

Interior studio lighting for both film and television studios use bulbs that burn at 3,200° K as professional films and color electronics are balanced for this rating. On stage, the color temperature of the lights is not important for general viewing, only the intensity of the lamps, because playing areas are lit with lights with various filters that soften or tint the white aspect of the lights.

Daylight-type films (most professional still films are rated for daylight exposure or lighting color temperature) are balanced to be exposed between 5,400° K and 5,800° K. Curiously, although a variance of just 100° K in the area of 3,200° K for interior lighting can produce an effect that is quite noticeable in the overall picture, appearing as warmer (or yellower) at 3,100° K and bluer (or colder) at 3,300° K, a variance of 200 or 300° of color temperature under daylight conditions (5,400° K to 5,800° K) produces little noticeable difference in overall tonal value in a color determination sense.

As such, the color of the previously mentioned green material will reproduce more faithfully by daylight-type film exposed in daylighting or by artificial light film exposed in tungsten light. Daylight-type films exposed under tungsten lights require the use of a light-balancing filter (that would be bluish) and not as efficient or as pleasing in results as exposing a tungsten 3,200° K rated film with an 85 filter (orangy tone) in daylighting. This is why professional color negative film for production work is rated at 3,200° K for interior shooting and an 85 filter is recommended for exterior filming with the same film in the camera. Coincidentally, television cameras operate in the same manner.

Some basic comparative color temperatures are:

Candle flame	1,850° K
150-watt incandescent tungsten lamp	2,900° K
3,200° K studio lamps	3,200° K
White flame carbon arc lamp	5,000° K

FIGURE 3.7 *Strong side lighting can create an unusual effect and add mystery and sensuality to a picture. As in Figure 3.6, the make-up is seriously affected by the lighting.*

High intensity Sun arc lamp	5,500° K
Sunlight (skylight plus sun)	5,400°–5,800° K
Daylight with overcast sky	6,800° K

As can be seen from the chart, the combination of sunlight and skylight in an unclouded sky produces normal daylight filming conditions, and when the sky is overcast, the light is bluer and the color temperature goes up. Although the television system can balance for this difference, Sun arcs are added to compensate for filming during overcast periods.

COLOR VALUES OF SETS, COSTUMES, AND MAKE-UP

Other factors to be considered are the reflective and color values of sets, wardrobe or costume materials, and make-up used for color television or film. Each affects the other to a certain degree in relation to skin tone. *Reflectance* is the term used to express the percentage of incident light that is reflected (or bounced off) from an object. A black velvet material has approximately 2 percent reflectance, while some white papers may be as high as 90 percent. This means that with a ratio of 1:45, the exposure for photography could be set in the center of the photographic range and no difficulty in reproduction would occur. Normally, it is extremely rare to have such a wide range in a picture, and the reflectances seldom exceed 1 to 16 even with dark clothing and white settings.

Reflectances in television approximate the same values, so these principles, in general, are applicable. However, the color television system can be balanced electronically to accept materials or settings of high reflective value. In this case, the make-up may have to be commensurately lighter in tone to standardize this balance and bring the skin tones down in coloration overall.

Correctness of flesh tone is recognizable immediately in color mediums, and while any other tint or shade may be varied lighter or darker, skin coloration must appear normal in relation to the time of day or created lighting effect. In ratio to the other colors appearing on the screen, the flesh or facial tones should register above the middle of the ratio scale on a meter reading. However, ratio values for proper skin tones are subjective and the judgment of percentage values depends upon how the final arbiters (the director, producer, or such) determine the amount of pink or tan-beige desired and the density or saturation of the skin tone on the screen in the finished print or out of master control in television that they want as the perfect skin tone values. To expand on this statement, the British have always selected a lighter make-up tonality (providing a higher reflectance value for the skin tone) than most producers in the United States. As well, a production make-up done in California studios may be slightly deeper and more saturated in shade (for both men and women) than a make-up done in the New York area, both in production and commercial make-ups.

Another property of light is its propensity to shift its value of reflectivity in film or amount of luminance in television. *Luminance shift* means that the average brightness of the scene is affected when very light and very dark (or black) wardrobe, scenery, or make-ups are placed in view of the camera. These conditions are monitored in the television studio control room by a video engineer who can balance the signal relative to the darkest, lightest, and skin tone values of the gray or color scale. The home viewer in television does have a brightness knob on the receiver, but this controls only the received signal from the transmitter and not the *average brightness* of the scene. In film, one bright object can bloom in the same manner because of its high reflectance of light at a certain exposure, but the average brightness is not affected in the same manner as it is on television.

A *color shift* from the scenery can change the aspect of facial color of the make-up when a performer is placed too close to a set. For example, a green wall will shift a greenish tinge to the make-up if the performer is too close to it for film or television. Also, with some special background projection changes in color for television, the facial tone can be affected. Some newsrooms formerly employed rather bright blue background for *Chroma key* effects,* and the newscaster's face would be grayed from the cold effect of the color shift of the blue. Replacement with more neutral backgrounds and advanced electronics corrected this.

Wardrobe

The old stage adage of the darker the wardrobe, the lighter the make-up appears, and vice versa, is quite true for films or television. Clothing affects the color and response of make-up to a great degree. Both black-and-white television and films reproduce clothing shades about the same except that, in most cases, television shows more contrast than film for a given material. Color television and films respond similarly, relative to contrast, but the reproduction of a certain color will vary as to the television camera system or the controllability of the film. In general, with a soft lighting scheme, light clothing, or a light background, a lighter make-up value is best; with darker clothing, darker backgrounds, or a harder or higher lighting arrangement, a darker make-up becomes preferable.

Spangles, sequins, rhinestones, and other brilliants

A good explanation of *Chroma key* is found in *Creating Special Effects* by Bernard Wilkie, "Chroma key {is} an electronic method of combining parts of television pictures received from separate cameras or other sources. Used chiefly for putting backgrounds behind action. Also known as color separation overlay." (page 155) or jewelry used on or with clothing are somewhat difficult to control on screen as they often have sharp reflective surfaces that give a kickback from the lights. Unless this is a desirable quality in the picture, these items should either be avoided or sprayed dull to reduce the glitter.

White and black wardrobe have a blooming or flattening effect respectively on the screen and must be carefully illuminated (as well as designed) to display any tonal value. The contrast ratio is affected in the overall scene by such and must be taken into consideration as to whether or not it disturbs the scene and/or affects the color saturation of the make-up.

Black and white, shades of yellow, pale blue, pale green, light tan, light pink, orchid, or very pale gray all picturize on the light gray scale side, while royal or air blue, green, deeper tan, and red will be a deeper gray, and deep blue, violet, maroon, and dark grays and browns will appear as a dark gray. Most grays and browns will picturize approximately as they appear in person and so are easy to judge in black and white, while the brighter colors must often be tested if the shade of gray is important to the scene. Remember, a pale blue and a pale tan or a royal blue and a deep red may not separate in gray scale at all.

Color wardrobe tests are always best for most film and television shows, and the dress rehearsal or costume parade on stage will settle any questionable qualities or shades for wear. Compatibility of wardrobe colors and make-up for women is discussed at length in the following chapters.

FILM TYPES

Although there are a number of manufacturers of professional film, the main U.S. supplier to studios is the Eastman Kodak Company. They supply a great variety of black-and-white and color films for both stills and motion picture use. Although there are a number of panchromatic black-and-white films, mainly color film is employed today as a good black-and-white print can be made from either still or motion picture color film.

Due to rapidly advancing technologies, new film types and improvements occur frequently. However, the current stocks include Eastman Color Negative II Type 5247 (35mm) and 7247 (16mm) E.I.*125 and Eastman Color High-Speed Negative Film Type 5294 (35mm) and 7294 (16mm) E.I. 400*. The newest ECN 5294/7294 has an E.I. of 400/320 and 7291 (E.I. 100). These are the principal types of professional motion picture films for production use.

Another category, called *video film,* is employed by newsfilm camera people or when a high-speed film is required:

Eastman Ektachrome Video News Film (Tungsten) Type 5240 (35mm) and 7240 (16mm),
Eastman Ektachrome Video New Film (High Speed) Tungsten Type 7250 (16mm) E.I. 400,
Eastman Ektachrome Video News Film (Daylight) Type 5239 (35mm) and 7239 (16mm) E.I. 160,
Eastman Ektachrome High Speed Daylight Film Type 7251 (16mm) E.I. 400.

Another professional film is the low contrast reversal film, Eastman Ektachrome Commercial Film Type 7252 (16mm).

Among the Kodachrome varieties available for amateur 16mm or the Super 8 films employed for test production or even some industrial use today are Kodachrome 40 Super 8 cartridge KMA 464 (silent) or KMA 594 (sound) and Kodak Ektachrome 160 cartridge EG464 (silent) or ELA594 (sound) Super 8. There are varieties of Kodachrome in both daylight and tungsten films in magazines or rolls that have an EI of 40 tungsten and 25 daylight for motion pictures, and the same in 35mm still film, which is widely used by professional still photographers, plus some newer 35mm still films with yet higher exposure ratings.

Production Filming and Procedures

When ECN II 5247 is employed as a camera film, it is developed and a negative image is produced. To be viewed, a positive image or print must be made from this negative. For the purposes of editing, a *work print* is made from the negative, and this is what is shown when the *dailies,* or *rushes,* are viewed by the film's director, producer, film editor, cinematographer, make-up artist, and other personnel technically concerned about what was filmed a day or so previously. From these dailies are selected the scenes that the film editor will splice together to form a *rough cut* of the production. After refining this rough cut to a point where the director, producer, and film editor are satisfied with the flow and story of the film, and the editor has marked where the effects such as the dissolves, freeze frames, fade outs, and so forth shall be employed, the negative is matched (by numbers on the edge of the film) to the work print and is cut and spliced. From this negative is made a *master positive print,* and then from it, a *dupe negative.* In turn, from the dupe negative are made any number of positive *release prints* that are employed for theater projection.

It can be seen that the original negative becomes the most valuable item in the chain, so it is well cared for and handled carefully. Each successive step taken from the original negative to the release positive is called a *generation.* Therefore, three generations take place from the original to the release print. Each step is controllable by the laboratory in exposure, color correction, and such, so the negative type of film is the most efficient method for professional film production use.

Color reversal film (the suffix *chrome* indicates that it is a reversal type) is processed to produce a positive from the camera film, and although most of the still and motion picture film so produced is for direct viewing, the low-contrast Eastman Ektachrome type 7252 (16mm) is often employed for some production use because additional prints for release can be made from it with some corrections possible. With the normal contrast reversal films, there is a degradation of the image when a copy is made from the original, and it is also not suitable when a number of release prints are to be made.

Printing flow charts (Figures 3.8 and 3.9) show the progression from negative or original to release positive prints for viewing.

Film Processing

In the Kodachrome and Kodak Ektachrome processes, the reversal method is used to produce positive color transparencies for projection, direct viewing, or photomechanical reproduction. After exposure, the film is first developed in a black-and-white developer. This produces a negative silver image of the subject in each of the three emulsion layers. The film is then re-exposed to fog the remaining silver halide deposit on the film and render it developable. By a process called

FIGURE 3.8   *Printing flow chart for color negative films. (Courtesy General Film Laboratories.)*

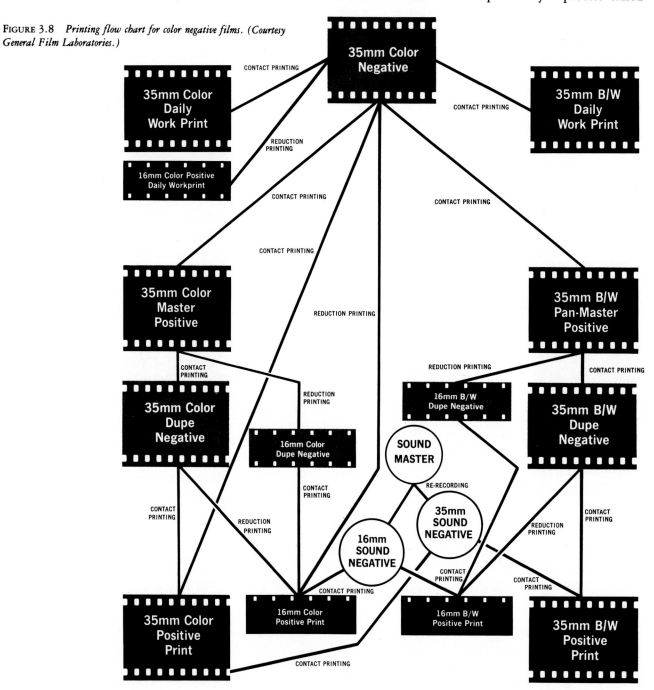

coupler development, the silver deposit is formed along with the three positive dye images: yellow, magenta, and cyan, one in each of the emulsion layers. Next the film is bleached and, without affecting the dyes, the silver is converted to salts soluble in hypo. The film is then fixed, washed, and dried.

In Eastman Color Negative II, Type 5247, the dye couplers are incorporated in the emulsion layers at the time of manufacture. After exposure in the camera, this film is developed in a color developer that produces a dye image along with the silver image in each layer. Each of the negative dyes serves to control the amounts of each of the primary colors—that is, cyan in the red-sensitive bottom layer, magenta in the green-sensitive

middle layer, and yellow in the blue-sensitive top layer. Then the silver images formed along with the dye images are removed by bleaching and fixing, and the remaining dye images are negative with respect to the tone graduations of the original subject. A positive film is then used to make the final print.

AUDIENCE VIEWPOINT

The judging of a production, whether it be on stage or screen or even printed on the pages of a magazine or book, is finally made by the eyes of the viewer or audience. This judgment is governed, aided, or restricted by the view afforded of the action, and there

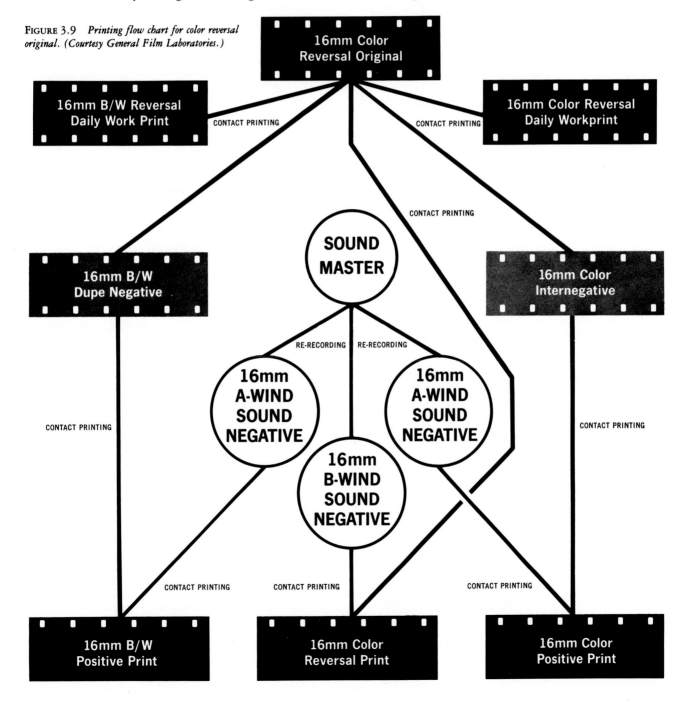

FIGURE 3.9   *Printing flow chart for color reversal original. (Courtesy General Film Laboratories.)*

are differences in the mediums of stage, screen, or print that affect the audience and its viewing position.

For stage productions, for example, the audience sits in a fixed position or seat in the orchestra section, the mezzanine, or the balcony, viewing the performers on stage directly through the medium of air. They see the entire production from this one spot and aspect of line of sight.

Television and motion pictures are seen on a screen, in a direct or a reflected image respectively, but have the facility of moving the audience to positions of viewing not possible in stage productions by the means of the movement of the camera or the lens that can change the aspect of the scene. As such, a camera can be moved in and out to show a close-up or a long shot, a lens changed to widen or narrow the size of the image. The camera may be mobile—that is, moved on studio dolly or crane or in an automobile, helicopter, or plane to place the audience in varying positions of viewing. Such is the extreme versatility of the screen medium, and due to this, make-up for the screen has been refined to an art of extreme naturalness so that even in the tightest of close-ups, the make-up must appear creditable and perfectly natural, whether it be straight or complicated character make-up.

Camera framing definitions for both film and television are:

Extreme long shot Overall establishing scene of an area;
Long shot Normally, a full length of a person (or persons),
Medium shot Waist up of people;
Close shot Head and possibly shoulders of a person;
Close-up Although employed in various ways, really a shot that shows only a portion of the face or body, like an eye.

Still photo reproduction or printing on a page has the additional advantage of *retouching,* which serves to remove or change facial areas by drawing in any item area, or color for the final print. The skill of the retoucher can take a badly applied make-up and make it appear perfect in color or black and white. Nevertheless, make-ups done for stills should not rely on the retoucher to correct faults such as incorrect skin color, badly drawn liplines, and so forth, but always should make the subject appear as perfect as possible.

Today, it is even possible to change each and every frame of a motion picture make-up to correct it in the same manner, with certain special visual effects added by laboratories that specialize in electronic or computer changes. An example of this can be seen in the film, *Altered States,* released in 1980 in which electronically created abberations were superimposed over the make-ups in some scenes to indicate body changes taking place that were not possible with make-up.

Diffusion

On stage, the distance the viewer is from the performer sets the amount or degree of diffusion or unclarity in a make-up. In film or television this diffusion is somewhat inherent in the system. When a film is projected on a screen, the distance it is projected as well as the size of the screen it must fill and the size of the film being projected produce a diffusion to the eye by this enlargement. In television, this diffusion occurs in the fidelity of the system to reproduce a picture. This phenomenon is often referred to as the *image structure.* A basic comparative image structure between television systems and motion picture films is shown in Table 3.2. This means that the image structure product of resolution, which is an expression of the clarity of the viewed image, is less in television than on film (professional 16mm and 35mm film, that is). In the assessment of general picture quality in television, the horizontal resolution is the critical measurement, and this capability is constantly being improved. *High Definition* is expected to improve the quality, clarity, and overall resolution of all television reception and is the direction of the future of video.

Suppose the proper quality and amount of light are provided in the make-up room (Figure 4.1). A good way to judge a screen make-up prior to shooting is to view the completed make-up in the mirror from the image reflected from the subject sitting in the make-up chair (3 to 4 feet from mirror and lights approximately). What is seen by the make-up artist is approximately what will be seen by the camera relative to the blending of colors, distinction or pronouncement of lines, shadows, highlights, or character make-up effects such as beards or the edges of prosthetic pieces.

Stills taken of a make-up and enlargements made thereof may well show a higher degree of resolution

Table 3.2   Resolution in Lines (Screen ratio 4:3)

	HOME VCR'S	BROADCAST TELEVISION	COLOR TELEVISION RECEIVERS	8 MM FILM	16 MM FILM	35 MM FILM
Horizontal (H)	240	330	350–400	230	490	1,000
Vertical (V)	—	—	370	230	490	1,000
Approximate Product H × V			129,500–148,000	52,900	240,000	1,000,000

and image structure than for motion picture film or television due to the fact that the image is stopped in a fraction of a second. As such, some studio stills taken for distribution for publicity may show up the lace on hair goods, the edges on prostheses, and so forth (requiring some retouching), whereas these things would not show up on the motion picture or television screen. Photos taken by Polaroid or Kodak Instant Film in the make-up room are useful for matching of design or shape but should not be considered to be of sufficient color accuracy for matching or testing because too many conditions can affect the print.

Still photos can be taken of a television screen (United States television has 525 lines per picture at a frequency of 30 pictures per second) by setting the picture to normal color and brightness and exposing the film at one-thirtieth of second with the lens opening set according to the exposure index of the film employed.

The Make-up Department

IN THE LARGE HOLLYWOOD STUDIOS, THE MAKE-UP DIRECtor or head of the make-up department usually had the best make-up room with the best facilities of lighting, chair, and so forth. It is here that the publicity department of the studio took the pictures of the complicated character make-ups or when the stars were to be shown in a make-up room. With the walls covered with photos, award plaques, and facial casts of the famous, the make-up room provided the background for public relations and imagery as well as displayed to any visitors what a make-up room should be.

However, today, depending upon the studio and in most other places around the country, the everyday make-up work is more often done under less than perfect conditions and comfort. Some studios do furnish proper chairs, mirrors, and other appurtenances in either boothlike setups or rooms or a large area with a number of chairs and mirrors in a row, while some others (often the small or individual studios that do commercials) assign the least desirable area for make-up. An ordinary low, nonswivel chair, a mirror surrounded by 25-watt lamps, and a fluorescent fixture high on the ceiling may be seen as the norm in many local television and film studios. Under such conditions, however, straight make-up is difficult to apply and judge, and most character make-up is impossible to accomplish.

A basic booth or room area for professional make-up application (Figure 4.1) consists of:

Proper and sufficient lighting,
A good mirror of adequate size,
A comfortable chair that can be adjusted to height,
Space for the make-up kit and items necessary for work,
A degree of privacy away from activities in the studio.

A similar setup for hairdressing, but with a sink with running water and a shampoo table, is a necessity.

LIGHTING THE MAKE-UP ROOM

Too often studio designers will place 25-to-40-watt incandescent bulbs around a mirror in a make-up area. Not only do they provide insufficient light, but also their color temperature is too low for proper color judging. They also produce excessive radiant heat, thereby warming the area unnecessarily.

A far better system is that illustrated in Figure 4.1, employing special fluorescent bulbs. Now although color temperature is the measure of the energy distributed over the spectral range—that is, *color quality* of a light source with a *continuous* spectrum (such as tungsten incandescence)—an *apparent* color temperature can be assigned to lights like *fluorescents* for comparative values.

Fluorescent bulbs diffuse light with a scattering or criss-crossing of light rays that produces a general illumination rather than a direct radiation. Thus, there is no filament that burns or incandesces under the application of electrical excitation but only the fluorescence of certain gases that produces the light values. Fluorescents cannot have their light directed or focused as can incandescent lighting. Many people have a poor knowledge of the varieties of fluorescent bulbs that are available as most hardware and even electrical supply stores stock and sell only one type of bulb—normally, a cool white type of bulb, which has little red in its spectrum so the yellow and green portions predominate (Figure 4.2).

FIGURE 4.1 *Basic booth area for make-up.*

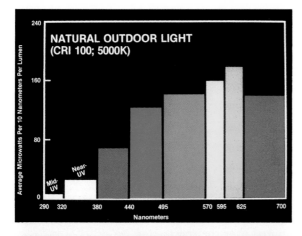

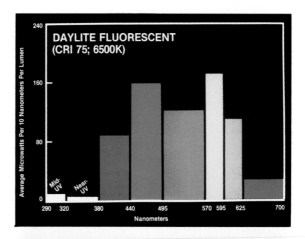

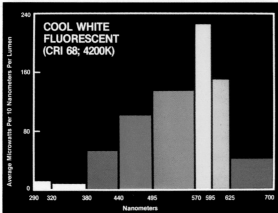

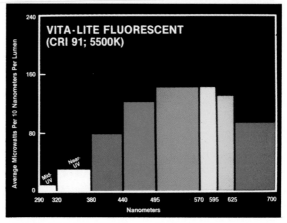

FIGURE 4.2 *Spectral energy distribution charts show average amount of light generated in each color band by the light source being measured. These are the color ingredients of each type of light.* TOP LEFT: *The chart is specified by the International Commission on Illumination as representative of natural outdoor light. The other charts represent the two common fluorescent lamps and Luxor VITA-LITE* (BOTTOM RIGHT). *(Courtesy Luxor Lighting Products, Inc.)*

Table 4.1 indicates the apparent color temperature of available fluorescent bulbs and adds a better judging rating known as the Color Rendering Index (CRI) (which has to do with the way that color materials appear under a specific light source). A comparative listing of natural sunlight shows the optimum rating of CRI 100 for comparison.

As can readily be seen, the VITA-LITE has the closest CRI to natural sunlight and an apparent color temperature within its exact range. As such, these bulbs are the best type to place in a make-up room to judge the color of make-up and hair as well as to provide a proper amount of light with the least amount of heat. For example, a mirror 4 feet high and 3 feet wide can be placed between two 4-foot and one 2-foot fixtures. With the make-up chair some 4 feet away, the amount of wattage is only 200 and rates about 115 foot-candles, which is quite sufficient for judging and applying most make-ups.

As the color temperature of sunlight is approxi-

mated by these lamps, they provide the optimum lighting conditions in color quality and color matching for make-up, hair, or wardrobe. Any make-up done under regular incandescent bulbs (2,900° K) will not provide the proper color temperature to judge red or pink tones, such as with lip- and cheekcolors, so when the subject is taken out of doors after having make-up applied under incandescent lighting, the color match can be off or the cheekcolor might appear streaked.

ROOM FURNITURE

The make-up chair should be comfortable and have a removable headrest and a hydraulic or swivel-up rise to a height that ensures the make-up artist does not have to bend over the subject in the chair. If the floor is carpeted, it is a good idea to surround the chair

Table 4.1 Kelvin and CRI Scale

LIGHT SOURCE	°KELVIN	CRI
Sunlight	5,400–5,800	100
VITA-LITE*	5,500	91
Cool White	4,500	68
Warm White	3,000	56
Daylite	6,500	75
Spectra 32*	3,200	82

*Luxor Lighting Corp. (see Appendix B)

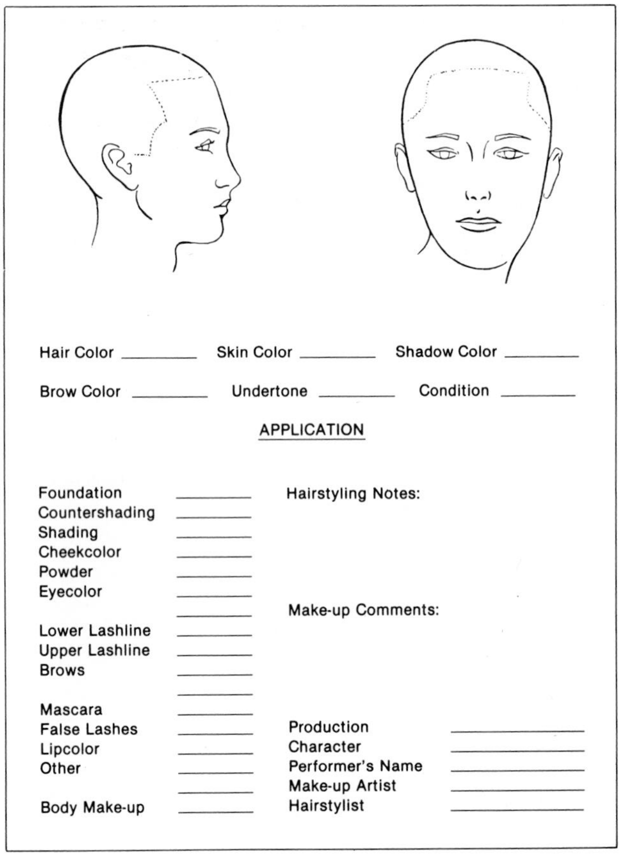

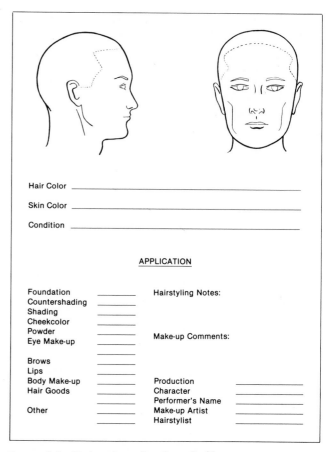

Hair Color _____   Skin Color _____   Shadow Color _____

Brow Color _____   Undertone _____   Condition _____

APPLICATION

Foundation _____   Hairstyling Notes:
Countershading _____
Shading _____
Cheekcolor _____
Powder _____
Eyecolor _____
                             Make-up Comments:
Lower Lashline _____
Upper Lashline _____
Brows _____

Mascara _____
False Lashes _____   Production _____
Lipcolor _____   Character _____
Other _____   Performer's Name _____
                             Make-up Artist _____
Body Make-up _____   Hairstylist _____

FIGURE 4.3   *Women's make-up chart for studio file use.*

Hair Color _____

Skin Color _____

Condition _____

APPLICATION

Foundation _____   Hairstyling Notes:
Countershading _____
Shading _____
Cheekcolor _____
Powder _____
Eye Make-up _____   Make-up Comments:

Brows _____
Lips _____
Body Make-up _____   Production _____
Hair Goods _____   Character _____
                             Performer's Name _____
Other _____   Make-up Artist _____
                             Hairstylist _____

FIGURE 4.4   *Men's make-up chart for studio file use.*

with a plastic mat similar to those used in barber shops so the area can be easily cleaned.

The make-up table should not be so deep that the mirror is too far from the subject, but it should be wide enough to accommodate the make-up kit and allow space to place the tools and make-up items. Formica tops are best for cleaning facility and cleanliness. A good practice is to clean off the top of the table with alcohol after each use. Wall and furniture colors should always be neutral in tone, such as white, off-white, very pale gray, beige, and so on so that no color shift will occur to the make-up. Extra shelving for reference materials and supplies are helpful, but a steel, lockable storage cabinet is best.

ON-SET FURNITURE

On-set furniture for the make-up department sometimes consists of a movable table with a mirror and a number of incandescent bulbs around it, again providing an inadequate setup that would be improved greatly with the addition of the proper fluorescent bulbs. A regular director's chair (folding, with a canvas back) is far too low for make-up, but some have long legs, often called a *Captain's Chair,* and while they do not provide a headrest, they are tall enough to facilitate make-up retouching. Some make-up artists have con-

structed their own wooden folding chair designs that incorporate a removable headrest for location or on-set use. The more comfortable a performer is in the make-up chair, the easier it is for the make-up artist to work.

LOCATION MAKE-UP ROOM

When on an out-of-town location, the production company should have a room set aside for make-up, possibly close to the performers' rooms in the hotel or motel if practical. Most rooms do have some form of adequate mirror, but the lighting must be augmented. A portable fluorescent setup can be constructed for the best lighting, but in lieu of that, portable clip-on light fixtures with reflectors holding 3,200° tungsten lamps (or similar types) can be utilized. Two such clip-on lights should be part of any make-up artist's extra supplies. It is also more convenient to have separate make-up and hair rooms, either in the studio or on location, whenever possible, for the convenience of the performers as well as for adequate work space.

MAKE-UP RECORDS

Record sheets or cards for both male and female performers should be kept to ascertain results from tests or for matching a specific make-up when retakes occur (often some weeks later). These records should have all the pertinent facts of the individual listed as well as all the make-up materials and special procedures. A blank drawing of a face allows the make-up artist to make visual notes of the make-up for reference or comparison (Figures 4.3 and 4.4). Attaching a Polaroid picture can add to the record, or a number of them can show progression. When the lighting is VITA-LITE fluorescents, daylight film can be used without flash most times.

In addition to the normal make-up department, a laboratory for manufacturing special materials and prostheses, taking casts, preparing hair goods, and such is often found in large studios. However, a growing trend is to contract out some or all of these services to special effects make-up artists who have built up specific facilities to deal with all forms of latex and plastic appliances, maintaining a staff of sculptors, mechanical designers, and other specialists. Even the unions are creating new categories for these artisans in our changing world of make-up procedures. Much of this special equipment and these materials are discussed in Part IV.

CHAPTER 5

Make-up Products and the Kit

TODAY'S PROFESSIONAL MAKE-UP ARTISTS CARRY MANY more new and different items than they did some 10 or 15 years ago, so one can always tell the difference between an outmoded or amateur kit and that of the currently working professional. Products can be divided now into three basic categories: commercial, theatrical, and professional items.

Commercial products are generally sold in department or drug stores for street make-up to be self-applied, and fashion dictates govern their sale, variety, and promotion. They are often merchandized with voluminous advertising and sales promotion, with the buy-and-try adage being the main sales pitch. Competition between commercial cosmetic lines does not consist so much in product difference in the main but in how ostentatiously the item is packaged, how loudly the brand or product is touted with advertising, along with the invention of various phrases and semantics spiced with juicy brand names. Red is never just *red* but must be Witch's Blood, Roaring Red, or the like, and many directions are strongly overworked by the employment of terms such as *natural, organic,* and so forth, without more than a that's-what-sells attitude beneath it all.

Such products are designed not for use by professional make-up artists but as self-application items by the purchaser. Low-pigment liquid foundations (that provide little or no skin coverage), supposedly long-lasting (and possibly staining) lipsticks, soft pencil or cream-style eye make-ups (that smear but that are easily finger applied, nevertheless), lotions and skin care creams with exotic ingredients such as mink or turtle oil, fruit or vegetable additives (strawberries, cucumbers, and so on), and pH value ratings (with little or no real meaning of what they actually mean or why) are but a few of the directions of sales jargon and ingredients for commercial cosmetic products. In general, these are neither designed for coordinated use for screen or stage make-up nor really recommended as such by their manufacturers.

Theatrical make-up lines are mostly defined today as those advertised for sale in student kits or for amateur school and college use. They still have the archaic foundation names such as Juvenile Flesh, Sallow, Hero, and so forth, as well as many colored powders for the face, old nose putty, and such, all of which have seen little or no use for some years by make-up artists.

Professional make-up products, however, are manufactured for use by make-up artists and are not usually sold on a retail level (seldom, if ever, in department stores or the like). In the main, they are rather plainly packaged and designed to fit in the make-up artist's kit or in larger stock sizes. An enormous variety of products, sizes, special materials, and items are supplied in the professional category that may never be found in any of the commercial lines and are often more advanced products than those for similar use in the theatrical lines.

Note: Most of the cosmetics that are made in the United States are quite reliable overall (due to stringent pure food and drug rules that govern the ingredients), and reactions of an allergenic nature may occur only to the perfumes, stains, or to sensitivity to a particular ingredient by a particular person. However, all the theatrical and the professional make-up companies (not the commercial companies whose products are sold in department or drug stores) generally sell certain items that are not in general use by the public. For example, resin-based adhesives, vinyl-based sealers, plastic waxes, latices, or plastics that are in common, everyday use for character make-up by professional make-up artists are not sold except for this use. Since the solvents for these items may be alcohol, acetone, and such, there are possibilities that some performers will have a sensitivity to these products, resulting in an irritation factor. If the make-up artist finds that anyone does have a reaction to any such material, it should be removed and not further employed for that person. Such cases *are* rare because most human skin can stand more than is imagined. A face covered all day, for long hours, with a number of latex or plastic pieces will not breathe as well as one with just a coat of foundation, but such character make-up is necessary to the work of the professional make-up artist to create the necessary effects.

Many schools and colleges that have a make-up course as part of an audiovisual, television, film, or stage program or curriculum or that have drama clubs as extra activities for some years always purchased make-up products for the group and simply left them out on the make-up tables for use by the performers. As such, the destruction level of the products was high, and cleanup after one or two performances was minimal. Most often the make-up kit or box was a veritable mess. Only in those institutions or groups where one person (who was often more interested in make-up than in performing) had charge of the kit, show after show, was it kept in any semblance of cleanliness or completeness. Eventually, instructors of make-up classes found that small individual kits served the purpose better and were sold to the students or included in their lab fee. This may be fine for a performer level but is quite inadequate for one who wishes to study in make-up artistry or design. Here, a full make-up kit, on the level of the professional make-up artist, is really a necessity.

THE MAKE-UP KIT

Make-up materials consist of foundations, shadings and countershadings, cheekcolors, lipcolors, eyecolors, powders, pencils, mascaras, skin care products, tools of the trade, special materials (such as adhesives, sealers, latices, and plastics), cleansers, and many other additional items such as prostheses, hair goods, and so forth. The stock of the products in a professional make-up artist's kit changes as technological advances are made in the color response of film and television systems or advanced stage lighting techniques as well as when new products and materials become available through research. As such, the make-up kit should go through constant change and improvement.

The old theatrical make-up manufacturers originally serviced their foundations in paper tubes, and when films came into production use, a new form of soft paint in a metal squeeze tube was devised. This spread more easily and did not require an undercoating of cold cream to spread it properly. The next major advance was the introduction by Max Factor of a water-applied cake (Pan-Cake) make-up, which was basically a heavily pigmented mixture with clays and binders. This form applied very rapidly with a wet, natural sea sponge and was the mainstay of the make-up for black-and-white films and television production. As this type of foundation was often drying to some skins and appeared rather flat and dull looking on the face, a newer type of cream make-up in a swivel-up tube came next, then a creme-cake formula was introduced, both of which were applied with a dry, foamed sponge. Attempts were made at various times to produce a liquid or semifluid professional foundation, but they did not achieve acceptance. Today, the main foundation types are the creme cakes (in a variety of sizes) and the swivel-up in the 1/2-ounce size.

Since professional foundations produce the basic skin tone, they become the main criterion of film and TV make-up, and the carrying cases or kits of each artist are often made to contain either one or the other variety as a basic direction of design of the kit, with those employing the swivel-up foundations often using the

FIGURE 5.1 *Creme-cake foundations in various sizes.*

multidrawer variety with the top portion opening up to reveal a deep tray where the foundations are placed label side up for selection.

There are a number of varieties and styles of these kits, in wooden construction generally. A newer and quite efficient case of the accordion variety works very well with the flat, round, creme-cake foundations, and the top section will hold up to 30 of the 2/5-ounce size or 24 of the 1-ounce size in the same space, thus giving a variety of shades to choose from (Figure 5.1). Most make-up artists also have a *hair kit* (not to be confused with a hairdresser's kit), which normally consists of a soft canvas or leather bag in which are stored the items employed by the make-up artist for character work, plus extras such as tissues, towels, and other supplies for the make-up kit. For complete list of items recommended for make-up and hair kits, see Appendix E.

HISTORICAL PROGRESSION OF MODERN MAKE-UP

The first technical study of film make-up was published in the *Journal of the Society of Motion Picture Engineers* under the title of the "Standardization of Motion Picture Make-up," by Max Factor, in January 1937. It discussed in detail the tests and results of designing make-up bases or foundations for black-and-white film rather than using the old theater make-up shades. A basic series of shades, ranging from a light pink to a deep orange-tan and numbered from Panchromatic 21 to 31, became the standard for many years for all black-and-white screen use. This report discussed the history of the use of make-up on the stage and the evolution of film from the orthochromatic to the panchromatic types.

Some attempt was made in manufacuring the make-up shades so that psychologically the performers were not faced with highly unnatural colors; nevertheless, *panchromatic* make-up did appear rather yellowish to orange-tan in skin tone, and even though a cheek rouge of a very pale shade (Light Technicolor) was employed later to remove the flat look from the cheeks of women, the overall effect still had a made-up look offscreen. Incidentally, the cheek rouge was so light that it did not register on the black-and-white screen with any tonal difference and was solely for performer psychology. Unfortunately, for some years performers became so accustomed to seeing themselves in panchromatic shade make-up for films or television that they often attempted to employ it on stage—where they looked as if an advanced stage of jaundice had set in, especially in the men's shades, rather than the pink-ruddy-tan shades correct stage make-up required.

In the Max Factor tests, Lovibond tintometer determination was employed to carefully design the shades of make-up that would translate into comparative natural skin tones for performers in a black-and-white medium. Pan 25 and 26 became the women's shades, and Pan 28 and 29 were used for men, with a shade that was three numbers below the base shade (say, Pan 22 for women) employed to create *highlights* that would photograph three shades lighter on the gray scale, while one that was three numbers above (such as Pan 28) appeared as a *shading* that photographed three shades darker on the gray scale. All this information worked out perfectly well for black-and-white television too, but when film and television went to color, these principles became ineffective and basically incorrect.

The first technical paper on make-up for color mediums was published in the *Journal of the Society of Motion Picture and Television Engineers* under the title of "New Make-up Materials and Procedures for Color Mediums," by Vincent J-R Kehoe in November 1966, with a revision in the April 1970 issue and an entirely new set of recommendations in the November 1979 issue of this same journal. Here were set up the newest principles and methods, as well as new products, for color—all of which could be employed compatibly when a black-and-white print was made from a color film or a television colorcast was to be viewed on a black-and-white receiver. As such, a complete change of what the professional make-up artist should carry in his or her kit was made, and it became no longer necessary to carry any black-and-white or panchromatic shades.

Over the years, with the improvement in film types as well as the technological advances in color television, newer foundation shades were devised so that now, only one basic series of pink-beige to ruddy, natural tan that is very natural looking even offscreen or offstage is necessary for most professional use for Caucasoids. The design of this new series was made by the author after exhaustive comparative tests on stage and screen, with some of the leading film manufacturers and television stations acting as test areas.

FOUNDATION SYSTEMS

There are essentially four different *foundation systems,* each designed for its own mediums, along with some interim ideas and experiments in between. The old stage make-up system divided the shades of foundation or base color (with powders to match) into the skin tone effect required for the character to be portrayed:

Very Light Pink	Ivory	American
Pink	Juvenile	Indian
Deeper Pink	Gypsy	East Indian
Flesh	Othello	Negro
Deeper Flesh	Chinese	Mulatto
Cream	Japanese	Lavender
Sunburn	Arab	Vermilion
Sallow Old Age	Spanish	White
Robust Old Age	Mexican	Black

Next came the panchromatic or black-and-white television make-up that was numbered as Pan 21 to 31 and later, when some of the orange tone was removed, 1N to 11N. When color film make-up and later color television make-up was being experimented with and devised, much emphasis was placed on the development of this same type of rising skin tone value idea. However, later research proved that this was not the best manner in which to design color medium make-up since many of the in-between numbers saw no use at all.

It is interesting to note that during the 1930s to 1950s, when this era of professional make-up was taking place, commercial cosmetic companies seldom produced more than six shades of foundation in their lines, and most had a low percentage of pigment to the amount of vehicle in the formulations. However, many spoke of *color correcting* the natural skin with such products. For example, pinkish areas were said to be correctable with beige tones and sallow skins could be heightened with pink foundations. Some went as far as selling a green-tinted foundation to correct redness and a lilac-tinted one to correct sallowness. These were to be applied before the normal foundation but did little else but put a rather heavy coating on the face—all of which seldom, if ever, appeared compatible in color to the skin on the neck, producing a very mask-like look.

For some years commercial companies also recommended three shades lighter for highlights and three shades darker for shading without having such colors in their lines. Erroneously, these principles were effective only for black-and-white mediums, while street make-up has always been a color medium. Certainly, shades such as Ivory, Beige, Peach, Natural, Rose, and Suntan did not lend themselves to this black-and-white theory, but being so lightly pigmented, little harm could be done—and little effect realized!

In 1962, the Research Council of Make-up Artists (RCMA) was formed. This group devised the system known as *Color Process,* which provided a truly modern method of skin tone and undertone matching for professional foundation products. Ethnological studies showed that Caucasoid skins could be defined or separated by the terms *Northern European, Middle European,* and *Southern European* and as going from a pink to ruddy to an olive to olive-brown undertone value. Orientals or Mongoloids had a rather yellowish cast to their undertone value, while Negroids had a gray undertone. Skins could be divided into light, medium, and deep plus the value of the undertone color, and this rating better defined true skin tone. The make-up shades required then were light, medium, and deep with undertone values of pink, olive, yellow, and gray for both men and women. In addition, some suntan shades for Caucasoids could be added and a series of spectrum colors to change any foundation shade one way or another.

Compatibility to the various mediums (street, stage, and screen) then had to be taken into consideration, and the color response required for each film type, television system, or the stage and street had to be built into the shades finally selected. To achieve a natural skin color look, the foundations would have to be very highly pigmented so that a little, spread properly on the skin, gave it the appearance of a skin color rather than a coat of make-up.

To eliminate allergy problems, the foundations should not contain lanolin or perfume. Also, vegetable oils rather than mineral oil should be used as the vehicle to lessen facial shine and the constant retouching that was required with old greasepaints and swivelsticks that used mineral oil in their formulations.

In the RCMA Color Process line, all the foundations, shadings, countershadings, and cheekcolors are made with the same vegetable oils and waxes with only the color pigments varied. Although most commercial brands of liquid make-up contain only 18 to 23 percent of pigment to vehicle and theatrical brands are in the 35 to 40 percent range, the RCMA shades are made with 50 percent pigment to vehicle, giving them a very high degree of coverage on the skin. Manufacturing controls of RCMA call for every batch to be skin tested so that no variance in color is allowable. Most commercial and theatrical batches have a tolerance of plus or minus 5 percent to up to 15 percent, but with lower pigmentation the difference on the skin may not be objectionable. Nevertheless, most professional studio make-up artists prefer their colors to remain constant.

William Tuttle's line of foundations has strict manufacturing controls as well, and he personally tests batches to standards to maintain accuracy of shade. In addition, good standardization has been noted in the Ben Nye professional line.

Color Matching

Matching a color or shade from one manufacturer to another is a very difficult procedure because of the infinite variety of shades that can be produced by the combination of the various earth colors available as well as the dilution of these colors with the waxes, oils, and tints of the colors that can be made by varying the amount of white. Exact shades are more difficult to reproduce in the light than in the dark skin shades since sometimes the deepness of the color will obscure some minor addition of a weaker color element.

As such, it is most generally accepted that one manufacturer's shade of olive and another manufacturer's olive, although semantically identical, may not only be a different concept of the shade but also, even though a match is intended, will show some variety when made in a professional foundation with a high percentage of color additives. In addition, to attempt to classify all the olives, reds, greens, and so forth into one master chart of everything made is a fool's paradise at best and a useless exercise in the least, due to the incompatibility of foundation materials such as waxes and oils from one manufacturer to another. For example, to say that the olive shade of one line is deeper in tone than that of another line, and so to number them as Olive #1 and Olive #2, implies that if one wants a deeper shade of Olive #1, then Olive #2 could be employed. However, Olive #1 may be a tube greasepaint with a mineral oil base, while Olive #2 may be a creme-cake variety with a purely vegetable oil and wax base, which on the skin may appear, as well as powder down, quite differently and not be at all compatible with other products used on the face from the same line of make-up. It is, therefore, considered good advice to select and employ the foundations, shadings, countershadings, and cheekcolors of a single manufacturer of professional products to complete this portion of a make-up so that they all accept powder with the same degree of absorption to make an even-surfaced skin texture. This advice is given, of course, with the premise in mind that all the named products such as the foundation, shading, countershading, and cheekcolor are made with the same wax-oil ingredient combination to insure compatibility on the skin.

Color Perception

It is essential that the make-up artist develop a keen sense of color perception and have a knowledge of the basic color ingredients that are utilized to manufacture the shades and colors of current make-up products. Color film and television systems today are extremely sensitive to nuances in the tonal values of foundations, lipcolors, cheekcolors, and so forth, so that identical batch matches during manufacture are essential for obtaining a correct color continuity from day to day on a performer. Basically, some eight earth colors are utilized for compounding skin tone shades to which are added white (titanium dioxide) and sometimes a bit of pink or red certified color to add a rosy tone. With careful combinations of these, professional make-up shades are manufactured.

The earth colors range from a light ochre, ochre, warm ochre, burnt sienna, red-brown, red oxide, burnt umber, to umber. Pure shades and dilutions in titanium dioxide can be seen in the RCMA Prosthetic Base Series 1 to 8, from A to C, in some 24 shades (three dilutions of each color) to show the effect of the white pigment on these basic colors. Note that judgment of color should always be made on the skin and never in the container since what appears to be the same shade may be quite different on the skin.

Foundation Shades

In the RCMA line the basic number shades are the F, KW, and KM series for Caucasoids, the KT series and Shinto for tan or yellow undertone skins, and the KN series for dark skins with gray undertones. Other special shades are sometimes named rather than numbered. The *shade* descriptions are given in the various lists for comparative purposes only because the key letter-number designations are all that are usually shown on the containers and that is generally how make-up artists delineate the shades.

In all cases in the foregoing lists, the light shades are listed first or with lower numbers, and the following ones increase the tonal value and color saturation in each particular series. Combination shade numbers are equal mixtures—for example, KW-36 is an equal compounding of KW-3 and KW-6 to produce the in-between shade.

In the basic shades, KW-1 to KW-4 show a cream to olive gradation, and KW-5 to KW-8 are a light to tan-pink for women. The men's shades, KM-1 to KM-3, are rising olive-tan tones, while the KM-4 to KM-6 are pink-tan to reddish tan and the KM-7 and KM-8 are red-brown skin tones:

WOMEN		MEN	
KW-1	Pale Cream	KM-1	Olive-Tan
KW-2	Beige	KM-2	Deeper Olive-Tan
KW-3	Warm Beige	KM-3	Dark Olive-Tan
KW-4	Olive-Beige	KM-4	Pink-Tan
KW-5	Light Pink	KM-5	Reddish Tan
KW-6	Warm Pink	KM-6	Dark Reddish Tan
KW-7	Beige-Tan-Pink	KM-7	Red-Brown
KW-8	Tan-Pink	KM-8	Deep Red-Brown

However, today's current recommendations for a basic series of foundation shades for screen and stage make-up are a combination of those mentioned here.

WOMEN		MEN	
KW-36	Light skin tone	KM-36	Lighter skin tone
KW-37	Normal or basic shade	KM-37	Normal or basic shade
KW-38	Deeper shade	KM-38	Deeper shade

For still photography and street make-up in general, the following shades are recommended for women (according to skin type):

PINK UNDERTONES		OLIVE UNDERTONES	
F-1	Very pale shade	F-2	Very light Cream shade
F-3	Pink Cream	F-4	Light Cream
KW-13	Medium (or basic shade)	KW-14	Medium (or basic shade)
KW-23	Deep	KW-24	Deep
KW-67	Pink tone	KW-34	Deeper

There is also a shade called Gena Beige that is a color sometimes used on light-blond women that have rather rosy skin undertones. This particular shade will mute the redness and even out the overall skintone.

NATURAL SUNTANS			
WARM TAN		YELLOW TAN	
KW4M2	Light	KT-1	Light yellow-tan
KW4M3	Medium	KT-2	Yellow-tan
KM-23	Deep	KT-3	Deep yellow-tan
		KT-34	Dark yellow-tan
		KT-4	Dark brown-tan (Tahitian)

Although the KT series is also suitable for warm tone oriental shades, another series, called *Shinto,* has a fully yellow undertone in four shades: Shinto I, Pale (yellow undertone); Shinto II, Medium Oriental; Shinto III, Deep Oriental; and Shinto IV, Dark Oriental.

A series for Negroid skins serves both men and women as there is no male-female undertone distinction as for Caucasoid skins, with some men being considerably lighter than some women and vice versa:

NEGRO SHADES (all with grayed undertones)			
KN-1	Very Pale Brown	KN-4	Deep Brown
KN-2	Pale Brown	KN-5	Dark Brown
KN-3	Brown		

Also

	(with gray-yellow undertones)	
KN-6	Yellowed Brown	KN-7 Deep Yellowed Brown
	(with gray-blue undertones)	
KN-8	Blued Brown	KN-9 Deep Blued Brown

while the KN series from 1 to 5 are ascending brown tones on the warm side, the KN-6 and KN-7 shades have a distinct yellowed brown tonality (suitable for some South Sea Island skin colors), while the KN-8 and KN-9 shades have a blue complement in their coloration (good for some of the people in India and Southern Asia).

In the Ben Nye line, there are the following series of shades. The L, M, N, T, and Y series are all designed for male and female Caucasoids.

NATURAL	LIGHTER VALUES	MEDIUM WARM BROWN TONES
N-1	L-1 Juvenile Female	M-1 Juvenile Male
N-2		
N-3	L-2 Light Beige	M-2 Light Suntone
N-4		
N-5	L-3 Rose Beige	M-3 Medium Tan
N-6		
	L-4 Medium Beige	M-4 Deep Suntone
	L-5 Tan-Rose	M-5 Desert Tan

NATURAL TANS	OLIVE SKIN TONES
T-1 Golden Tan	Y-3 Medium Olive
T-2 Bronze Tan	Y-5 Deep Olive Tan and Deep Olive

The 20 series includes rising tones for Asians, Latins, and Negroids:

22	Golden Beige	25	Amber Lite	28	Cinnamon
23	Fawn	26	Amber	29	Blush Sable
24	Honey	27	Coco Tan	30	Ebony

There are also some special shades:

AMERICAN INDIAN SERIES	MEXICAN SERIES: OLIVE BROWN TONES
I-1 Golden Copper	MX-1 Olive Brown
I-2 Bronze	MX-3 Ruddy Brown
I-3 Dark Bronze	

Other special shades include Black, White, Ultra Fair, Old Age, Bronzetone, and Dark Coco. Of course, as in all professional lines, the color series designations may be different from others, and individual color descriptions may vary.

Custom Color Cosmetics by William Tuttle are mostly designated by names, but some numbered shades are color copies of some of the old Max Factor and other shades that are no longer available in some areas or requested by some make-up artists:

Ivory	Natural Tan	11019 (Dark Tawny Tan)
Truly Beige	Warm Tan	
Fair and Warmer	Tan-del-Ann	Tawn-Shee
	Tawny Tan	Dark Beige
Light Peach	Fern Tan	Toasted Honey Tan
Pink 'n Pretty	Bronze Tone	
Peach	Jan Tan	Cafe Olé
Rose Medium	TNT	Rahma
Light Beige	Light-Medium Peach	Natural Tan
Light-Medium Beige		Chocolate Cream
	Deb Tan	
Medium Beige	Suntone	Tan Tone
Medium Dark Beige	Medium Beige Tan II	CTV-8W
Bronze	Shibui	BT-5
Tan	Deep Olive	BT-6
Xtra Dark Tan	Chinese I	BT-7
Hot Chocolate	Chinese II	K-1
Sumatra	Western Indian	Hi-Yeller
N-1		Ebony
		Desert Tan

As the shades of the Tuttle make-up are not arranged in series, they are all listed here but not in any particular order. However, see the charts on page 000 for general recommendations for use.

Some make-up artists wish additional mixtures and shades for special or particular purposes, and these are often made available on order from some professional make-up manufacturers. Also, there are some RCMA foundations that will *minus* or *plus* the red content of facial coloration for certain corrective or character make-ups:

Minus Red   1667, a pale ochre; 1624 series, two shades of warm ochre.
Plus Red   6205, a russet tone.

When these are added over any regular RCMA foundation, they will change the appearance of the color from less red to more red content.

There is also a series of Color Wheel or Rainbow Colors:

Red	Midnite Green	Lilac
Maroon	Opal	Violet
Orange	Turquoise	Purple
Yellow	Ultrablue	Black
Lime	Blue	Superwhite
Green	Navy Blue	Light grey
		Dark grey

And there is a nonmetallic series (a combination of pearlescence and earth colors produce these colors) that includes gold, bronze, copper, pure pearl, and silver.

Foundation Thinner
RCMA makes a foundation thinner to dilute slightly any Color Process foundation to aid in spreading it on latex prosthetics or for rapid body make-up. There is another thinner for the (AF) Appliance Foundation series.

SHADING AND COUNTERSHADING
RCMA Color Process foundation base shades are limited to facial and body shades, and all *shadowing* and *highlighting* are done with shading and countershading colors, which are based on natural *grayed* or *countergrayed* skin tones plus light-created shadows for *natural contouring* or special tones for character make-up.

SHADING COLORS

S-1	Grayed Tan	S-5	Brown
S-2	Blue Gray	S-6	Red-Brown
S-2W	Light Gray	S-7	Freckle Brown
S-13	Equal Mix of S-1 and S-3	S-8	Black-Brown
S-3	Browned Gray	S-9	(Light) Lake #1
S-14	Equal mix of S-1 and S-4	S-10	(Dark) Lake #2
S-4	Grayed Brown	S-11	(7-34 equiv.) Brown and Gray

COUNTERSHADING COLORS

CS-1	Light Skin	CS-3	Dark Skin
CS-2	Medium Skin		

BEARDCOVERS
The special problem of men's beardlines may be corrected or concealed with a spread foundation called *beardcover:*

BC-1   Pink Tone (used for black and white only)
BC-2   Orange Tone
BC-3   Orange-Tan (best for color mediums)

Ben Nye uses the following designations for countershading, shading, and beardcovers:

CREME HIGHLIGHT	CREME BROWN SHADOW		FIVE O'SHARP
Extra Lite	#40	Character (Purple-Brown)	Olive
Medium			Ruddy
Deep	#42	Medium (Brown)	
	#43	Dark (Warm Dark Brown)	
	#44	Extra Dark (Rich Dark Brown)	

In addition, a series of creme lining colors covers a number of uses, such as a Sunburn Stipple and a Beard Stipple, plus the brighter colors of Forest Green, Green, Yellow, Orange Fire Red, blood Red, Maroon, Misty Violet, Purple, Blue, Sky Blue, Blue-Gray, Gray, Black, and White. The following are some Ben Nye Pearl Sheen liners:

White	Shimmering	Copper
Green	Lilac	Mango
Turquoise	Ultra Violet	Rusty Rose
Royal Blue	Amethyst	Cabernet
Sapphire	Gold	Rose
Brown	Bronze	Charcoal
	Walnut	

William Tuttle recommends Special Hi-Lite for erasing circles under the eyes, and there is also Hi-Lite, while shading colors are Shadow I and Shadow II. For corrective contouring there is also Red-Out for erasing darkened areas or Sunburn Stipple for an outdoor look.

CHEEKCOLORS
Although cheek make-up has been made in both wet and dry forms for many years, currently the term *cheekcolor* denotes a compatible product to the foundation, while *dry blush* is applied as an additive color only after the foundation and other facial make-up has been powdered.

CHEEKCOLOR

Regular	Pink-Orange (similar to a Light Technicolor Blush)
Special	Muted shade of above (with a pinkish foundation)
Flame	Coral
Red	Stage red
Dark	Burnt Orange (similar to Dark Technicolor Blush)
Raspberry	Deep Pink (similar to old Natural Blush)
Lilac	Deep Lilac
Grape	Bright Grape
Genacolor Pink	Natural Pink (*Genacolor* is a trade name of RCMA for natural cheek make-up colors.)
Genacolor Rose	Natural Rose
Genacolor Plum	Pale Plum
CTV	Brick Red

DRY BLUSH

Plum	Peach Pearl	Brown Pearl
Red	Bronze Pearl	Golden
Mocha	Rust Pearl	Brandy
Pink Pearl	Raspberry Pearl	Blush Pink
Candy Pink	Claret	Flame
Garnet		Lilac

Cheekcolor and dry blush shades are added to, depending upon fashion dictates, but the Genacolors Pink and Rose see the most use for studio production make-ups.

Ben Nye lists the following cheekcolors:

CREME CHEEK ROUGE	DRY CHEEK ROUGE	CHEEK BLUSHERS
Red	Red	Dusty Pink
Dusty Rose	Raspberry	Nectar Peach
Raspberry	Coral	Golden
Dark Tech	Dark Tech	Amber
Coral		
Blush Coral		

William Tuttle has Blusher, Tan Rouge, CTV Rouge, Mauve Blusher, Persimmon Piquant, and 007.

LIPCOLORS

Most commercial and theatrical lipsticks contain dibromo or tetrabromo fluorescein dye pigments (different from the more inert and nonstaining lipcolor pigments) to provide a long-lasting look to the lips. These act, however, to stain the lips (due to their reaction with the pH value of the saliva) with a pinkish hue, often destroying any color match with the wardrobe that may be red, peach, or orange colored. It was also stated in Edward Sagarin's comprehensive book, *Cosmetics—Science and Technology,* in an article by Sylvia Kramer, that "A possible increase in the incidence of reactions to lipsticks was noted with the introduction of the 'indelible type' lipsticks which use higher concentrations of these dyes than had been used previously" (p. 885). Although reactions to inorganic colors seem to be practically nonexistent, it was shown in a comparison chart in Sagarin's book that 48 percent of all cosmetic allergenic reactions were due to such dyes in lipsticks (p. 881).

RCMA designs and manufactures all their lip products including lipcolors, lipliners, and lipglosses without these dye stains to eliminate these problems. Although there are many shades of lipcolors, most make-up artists carry only a basic number of shades in their kits, varying them as fashion introduces new or revives an old color direction. In addition to the pigment colors, pearls, nonmetallic golds, bronzes, silvers, and coppers are employed for new color concepts (for a complete list, see Appendix A).

EYECOLORS

Years ago, brightly colored sticks called *liners* were used for eye make-up. On the advent of black-and-white film where colors became grays, the term *eye shadow* came into use to denote the application of a deep blue-gray or a deep brown paste employed to shadow the eye areas. The change to color film and color television brought back color into the make-up artist's vocabulary and kit, and once again bright colors were introduced for eye make-up, cheek make-up, and lip make-up, so the terms *eyecolor, cheekcolor,* and *lipcolor* described the placement effect with more exactness.

Eyecolors are now made in many forms including the following:

Pressed cake type, applied with a moistened flat brush;
Pressed cake type, applied with a dry soft brush or Q-Tip;
Liquid variety, brush applied, mostly in pearled colors;
Creme-cake style that requires powdering after application;
Large pencil or even some stick forms (similar in shape to old liners) of both hard and soft varieties;
Professional 7-inch pencils in bright colors.

The most efficient type for professional use is the water-applied cake variety because it is easy to blend and is long lasting. Most of the creme-cake, large pencil, or stick types are sold commercially and are easy to self-apply, but they run or gather into the folds around the eyes and therefore require constant attention to keep the area fresh appearing, unsmeared, and matching from shot to shot for any screen make-up.

PENCILS

The 7-inch, professional-type pencil comes in a variety of hair colors and bright colors, as well as lipcolor shades. They can be sharpened with a make-up pencil sharpener or a razor blade.

HAIR COLORS	BRIGHT COLORS	LIPLINER COLORS
Black	Blue	Light Red
Midnite Brown	Navy Blue	Red-Red
Dark Brown	Green	Dark Red
Medium Brown	Kelly Green	Maroon Lake
Light Brown	Gray Green	
Frosted Brown	Silver Green	
Auburn	Lilac	
Blonde (Silver Beige)	Turquoise	
Ash	Gold	
Deep Ash	Silver	
Gray	White	
Dark Silver Gray		

Just as with lipcolors, take care that only those pencils that do not contain dye colors should be used or they will stain the lip outline pinkish.

MASCARAS

Different forms of mascaras appear from time to time, but the most common varieties are the cake, which is applied with a wetted brush, and the semiliquid, which is in a tube with a wand applicator. Although they all come in a number of shades, the black and dark brown colors are the most useful and most natural. While the wand variety may be acceptable for personal use, make-up artists more often prefer the cake type as the brushes can be easily cleaned and sterilized between uses while the wand cannot.

FALSE LASHES

As an adjunct to or often combined with mascara are false eyelashes. These may be applied either singly or in strips. Most are supplied in black or brown, but colors are available on special orders from certain manufacturers. They are best adhered with a latex-type *eyelash adhesive* for the safest and easiest method. A type of strip lash with a transparent filament base is available when eyelining is to be kept to a minimum but when some emphasis is desired in the lash area. Take care in applying the single or tuft-type lashes (singles with three or four hairs attached) with any of the plastic glues or adhesives because these may cause damage to the natural lashes to which they must be attached. (See pages 65–66 for further information on false eyelashes.)

POWDER

Years ago, paint and powder were the standbys of theatrical make-up, and in fact, early film make-up used the same system of a shade of powder for every shade of foundation color. With the advent of color film and color television, it was found that the heavily pigmented powders actually deepened the foundation shade and changed the make-up color on the face during the constant retouching required from early morning to late afternoon. For black-and-white film this change was not important since such tonal variations were not that apparent in the gray scale, but in color mediums, the effect was quite noticeable and undesirable. Therefore, a lightly pigmented *neutral* powder was first devised, and then later, a *translucent* (less filler and pigments) powder was recommended for all screen make-ups.

RCMA introduced a new concept of a transparent No-Color Powder for use with Color Process materials that did not change color on the face whatsoever, as it neither contained any pigments or fillers nor caked with constant use. As such, No-Color RCMA Powder can be applied to any shade of foundation, from the lightest to the darkest, with no color change or caking, so that only one powder need be carried in the make-up kit for general use. With RCMA Appliance Foundations a more heavily filtered (but not pigmented) powder called *PB Powder* provides a powder with more absorption power for these more heavily oiled foundations.

There are also some *overpowders* with gold and pearl pigments added, which are used for special facial and body shine effects (such as on the shoulders or cheeks). William Tuttle makes an Extra Fine Translucent Powder, while Ben Nye has Translucent, Coco Tan (for dark skins), and Special White (for clowns).

CLEANSERS

Because there are basically three types of skin conditions—namely, dry, oily, and normal—each requires a different variety of cleanser for best results. Dry skin needs a highly emollient and penetrating cleanser that has no drying aftereffect. A mineral oil-water lotion type like RCMA Deep Cleansing Lotion is recommended rather than the heavier solidified mineral oil creams.

Oily skin is best cleansed with a liquid that is non-soapy but that has a foaming action so that it penetrates the make-up. It should not leave any oily residue or feel on the skin. This type is often preferred by men, and RCMA makes one called Actor's Special, which acts as a cleanser and a skin freshener with a slight lemon odor.

Normal or combination skins (with both dry and oily patches) may use one of the water-based cleansers that can be rinsed off after use with warm water. RCMA Hydrocleanser is one of this type.

Special Cleansers

RCMA Studio Make-up Remover and Studio Brush Cleaner are essentially the same product and are excellent for removing heavy make-up applications or for cleaning brushes after use.

Klenzer, RCMA's Appliance Foundation Remover, has been especially designed for removing RCMA AF make-ups or any other type of *rubber grease* foundations.

Mascara remover pads are oil-soaked cotton pads for removing mascara or other eye make-up without irritating the eyes. It is recommended that after removing make-up products from the face with a cleanser, a wash with warm water and a neutral soap (like Neutrogena) is best to clean the skin completely.

Ben Nye makes a make-up remover and a brush cleaner, while William Tuttle produces a more extensive cleanser and skin care line with:

Cleansing Lotion (for dry and normal skin),
Freshener (for dry and normal skin),
Oily Skin Cleanser,
Oily Skin Astringent,
Moisturizer (dry and normal skin),

Skin Conditioner (very dry skin),
Facial Scrub,
Sun Screen Moisturizing Lotion.

SKIN CARE PRODUCTS

Before applying any foundation to the skin, performers with dry skin should have a coating of RCMA Prefoundation Moisture Lotion spread on the face with a dry foamed sponge. This emollient will sink into the skin and lubricate the surface of the face so that the foundation will apply more easily and smoothly. It is always best to use the same manufacturer's moisture lotion as foundation to ensure that they are compatible. Other brands may break up or clot when applied.

Oily skins often require the application of a freshener (with no more than 9 to 10 percent of alcohol) to eliminate some of the surface oil on the skin. Astringents or after-shave lotions often have up to 35 percent alcohol content and are quite strong for some skins—even though the skin is oily.

Hand and Body Lotion is recommended for use on the hands and body, if the skin is dry, prior to any make-up application. This product is especially designed for the body rather than the face area. For actresses with very dry skin, it is well to recommend proper skin care after removing make-up at the completion of the day on camera or stage by applying a coating of RCMA Overnight Skin Lotion or some similar product to combat skin dryness overnight. These are made with a high vegetable oil content and do not leave a greasy feeling on the face.

TOOLS AND EQUIPMENT

Only commercial cosmetic companies and amateurs advocate the application of any make-up product with the fingers for a beauty make-up. The professional make-up artist always employs the correct tools such as brushes, sponges, puffs, tweezers, Q-Tips, and so forth for applying make-up. Many application tools have changed over the years and, with advancing technology, will continue to do so.

Sponges

Foamed sponges are the mainstay of foundation application. However, it has come to the attention of dermatologists and skin specialists that certain *pseudocosmetic* reactions may cause skin irritations that are similar to an allergenic cosmetic reaction but not due to the cosmetics themselves. One of these is the use of foamed *rubber* or *latex* sponges for applying make-up. It has been discovered that a chemical employed in the manufacture of the foam, *mercaptobenzothiozole,* may produce a dermatitis effect on human skin. By using nonrubber sponges (such as polyurethane types), this condition or its possibility is completely avoided or alleviated. It is strongly recommended that only fine-foamed polyurethane sponges be employed for professional work and a new sponge be used on each person to ensure cleanliness and sterility.

There are also *stippling sponges* in both red rubber and foamed plastic with a variety of pores-per-inch styles that are utilized for applying various character effects. They should be cut into small pieces for facility in use.

Puffs

Three-inch, cotton-filled, velour powder puffs are best for applying powder. Note: Avoid using the so-called powder brushes as they are not only difficult to clean and sterilize, but also may streak the facial make-up.

Brushes

Brushes are made from many hair materials such as red sable (from the tails of the Kolinsky or Red Tartar marten); black sable (wood marten or stone marten); ox hair (from the ears of oxen); camel hair (from squirrels, ponies, or goats—not camels!); fitch hair (Russian fitch or skunks); badger hair (from Turkish or Russian badgers); goat hair (back and whisker hair); and many varieties of artificial hair made from plastic filaments. The best type for make-up brushes for most uses are those made from the best of the red sable hairs because they have better spring and workability (Figure 5.2).

Bristle brushes are different from hair in that hair has a single, individual natural point, while bristle has multiple natural tips or flags and also has a taper. This taper gives natural bristle brushes (such as those used for eyelash or eyebrow brushes) certain working qualities over any of the plastic bristle brushes. Bristle comes from hogs or boars, and the best comes from the back strip of older animals. Good real bristle brushes are difficult to find in today's plastic world.

Proper construction of a make-up brush from red sable hair for professional use is also important. The length-of-the-hair out of the ferrule (LOOF) must be combined by the best ferrule metal, and the hairs must be secured tightly in the ferrule. Handle length is also important so that brushes can be easily stored in a make-up kit. Although steel, copper, and aluminum are often used for making ferrules, seamless, nickleplated brass is best because the hairs can be permanently heat sealed in a material known as Nylox, which is quite insoluble in solvents such as acetone, alcohol, and so forth. As such, the hairs stand up to repeated cleanings in RCMA Studio Brush Cleaner without falling out.

The length of the hair out of the ferrule controls the flexibility, snap, and painting quality of the brush that has been dipped in a material for use. Those for oil-wax products should be of shorter hair length, while those for spreading water-based products should be somewhat longer. A slight fraction of an inch too much

FIGURE 5.2   *Make-up brushes.* TOP TO BOTTOM: *Eyebrow brush; shading and countershading brush #7, flat; lipcolor brush #4, flat; eyecolor brush #3, flat; eyelining brush #1, round.*

out of the ferrule will make a lip brush too pliant or an eyecolor brush too stiff for best application.

Brushes designed for the fine arts where artists paint with heavy oils or with watercolors are not good for applying the waxy cremes or heavily concentrated wet or dry colors of make-up materials. A brush that is about 7 inches overall is best for make-up use and fit in the kit, and a walnut-finished handle adds to the professional look of the brush.

Both round and flat brush styles are used, and the following chart provides the brush sizes employed by many make-up people. Incidentally, the higher the number, the larger the brush size in number of hairs employed in its manufacture.

SIZE		USE	BEST LOOF
#1	Round	Eyelining	1/4 inch
#3	Round	Character lining	3/8 inch
#10	Round	Adhesives, sealers, and character work	11/16 inch
#3	Flat	Eyecolor application	9/32 inch
#4	Flat	Lipcolor application	1/4 inch
#7	Flat	Shading, countershading, blending	3/8 inch
#12	Flat	Blending	9/16 inch

The #10 round brush should have an unfinished wood handle so that it can be left standing in a bottle of adhesive or sealer without danger of the handle finish being removed by the solvent during use.

A camel hair brush for applying dry cheekcolors or powder to deep areas of prosthetic appliances with hairs about 7/8 inch out of ferrule and about 5½ to 6 inches in overall length is also useful.

Bristle: Natural or Plastic

Eyelash   One row of seven tufts in a 4-inch handle is best. However, many artists cut off three rows nearest the hand-holding area to provide a smaller and more controllable applicator for mascara.

Eyebrow   Two rows with a 4-inch handle.

Hair Whitening or Wig Cleaning   Two rows in an unfinished bamboo handle.

Although artificial plastic bristle brushes (for brows and lashes) work almost as well as natural bristle brushes, the trend toward making application brushes from cheaper plastic fibers or filaments by some of the theatrical and commercial firms provides brushes that are less than suitable for fine blending or exacting professional work. There are also some foamed sponge tip applicators for eye products, but these are for personal use only and cannot be cleaned or sterilized for make-up artist service.

It is also well to avoid the large 1½- to 2-inch wide, long-handled soft camel hair brushes that are merchandized for dusting on blush or powder products. First, they are difficult to clean between uses, and next, a smaller brush for blush and a puff for powder are far better for control. Many pseudo-make-up personnel appear to be making fancy incantations and performing miracles with the flying camel hair dusters, but the overall effect in the end is minimal—often matching their talent.

Special Tools

Certain medical and dental tools have found use by make-up artists (Figure 5.3):

College Pliers   Six-inch, stainless steel, curved tip dental pliers (large tweezers) are best for handling false lashes, small prostheses, and other items.

Dental Spatulas   The stainless steel, 2½-inch blade spatula is excellent for mixing colors, stirring liquids (like stipples), taking a bit of make-up out of a container to be placed on a plastic tray or disposable butter chip for individual portions of lipcolor, and so on.

A variety of scissors are useful:

Hair Scissors   Get the best available, keep them sharp, and cut *only* hair with them. The 3-inch barber's style is most useful.

Straight Scissors   A good pair with short blades for cutting all other materials such as plastics, fabric, and so forth is best.

Curved Scissors   Surgical, stainless steel with a 1½-inch blade length are best for cutting curves on latex or plastic appliances.

Pinking Shears   A small pair of these for trimming lace on prepared hair goods should be part of a make-up artist's hair kit.

Comb   An aluminum tail-comb with widespread, unserrated teeth for hair work.

Eyecare items would include *tweezers*. A good pair of slant-cut tweezers for brow care is important. Many complicated varieties of plucking tweezers are made as well.

Although some female performers like to curl their lashes, it is not a good practice to carry or offer an eyelash curler for general use. Not only are the rubber or plastic pads on their curlers hard to clean and sterilize, but also new research has shown that some individuals are allergic to the plating on the finish of some of the curlers when they are pressed on the eye area. Also, if the pads inadvertently slip or fall off, one can clip off all the lashes at the roots. This, of course, is not a pleasant prospect, but it has happened.

CHARACTER MAKE-UP MATERIALS

Many special materials that a make-up artist must have to do various character make-ups see little or no use for straight or ordinary corrective make-ups. These are listed here in alphabetical order rather than in importance, and their use is briefly explained. More information as to specific uses will be found in the chapters on character make-up. It is not necessary to carry all these items in the kit all the time, but in most cases, a bit of each that the artist may find important should be part of the everyday make-up kit.

We reiterate the statement that these are not ordinary materials and that some performers may have allergies to some of the products or solvents. If this occurs, discontinue their use on that particular performer.

ACETONE   A highly volatile liquid that is one of the main solvents of adhesives, sealers, and prosthetic plastics. Also used for cleaning the lace portion of hair goods after use.

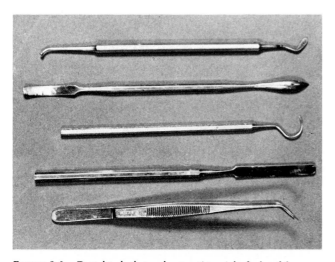

FIGURE 5.3   *Dental tools that make-up artists might find useful.* TOP TO BOTTOM: *A tool used for filling small indentations in a plaster mold; another tool for larger filling; a probe for picking up small items; a dental spatula for handling wax, and so on; and dental college pliers.*

ADHESIVE TAPE   Some new varieties of this old material are translucent and porous and have excellent adhesion. The 1-inch width is best.

ADHESIVES   This describes a very wide category of make-up materials and changes often as new items are researched and discovered. The original adhesive material for make-up was called *spirit gum,* which was nothing more than a solution of rosin in alcohol or other solvents (still made by many theatrical supply companies). Although it had fairly good adhesion, after a short while the dried gum took on an unwanted shine and, in the presence of excess perspiration, lost much of its adhesiveness and cracked off the skin. Today, although rosin is still one of the ingredients of most hair goods adhesives of that type, other materials have been added and combined in the manufacture to make an adhesive that not only has better stick but also does not have unwanted shine and, in some cases, stands up to the ravages of perspiration.

To disguise the lace portion of a hairpiece or beard, a firm bond must be made to the skin, and in most cases, old spirit gum darkened when any foundation make-up was placed on the lace or even close to it. A strong line of demarcation often occurred, necessitating removing the piece, cleaning it thoroughly, and re-adhering it. When the foundation to be applied was to be deeper than the color of the subject's skin, this presented a constant problem of maintenance. However, some of the new plastic adhesives have solved this problem (See "PMA Matte Lace Adhesive").

One way to take some of the shine out of spirit gum is to add some clay material (such as Kaolin or Attapulgus Clay) into it and stir thoroughly until the powder is well suspended in the gum. Although this will provide a gum with little shine, under some conditions it will whiten (or gray, depending on the shade of the clay) and therefore show—particularly on lace hair goods.

Matte, or nonshining spirit gum, was originally made by adding a very fine silica material to it and mechanically mixing the powder in. This produced a thicker gum and provided less shine but was not as flat as clay in matting. RCMA researched all these gums and matting materials and provides a series of matte adhesives for various uses and conditions.

RCMA MATTE ADHESIVE   This material was introduced in 1965 and is made by combining microsilica materials of various micron sizes with the resin mix under high shear. In this way, the smaller-sized silica particles hold the larger ones (the latter provide better matting) in solution and, at the same time, adds considerably more adhesion to the final mix due to the molecular structure of the finished material. RCMA Matte Adhesive can be employed both as an adhesive for hair goods or lace to the skin as well as a skin sealer with residual tack.

RCMA MATTE ADHESIVE #16   A superstick adhesive for hair goods that contains additional solids plus a plastic material to aid adhesion even when a subject perspires more than usual.

RCMA MATTE PLASTICIZED ADHESIVE   An adhesive designed for use with latex appliances that provides more all-ways stretch capability and the tackiness required for holding slush cast or foamed appliances to the skin.

RCMA SPECIAL ADHESIVE #1   Specifically designed to adhere any lace goods or hair to plastic bald caps. Normally, the bald plastic cap or front is attached to the subject with RCMA Matte Plasticized Adhesive or one of RCMA's Prosthetic Adhesives and then the hair goods are attached. Special Adhesive #1 is a very fast-drying heavily matted adhesive/sealer that will form a film that incorporates itself into the plastic bald cap while adhering the hair or lace to it. The product must be thoroughly shaken before use and applied *over* the lace as the hairpiece is held in place. Foundation make-up may be applied directly over RCMA Special Adhesive #1 without darkening the surface of the lace (also see pages 213–215).

RCMA SPECIAL ADHESIVE #2   A neoprene-based adhesive for attaching velcro or other items to appliances.

RCMA PROSTHETIC ADHESIVES   These are adhesives from the medical profession that have found various uses by make-up artists.

Prosthetic Adhesive A   This is a solvent-based, clear, quick-setting contact adhesive that has a very low irritation factor to the skin. It is less affected by perspiration and water than other adhesives, so it can be used for scenes in the rain or water. Diluted with Prosthetic Adhesive A Thinner, it can be sprayed on body surfaces to attach large appliances or hair goods (also see pages 256 and 257).

Prosthetic Adhesive B   This is a water-based, milky white (but dries clear) acrylic emulsion adhesive that sets less rapidly than Prosthetic Adhesive A and also has a low irritant factor because it does not contain any strong solvents. Although the dried film is insoluble in water, the liquid can be diluted with a few drops of a mixture of isopropyl alcohol and water.

A word of caution: Take great care in testing or using some of the new superadhesives made for industrial use as they are extremely difficult to remove from the human skin without serious damage.

Ben Nye, Stein's, and some of the other theatrical make-up manufacturers produce various adhesives of the spirit gum style but do not furnish any matte type of sealers or adhesives. They also have stage bloods, nose and scar waxes, and so on for character stage work. Also see Appendix A for full listing of these and other items of a similar nature.

Adhesive Remover   RCMA makes a cleanser that will dissolve and remove from the skin any type of plastic sealer, adhesive, scar material, or such made for make-up use, leaving the skin soft and lubricated. It is not for use in cleaning adhesives from lace goods or in cleansing the face *before* applying hair goods as it contains a moisturizing agent that prevents proper adhesion.

Alcohol   Ordinary drug store rubbing alcohol is only a 70 percent type and is not a suitable solvent for make-up use. A 99 percent isopropyl alcohol should be stocked because many adhesives and sealers have a mixture of 60 percent acetone and 40 percent isopropyl alcohol as solvents. Alcohol is also a good sterilizing agent to clean tools and table tops.

Artificial Bloods   The search for realistic-appearing human blood has led to many products from casein paints to food-colored syrups. RCMA makes the following types:

COLOR PROCESS TYPE A   A water-washable material that is very realistic in appearance and flow. It is nontoxic and does not cake and dry but appears fresh looking for some time.

COLOR PROCESS TYPE B   A resin plastic formulation designed for use where a blood effect must remain in place and not run during a long scene. It is solvent based and sets quickly, remaining shiny and fresh flowing in appearance. It is removed with RCMA Adhesive Remover.

COLOR PROCESS TYPE C   A soft creme variety serviced in a tube to make bloody areas for an effect where the blood does not have to run.

COLOR PROCESS TYPE D   A rapid-drying liquid suspension of a brownish tone employed to simulate dried blood on bandages. Not for any fresh blood effect or skin use.

Artificial Tears and Perspiration   A clear liquid that can be used to simulate tears when placed in the corner of the eye. It can also be stippled or sprayed on the skin to simulate perspiration.

Beard-setting Spray   A solvent-based artificial latex material that is used to set pre-made laid hair beards on forms. Not for facial use as a spray (also see pages 198–200, 252–253).

Gelatin Capsules These capsules are obtainable in various sizes from most drug stores. They can be filled with RCMA Color Process Type A Blood and then crushed with the fingers or in the mouth for blood flow effects. Don't prefill these for future use because the artificial bloods will soften them too much.

Hair Whiteners RCMA makes four shades of cream-style hair whiteners: HW-1, Grayed White; HW-2, Pinked White; HW-3, Ochre White; and HW-4, Yellow White. In addition, there is a Superwhite that can be used for highlighting. These are for small areas of whitening only, and full head graying or whitening should be done with sprays of liquid for this use. Nestle-LaMaur Company makes the following shades of liquid sprays in cans: White, Beige, Silver, and Gray. It also makes other hair colors such as Brown, Black, Blonde, Auburn, Light Brown, and Gold as well as Pink and Green for special effects. Very realistic hair changes can be done with these sprays (see page 258).

Latex There are many grades of latices for specific uses:

PURE GUM LATEX An unfilled pure gum rubber that air dries to a tough elastic coating. It is not suitable for casting but is excellent for making inflatable bladders (see pages 187 and 191).

CASTING LATEX A latex compound employed for slush or paint-in appliance making. Can be tinted to any shade with colors (see page 187). Casting Filler can be added to this product to control buildup density and stiffness of the finished item.

FOAM LATEX A three- or four-part combination of materials used to produce foamed latex appliances, the actual latex portion being a heavy gum mixture.

EYELASH ADHESIVE A special latex form that has excellent adhesive qualities for attaching strip and individual lashes as well as for an edge stipple for latex appliances (see pages 65 and 66).

RCMA OLD AGE STIPPLE A compound containing latex made specifically for wrinkling the skin. Not just any latex material will act in the same manner. RCMA Old Age Stipple is made in four regular shades: KW-2, KW-4, KM-2, and KN-5, and special colors are available on order (see page 109ff).

PMA Molding Material PMA (Professional Make-up Artist) materials are made by RCMA for professional use and encompass some interesting new special materials. PMA Molding Material is a paint-in type of plastic for making small or flat appliances. It dries rapidly and builds up well. It is supplied in three shades: Light (KW), Deep (KM), and Dark (KN) colors (also see pages 204–205).

Plastic Cap Material The lightly tinted variety is used for making plastic bald caps and fronts. It is also available in Clear for coating plastalene sculpture (see pages 173, 203–204).

PMA Press Molding Material A clear, heavy liquid employed to make press molded appliances (see pages 205–207).

Appliance Foundations or Prosthetic Bases RCMA makes a series of Appliance Foundations (AF series) for use with latex or plastic foamed or non-foamed appliances in a number of shades (see page 272).

Scar Material A slightly matte scar-making material with a tinge of pink color that dries on application to form very realistic incised scars. Can be removed with RCMA Adhesive Remover (see pages 235–236).

Scar- or Blister-Making Material A molding plastic type that can be formed into scar tissue or dropped on the skin to simulate second degree burns or other blister effects (see pages 234–236). Serviced in a tube for easy application.

Sealers One of the first sealers used by make-up artists was flexible collodion that was gun-cotton dissolved in ether with castor oil as a plasticizer. Employing build-ups on the face with successive layers of spirit gum, cotton batting, and a cover of collodion was the method employed by Jack Pierce to do the first Frankenstein Monster on Boris Karloff. This was a laborious method and did not guarantee a fully controllable surface or buildup. Most theatrical make-up books employed this procedure for many years, and unfortunately, some actors still thought it was the only way to change features (see story on Lon Chaney's Frankenstein Monster on pages 142–143). The resultant film over the cotton did not have much flexibility and hardened as the day went on. The surface could also be easily marred if pressed.

Next came the vinyl plastics, one of which was called *Sealskin*, a medical sealer made from a polyvinyl butyral that was too slow in drying. A similar type, but faster drying, was George Bau's Sealer #225. Unfortunately, both these sealers (and many of this type on the market) dry with a glossy, objectionably shiny surface, and foundation make-up slides off it easily.

The RCMA sealers combined various varieties of polyvinyls and added matting materials to produce sealers that had not only good adhesion and no shine but also sufficient tooth to hold make-up foundations better.

RCMA MATTE PLASTIC SEALER   Can be employed both as a surface sealer for wax buildups or as an adhesive for lifts (see pages 110 and 232). It can also be used in conjunction with other materials to cover eyebrows and seal latex pieces and wherever a film former is required.

PMA MATTE MOLDING SEALER   A sealer for the edges of appliances made from PMA Molding Material. Dries matte.

PMA MATTE LACE ADHESIVE   A sealer/adhesive that is very fast drying and suitable for use with hairpieces, blocking brows, and so forth (see page 255).

Toupee Tape   A number of varieties are available but a product called *Secure* is a colorless, two-sided, very sticky tape that is excellent for holding down toupee tops (see page 255).

Waxes

DENTAL WAXES   (See page 234.)

Black Carding Wax   Useful for blocking out teeth for a toothless effect,

Red   For simulating gum tissue,

Ivory   A hard wax that can be used to form temporary teeth.

MOLDING WAXES   Although some grades of mortician's wax may see some make-up use, old nose putty is seldom used today. A new type of microsynthetic-wax material is made by RCMA and comes in various shades and is less affected by body warmth than the mortician variety (see page 232).

RCMA PMA PLASTIC WAX MATERIAL

Light	Pale shade
Women	KW-3 color
Men	KT-3 color
Negro	KN-5 color
No-Color	A clear wax that can be tinted with RCMA Color Process foundations.
Violet	Matches RCMA Color Process Violet and is used to make raised bruises.

In addition to RCMA's regular Color Process foundations (which can be employed with foundation thinner for foamed appliances), RCMA makes an Appliance Foundation series and an Appliance Paint series especially for appliances of all kinds. See Part IV under *Coloration* for information on these. William Tuttle will make a prosthetic make-up on special order to match any listed color in his line.

DRESS AND APPEARANCE OF MAKE-UP ARTISTS

Most make-up departments do not set any dress standards, and the white wraparound barber's or lab coat is seldom seen except for make-up lab people who may use one to protect their clothing. In the main, a clean, professional look—hands washed, nails clean, breath fresh, no overuse of strong perfumes, colognes, or aftershave lotions—presents to a performer a pleasant and appealing look. The female make-up artist must be especially careful to be attractive and not over-make-up or look like a washerwoman. The female performer is apt to think, "Will I look like *that* when she finishes my make-up?" Dress on locations must often bow to the weather, but cleanliness of appearance must be paramount.

KIT MAINTENANCE

Just like the person, the make-up kit reflects the artist in many ways. Clean brushes, fresh sponges, sharpened pencils, and so forth are always necessary. Take time during lulls in work to keep the kit clean. Wipe off the tops and sides of foundation containers, and keep everything neatly labeled. Every once in a while, empty out the entire kit and clean and restock it. Do the same for the hair kit, and you'll always be proud to show off your working tools.

OBSOLETE TERMS AND ITEMS

The basic language of cosmetic products has changed over the years, and new terms are often required to bring up to date the materials or their usage and to correct old, confusing, and outmoded terms. Table 5.1 shows how the terms and products have changed.

Besides these basic materials, a number of character make-up items have changed or become obsolete as well (Table 5.2).

It is certain that some of today's make-up materials will also become obsolete as new forms and new advances in technology occur. Make-up artists should keep up with the new and avoid getting stuck in a rut of only using the old way.

Table 5.1 Comparative Make-up Descriptions

ARCHAIC (OLD STAGE TERMS)	OLD (BLACK-AND-WHITE TERMS OR PAST USE)	NEW
Base: A coat of cold cream for all skins before make-up.	Prefoundation Lotion: For dry skins. Freshener Lotion: For oily skins.	
Greasepaint: Stick or tube; heavy and greasy.	Foundation Base: Low to medium percentage in cake or liquid.	Creme Foundation: High percentage (50 percent) of pigment to vehicle.
Powder: Heavily colored shade for each greasepaint.	Neutral-Translucent: Low-pigment, lighter-weight powders.	No-Color Transparent: No pigments and fillers and never cakes.
Highlight: Yellow or white greasepaint.	Foundation: Three shades lighter than base tone.	Countershading: Tone opposite in color to natural shading.
Lowlight or Shadow: Brown or lake color liners.	Foundation: Three shades darker than base tone.	Shading: Natural gray-tan, browned gray, or grayed brown.
Under-Rouge, or Wet Rouge: Greasy, heavy, red on cheeks.	Cheek Rouge: Greasy colors to match lipstick shades.	Cheekcolor: Foundation bases in cheek shades; blend perfectly.
Cake Rouge: Heavily pigmented, usually dark shades.	Dry Rouge: Colors also matched to lip make-up (Exception: Dark rouges not used in black-and-white mediums.)	Dry Blush: Dry colors not heavily pigmented; great variety of shades.
Liners: Greasesticks in colors for eye and character lining and shadows.	Eyeshadows: Greasy stick or tube type used to deepen or shadow eyelids.	Eyecolors: Bright to muted colors used to color highlight lids or frontal bones; come in many forms.

continued

ARCHAIC (OLD STAGE TERMS)	OLD (BLACK-AND-WHITE TERMS OR PAST USE)	NEW
Lip Rouge: Greasy red colors in pot or jar.	Lipstick: The *stick* describes the shape; often contained the bromo-fluorescein dye stains.	Lipcolors: Stick or jar of nonstaining, non-color-changing, high-pigment type.
Cosmetique: A waxy black or brown substance that was heated in a spoon and applied to lashes with a toothpick.	Mascara: In cake or tube form.	

Table 5.2 Old and New Make-up Items

ITEM	REPLACEMENT
Bandoline: A hair stiffener.	Any good, firm hair-setting gel.
Blood: Old black-and-white medium blood ranged from a deep brown red to the use of chocolate syrup; later a type of red casein paint was popular.	The Color Process series by RCMA.
Brilliantine: A liquid to make the hair shine.	Hair spray or hair shine in aerosol cans.
Burnt Cork: Actually made from charred or burned cork; used for minstrel blackface.	As this phase of U.S. theater is seldom seen now, no replacement was made.
Carbon Tetrachloride: Often sold as *Energine* for a clothes cleaner; found to be toxic to humans, affecting the liver.	Modern cleaning fluids like TCTFE.

continued

Table 5.2   (continued)

ITEM	REPLACEMENT
Clown White: The old zinc oxide white grease used for clowns; still available.	RCMA Super White: Is whiter and has better coverage.
Collodion: *Flexible* formerly used as a sealer employed with cotton batting for buildups; *nonflexible* formerly used for incised scars.	RCMA Matte Plastic Sealer: A great advance over flexible material; RCMA Scar Material replaces nonflexible type.
Extender: A liquid formerly used to spread Pan-Cake make-up.	No longer required.
Fishskin: Actually the stomach lining of animals used as a film for various uses.	Matte plastic sealer, cap material, and molding plastics.
Lip Pomades: Very thinly pigmented lip make-up in light shades formerly required for early Eastman Color Negative (ECN) film (1950s) as reds filmed with a high saturation of color.	Lipgloss: In colors gives about the same effect with high shine; today's ECN uses regular lipcolor shades.
Lipstae (or -stay): A lacquer preparation for covering lipstick so the color would not come off; it dried the lips or peeled off.	Not used today.
Mortician's Wax: A too-soft product for building features; affected by normal body heat.	RCMA Molding Wax: In many shades.
Moustache Wax: Used to curl or set the ends of moustaches.	RCMA Molding Wax.
Nose Putty: A combination of clay and resin formerly used to make facial buildups; heavy in weight and difficult to cover and color.	Prosthetic appliances or RCMA Molding Waxes.

ITEM	REPLACEMENT
Scalp Masque: A dye type of liquid formerly used to stain the scalp to give the appearance of hair; had an objectionable shine and sometimes ran during the heat.	RCMA Eyecolor #1: In brown or black, applied with a sponge; use if scalp coloring must be done to hide a bald spot.
Sealers: Any sealer, plastic type, had much shine that could not be covered.	Matte Sealers: All good sealers now have matting material added to give the same reflective value as the skin.
Sponges (natural): Sea sponges used wetted to apply Pan-Cake-type products; not useful with cream products as they gum up immediately.	As most professional foundations today are of the cream-cake or stick variety, fine-foamed polyurethane sponges are best.
Spirit Gum: Old rosin-solvent type had much shine when overapplied and often lost its adhesive qualities under the effects of perspiration.	Matte Adhesives and other new Prosthetic Adhesives.
Tooth Enamel: Came in black, white, and ivory shades and was a lacquer to be painted on to cover discolored teeth or make them appear to be missing (black type); questionable to use on today's plastic-type dental work—could stain it.	Black Dental Wax or the teeth wiped dry and black pencil applied will serve as a temporary blacking out without the danger of staining.
Wig Cement: An adhesive with a clay matte material that was very difficult to remove from lace.	RCMA PMA Matte Lace Adhesive: Does the work without the mess.

Conclusion to Part I: A Lesson for Learning

WHEN DICK SMITH WAS THE HEAD OF THE MAKE-UP DE-partment for NBC-TV in New York City, his department was considered to be the standard of excellence. Some of the things he was adamant about were cleanliness of the make-up kit and studio setup, service to performers of make-up materials, adequate supplies stocked in each of the studios (NBC had some 14 or 15 locations from which television programs emanated in those days), and clean up after any make-up so the next make-up artist using the studio make-up room walked into a clean and well-kept place. In fact, the phrase "to Dick Smith the area" became a standard that meant clean everything—the kit, the setup, the materials, and the area—before you worked on any performer or left the studio where you had worked.

Following are a few comments made by Dick Smith to the Apprentice Training Program (a series of lectures and demonstrations that I designed and headed for the IATSE) on what he considered to be good and sound advice to apprentices or aspirants to professional make-up artistry. They are as valid now as they were then, and will continue to be, if one wants to be a professional.

Our union officers are trying to encourage more professional and conscientious behavior among all members. Some old-timers, myself included, have been asked to add our viewpoints. I believe that our craft would be much more respected if all our members followed the practices which I will outline, but I realize that individuals seldom change much just for the sake of their group. However, there is a chance that people will change if they realize that they could make more money. Actually, that is my main motive for doing these things: I get more jobs and better pay. It does not even take great talent. A make-up artist who acts professionally will seem more competent than a highly skilled man who goofs off. Acting professionally means the following to me:

1. When you receive an assignment, get as much information about it as possible so that you will be properly prepared (Color or B&W? Number of performers? Type of commercial or show? Type of make-up that will be appropriate? Time allowed for make-up? Any special make-up or hair problems? and so forth).

2. Keep your kit and materials clean and neat, and use sanitary procedures. A decent personal appearance is also essential. Nothing makes a worse impression on a performer than a slovenly make-up artist with a dirty kit who tops it off by offering to apply lipstick with a soiled brush and a pot of rouge that has served a thousand others. You might just as well spit on it! That couldn't make it worse. Dirty combs, sponges, puffs, and other items are almost as unpleasant to most performers, and the chances are that some will insist on putting on their own make-up rather than let you use yours. The rest will silently pray they never see you again. My system for lipstick is to scoop out a small portion from the lipstick jar (or tube) with a clean spatula, and put it on a disposable butter chip. I use a clean lipstick brush for each person, and make it obvious that I am picking up a clean one. If you want to save the lipstick for later use, you can fold the butter chip in two, and write the person's name on it with an eyebrow pencil. I carry plenty of clean sponges, puffs, brushes, and towels. The rest is just a matter of cleaning out my kit once in a while when I am waiting around on set.

3. A kit should not only be clean but also adequately stocked. Nowadays that means that everyone should carry color make-up even if he is not warned ahead of the assignment.

4. Reporting on time is obviously important. It's also a good practice to try to get your make-up done in the scheduled time. If you find that enough time hasn't been allowed for the job, tell the A.D. as soon as possible, and tell him how long you will take so that he can adjust the work plan whenever possible and explain to the director and so forth.

5. I believe one should do all the make-up he can. No make-up artist should sit around and watch a performer apply any kind of make-up, even hand-make-up, if the performer would permit him to do it. Nor should he tell any per-

former, no matter how unimportant, that he doesn't need any make-up if he has nothing else to do. Of course, this means more work, but that's what we are getting well paid to do. When a job is simple or dull, it's tempting to loaf, but if you do an easy job poorly, why should anyone believe that you could do a difficult job any better? You have to treat all work seriously if you want your work to be respected.

6. Handling performers with courtesy and understanding is obviously sensible. A happy actress may give you good publicity and get you more jobs. The opposite is also true. When your patience is at the breaking point, remember that a performer's face may literally be his or her fortune. How he or she looks is tremendously important both on the emotional and financial levels.

7. It is most important to apply make-up in a way that is suitable to the job. Find out as much as you can before you start. Go on the set and look at the background colors to see if they would affect the make-up. If the director isn't present when you start, keep the make-up natural until you are told otherwise. It is easier to add than to remove.

8. When you finish everyone, tell the A.D. so that he knows the performers are not being kept in the make-up room, and you won't get the blame if they are late getting to the set. If the crew isn't ready to start, tell the A.D. where to find you if you are going out for coffee, etc. In fact, anytime during the day that you want to leave the set, you must tell the A.D. or some other appropriate person.

9. At the earliest opportunity, check with the cameraman to see if the make-up is satisfactory to him. In some cases you should check with the director as well. It is particularly important if you haven't worked with these people before.

10. Putting the make-up on in the morning is only half the job, and often, the easiest half. Keeping it fresh throughout the day can be a lot of work. What makes it harder is that it is very tempting to take it easy and do as little retouching as you can get away with. But you are doing yourself harm if you wait for the cameraman to shout. Some of them are too occupied with other things or not observant to notice the make-up until it's half off. You should be more critical of

and more interested in your own work than anyone else. It's a good idea to do your retouching before the camera is ready to roll and the A.D. looks around and yells for make-up. You can usually tell when the time is near for a take and get your work done beforehand. Then you won't be in the position of making everyone else wait while you do a hasty, patch-job. Of course, there are some things that have to be left till the last minute, but keep them to a minimum. When there is a long break between scenes, tell the A.D. how many minutes you will need to get the cast ready so that he can warn you when it is time to do the retouching.

Retouching after lunch is important, and should be planned with the A.D. before lunch. Sometimes make-up should be done over instead of retouched. If there will not be enough time, explain the problem to the A.D. and perhaps, he will give only a half-hour lunch to certain performers and you. If he doesn't, you can't be blamed for holding up shooting after lunch.

What I have outlined above seems obviously business-like to me, and yet, I have found that many artists feel and act otherwise. Many times I have received warm praise for a simple job only because the make-up men who had preceded me had not observed these practices. They disappeared from the set, or they held up production, or they didn't get along with someone, or they failed to keep the make-up fresh. A frequent complaint of directors was that the artist did not follow his instructions. If the director says he wants only a touch of make-up, don't try to prove how great you are by using every trick you know. The chorus-girl make-up you give him will only make him wonder if you understand English. That situation is just one example of not following what is probably the most important rule: Adapt the make-up to the particular requirements. As I said before, always get as much information as you can, do your make-up accordingly, and the director will think you are tops!

I hardly think much can be added to these astute thoughts by Dick Smith. They completely express his philosophy and show the need to understand the pages in Part I of this book before attempting to do any make-up procedures.

Part
II
Beauty Make-up

Beauty is a harmonious combination of elements
that arouse aesthetic pleasure.

CHAPTER 6

Faces and Their Structures

FACIAL SHAPE THEORY FOR BLACK-AND-WHITE MEDIUMS

IN THE LATE 1950S, FEMALE BEAUTY CONCEPTS WENT through some revolutionary changes due to the rise in the use of color in films and television. For many years, the criteria of facial structures were based on the determinations and rules that were followed for black-and-white film and then television. In a system of facial shapes, with the *oval* shape (shaped like an inverted egg, that is) being considered the most ideal photographically (from a frontal standpoint), as well as the easiest to light, it was also the least common among most female faces. Faces were divided, in addition to the oval, into classes known as *round, square, oblong, triangle, inverted triangle,* and *diamond,* and many theories and concepts were employed or promoted to show the best make-up for each (Figure 6.1). Bone structure, cartilage and other tissue in addition to the placement of features served to define the general characteristics of each shape.

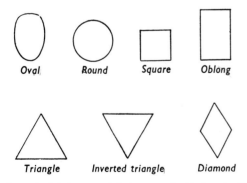

FIGURE 6.1 *Facial shapes, mainly important for black-and-white medium make-up or for hairstyling, can be divided into seven general types (as defined by Westmores in the 1930s). The appearance of the facial shape can be altered by shading or countershading to adjust to another type.* Oval: *This is the so-called ideal shape. The cheekbones are slightly wider than the forehead line, which is wider than the jawline.* Round: *Full cheeks and jawline with a rounded hairline.* Square: *Straight hairline, wide forehead and jawline, and a straight cheekline.* Oblong: *A long face with squarish but narrow construction and often a long nose.* Triangle: *Irregular hairline, narrow cheekbones, and wide jawline often with close-set eyes.* Inverted Triangle: *Wide broad forehead, wide-set eyes, narrow jawline, and high cheekbones.* Diamond: *A narrow forehead, high prominent cheekbones, and a small chin.*

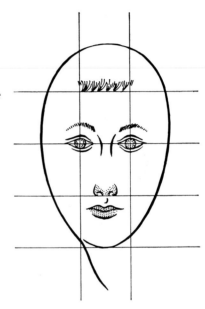

FIGURE 6.2 *Facial Balance. The face can be divided into approximately equal divisions, three of height and three of width, for the purpose of measurement for a balanced aspect.*

The *facial balance* from a frontal standpoint of the ideal, or oval, face was divided into three equal divisions of height and three of width for the purpose of measurement (Figure 6.2). Of course, the profile of a person could show another set of influences, with prominent features such as chins, noses, foreheads, or lips also being accountable in the overall aspect of beauty description, but these were seldom, if ever, equated with the frontal shapes.

In actuality, the human face is most unbalanced and asymmetrical, with the right side not matching the left side at all! So jawline shading on one side needed to be different as well as eye make-up and lip make-up to achieve a balanced result. By cutting a photo down the middle and reprinting it to show two left sides of the face and two right sides, we achieve three different faces (Figure 6.3).

FACIAL ANATOMY

The human skull plays an important role in defining the shape of the face. For the sake of anatomical bone structure, we can define the *dolichocephalic* as a long head type, the *brachycephalic* as the short head type, with the intervariety, or combination, as the *mesocephalic* type (Figure 6.4).

FIGURE 6.3   LEFT: *A normal face in a very tight close-up shows a facial imbalance and inequality of shape.* MIDDLE: *The same face with the photo printed to show two* right *sides.* RIGHT: *The same face with the photo printed to show two* left *sides.*

Faces can also be defined relative to their size, and some persons have larger features than others (regardless of facial shape). The more disproportionate the features, the less they fall into the category of the norms, and often, what some consider as a disproportionate face can be either ugly or unusually striking and, therefore, different enough to be noticed and considered by some to be beautiful. Certain actresses of fame have had large noses, jaws, or other features and have gained their individuality on such. Others have overemphasized with make-up application parts of the face such as eyes, brows, or lips and have created individual styles for themselves.

SKIN AND HAIR TYPES

For the purposes of record sheets (see pages 24–25), skin tones and colors of hair should be simply stated. For example, the following is a description of common hair shades:

Light Blonde: Platinum to light drab shades;
Golden Blonde: Bright to dull golden, yellowish shades;
Dark Blonde: Faded and deeper drab shades;
Light Warm Brown: Reddish tones in brown with blonde highlights;
Red: Carrot red to light auburn or reddened brown;
Dark Auburn: Dark reddish brown shades;
Brownette: Browntones darker than light warm brown with no blonde highlights;
Brunette: Dark brown, can have some reddish or drab brown;
Black: No brown overtones, but really black in shade;

White: White hair, can have yellow to bluish glint;
Light Gray: More white than other color, but a mixture;
Dark Gray: A dark shade of hair with some white in it.

Drab describes hair with a rather tannish color with no redness or warmth in it, while *golden* relates to a warm yellowness in shade. Warm tones can also indicate some reddish glints.

Changing a hair color with chemicals (from the natural shade) might make it appear that the skin tone changes, but neither the undertone nor the skin color changes at all. Recommendations for make-up shades for the street or still photography should be based on the *natural* hair color and undertone value and not the *created* hair shade. Therefore, a blonde normally has pink undertones, while the true brunette more likely exhibits olive ones. Redheads almost always fall into the blonde category, while dark blondes and brownettes can often have combination skins with a mixture of pink and some olive undertones or less pink or less olive than the more definite types. Most Caucasoids can be categorized into skins that are pale, light, medium, deep, and tanned (various densities), while Negroids (with gray undertones) may have a shade value of light coffee to very deep black-brown (often depending upon the intermarriage percentage with other races). Orientals, with their yellow undertones, can also run in pale, light, medium, and deep tones but exhibit no natural pink values whatsoever.

Eyes too, can show shades from very pale blue to the deepest of brown-black, with Negroids and Orientals having both black hair and brown-black eyes as predominating features, while Caucasoids can range from many shades of blue and greenish and grayish tints to pale and dark browns. See Chapter 10 for more on racial and national types.

COLOR SHAPING AND CONTOURING CONCEPTS

Today, the facial shape system is still employed by hairstylists, but make-up artists can no longer rely on the pictorial light and shade effects of black and white and must *think* in terms of color. The shading and highlighting principles of black-and-white make-up procedures have given way to judging make-up in terms of color and its relationship with the wardrobe and background. As such, color must be coordinated, compatible, and fashionable.

In the facial shape system, the round, square, and triangular faces had wide jawlines. In black-and-white mediums, the jawline could be minimized by applying a foundation that picturized three or four shades deeper

gray, thereby minimizing the actual jawline width by placing it in a shadow and making it less noticeable. Essentially, the jawline was the only area of the face in which such shading could attempt to change the facial shape (nose shading does not change the facial shape aspect). The width or height of the forehead area or the shape of the hairline was governed by hairdressing and the shape of the hair comb.

In color mediums, the effect was that the three-shades-darker-foundation principle picturized a different *color,* and the area of the jawline often looked like a beardline effect had been applied. As such, the black-and-white facial shape ideas had to be changed, and by the early 1960s much study and research had gone into the new concepts of color by both film and television make-up artists. Primarily, make-up for color should be designed to emphasize the best and most important features of the face with color, and in doing so, there appears to be less visual noticeability of the less desirable ones.

Skin tone and shades that comprise the complexion, the shape and color on the lips, and the design and color of coordinated eye make-up form a rising scale of accentuation and fundamentals for color make-up.

The eyes are usually the focal point of the face and are how individuals contact each other in conversation or communication; they thus maintain the most attention. Most beauty make-up products are made for eye emphasis. The lips are next in order with color, shine, shape, and style changing with the fashions. Both these elements can change drastically from decade to decade, but a good, clear, healthy, attractively toned complexion remains the basic element of necessity for the background of all the color splashed in various areas of the face. Combined with the hair dress and style, an overall compatible look must be achieved for each woman.

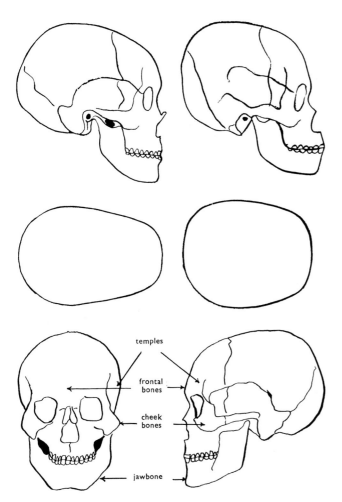

FIGURE 6.4 *Head shapes and facial bones.* TOP AND CENTER LEFT: *Dolichocephalic, or long, head type.* TOP AND CENTER RIGHT: *Brachycephalic, or short, head type.* BOTTOM: *The bone structures of the face that are most important to the make-up artist are the temples, frontal bones, cheekbones, and jawbone. Bone structure generally becomes more pronounced with age.*

COLOR PRINCIPLES

We can divide the subject of color principles for make-up into the categories of coordination, highlights, contouring, accent, framing, and response, or sensation.

Color coordination is the careful selection and matching of the various products and their relation to each other in the completed make-up; that is, the foundation to the skin color and undertone value, the pencils to the hair shade and so forth. It also embodies the coordination or matching of the eyecolors and/or lipcolors to the wardrobe colors.

Color highlights provide brightness and interest in the face. They are eyecolor, lipcolor, and cheekcolor. Any color—bright color—reflects in *color* with a stronger, brighter look than the skin tone shades of foundation, shading, or countershading. Consequently, color in make-up is neither a shadow nor does it produce one, but it color highlights any area to which it is applied. These products and their functions can also be referred to as *color-contouring* materials. *Color framing* is the term employed to describe the effects of eyecolor on the eyelid because it frames the eye in color. This *color framing* is part of a woman's overall wardrobe in a sense rather than just another make-up color. Eyecolor can be *color coordinated* into the main color of the costume, the accessories, or even as a separate *color accent* that coordinates itself into the overall *color response,* or *sensation,* of the complete view of the individual. Color accent is, therefore, an accentuation with the color highlight principle, and color response, or sensation, is the feeling that the eye receives when viewing the completed make-up harmonized with the hair dress and shade and the costume and its accessories. It is the total effect.

Such are the directions and principles we must employ for every medium where make-up is utilized today to enhance a woman's make-up potential in the fashion trends of the day. Although the directions will change from influences such as revival of old colors and designs, invention of new products, of new visual uses of color and accent and such, the basic tenets of this system of color emphasis will still be applicable to facial make-up.

COLOR CHARTS

Printed color charts of every shade of make-up are useless in the main since the printed page employs inks that, depending on the particular press run, can only be a combination of the number of rollers of basic color employed to produce the subtle shades of make-up colors of the multitude of gradients of foundation shades. The correlation of the tonal values of the make-up in the container or on the face to the printed ink on a page is often an exercise in futility for exact color matches.

Color chips in plastic have been produced to match nail and lipcolors quite well in some displays by the large commercial cosmetic firms (whose advertising budget can be a *high* percentage of the retail price of the items). Color photos can approximate closely make-up colors, but again, a slight variation in the developing process can change all the colors on a print.

One answer to color charts has been to provide extensive testers on drug and department store counters for prospective purchasers to try the products on their faces. However, the tester layout may be difficult to keep clean and maintain from customer to customer, and often looks very messy and unsanitary.

These color principles and concepts relate to make-up for women. See pages 70–73 for make-up for men.

CHAPTER 7

Straight Make-up

WOMEN

IN OUR SOCIETY, THE GREAT FEMALE BEAUTIES OF THE world can be observed in print, on stage, on the screen, and in person, advertising themselves by being seen and by the public being told over and over that these are the beautiful people. However, what is beautiful in some corners of this world is not in others, so environment and even country play a role in this judgment of beauty and fashion. We are, however, concerned herein with the role played by the make-up artist in our known societal structure and the effect of cosmetics, as we know them, on this premise. In this concept, make-up is an art form, with the make-up artist working on living flesh rather than lifeless canvas or board. As with all other art forms, the individual artist may conceptualize a woman's face in a certain aspect or trend, but this does not mean that this portrait is the *only* way to apply make-up to her to achieve that certain end. Each woman may present many looks and moods, such as natural or high fashion modes as interpreted by different make-up artists. In each aspect, nevertheless, the thought of coordination as to season, clothing, time of day, hairstyle, surroundings, and lighting effects must pervade the artistic concept of the make-up.

In general, street make-up (without doubt the greatest use of cosmetic products today) is, in a way, the most difficult medium for products as the make-up must stand inspection in sunlight or artificial light, from a distance or in close, with only the medium of air as a separation from the public's viewing position. However, improvements in film and television systems as well as the closeness of the theater-in-the-round technique require that make-up artists perform their applications in a manner that considers this intimacy of viewing and lack of monochromatic diffusion provided by black and white for all make-ups for color mediums.

Facial Coloration

The most basic premise of the application of make-up for use on the human face is to select a shade of foundation or base to establish a pleasing and natural skin tone to complement the person on which it is applied, as well as to know what shades will picturize properly in the medium (film, television, stage, or street). *Skin tone* has always been, and will continue to be, the most basic consideration of a correct make-up. In all mediums, if the skin tone is right, the other elements of make-up or wardrobe have a better chance to retain a balance. Therefore, the amount of lightness to darkness, redness to yellowness, greenness to brownness, or any other elements of color or shade that will make the skin tone appear as unobtrusively natural as possible for the person is the proper shade. Charts explaining the shades of foundation can be found on pages 30–31, as well as use charts for all mediums on page 55. Equivalency suggested charts will also be found in Appendix A.

The correct professional foundation must cover, with a thin coating, both area colorations (such as excess pink in cheeks) and blemish discolorations (freckles, spots, and so forth) and must be compounded so as not to require the use a a heavily pigmented powder to achieve coverage and to retain a natural skin *halation*. The world *halation* is very often employed by cinematographers to denote the correct reflectivity value of skin with make-up. To them, if a skin lacks halation, it has a make-up that is too flat (such as produced by most Pan-Cake make-ups). If the skin has too much halation, it is often due to the mineral oil seeping through the make-up and producing an unwanted shine. This shine must then be powdered to lessen it, but this might also flatten the make-up too much if heavy powder is used.

The foundation should be unperfumed because some people are allergic to certain fragrances, plus it should not contain any ingredients that might cause skin problems. The generalized and overused term *hypoallergenic* is considered by most cosmetic chemists as an approach to formulation rather than a rigid set of rules, so a good professional foundation need not have these words in its advertising scheme to engender the same approach. It should be noted that some oils, such as vegetable and seed oils, are more easily absorbed into the surface of the skin than mineral or many animal oils, so are apt to produce less surface shine in a foundation when used in their formulation. A waterless, vegetable oil-vegetable wax, cream-cake foundation, therefore, has been found to be easy to retouch and maintain in a fresh-appearing state for longer periods

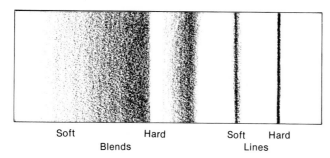

Soft   Hard   Soft   Hard
            Blends   Lines

FIGURE 7.1   *Blends: Soft and hard. Lines: Soft and hard.*

of time under production (or street, for that matter) make-up conditions.

Once the skin tone value has been established, next must be determined the *facial contours* and the *facial definitions*. The former are accomplished with blended placement of accenting or de-accenting materials and the latter with positive lines or accentuations.

Blends and Lines

The meeting point of two different shades of make-up can be blended into a soft or a hard edge or can be separated by a definite line (Figure 7.1). Blending, or fusing, the two shades at the juncture, with a sponge or a brush, of foundation with countershading, shading, or cheekcolor must be soft and subtle to produce the desired effect of a skin color area without producing a line or hard demarcation between them. These would be soft edges, while hard edges might find use in more impressionistic, fantasy, or special make-ups where the shade differences need be less subtle. Drawing lines is far simpler a procedure than learning how to make a good soft blend, and make-up requires a *soft* and not a *hard* hand for application techniques. The soft blend will have the longest demarcation scale, while the hard line will have the shortest.

Facial Contouring

A different approach and methodology than for black-and-white screen mediums is required for color mediums. Rather than *highlight* and *shadow* techniques, which served perfectly well for the gray scale, the expression *contouring* better describes this form of application as it not only can denote light and dark areas but also can imply (relative to make-up procedures) the emphasis derived from color and reflection.

Therefore, contouring may be divided into three basic categories for facial make-up purposes:

1. Natural: The use of shading and countershading colors to produce or correct certain natural or light-created shadows;
2. Color: Employing color—bright color—on the eyes, lips, cheeks or elsewhere to draw attention and to emphasize an area shape or accent.
3. Reflective: The employment of shining or light-

reflecting materials alone or in combination with colors to catch the available light and to utilize the reflections created as an accent.

Facial Definition

Although the lipline is normally placed within the area of color contour due to its prominence and size of coverage, we may also *outline* the very edge of the lipline with another color or shade for definition or accentuation. In addition, the eyes can be outlined with black, browns, blues, and so forth to delineate further the eyelashes or to define the shapes of the eyes, while the eyebrows can be colored and formed to stress the outer frame of the eye area. As well, freckles or moles (real or created) may be classified as facial definitions.

Sequence of Make-up Application

A sequence of make-up application that entails the use of all the aforementioned principles can be employed in this general order (Figure 7.2).

Foundation   Establishing a proper and correct skin tone.

Countershading   Natural contouring to lessen the shadows under the eyes or in the eyesocket near the nose.

Shading   To lessen the aspect of the frontal bone, to create seemingly more of a hollow under the cheekbone, to correct or minimize any excess weight of flesh under the chin by natural contouring.

Cheekcolor   A color contour to define the cheekbones. If a gloss or pearlescence is added to the cheeks, this area can also be reflectively contoured.

Powder   Only used to control some of the natural halation of make-up materials. In a sense, it is *anti*-reflective contouring.

Eyecolor   The use of color to frame or accent the eyelid, socket, or frontal bone. In this case, color should be construed to include only the bright colors since grays and browns more often fall into natural contours (unless the products used are compounded with reflective or shining materials). Both color contours and reflective contours are indicated here.

Lashlines   Eyelining made with water-based eyecolors or pencils may define the shape and size of the eyes and produce a facial definition.

Eyebrows   Again, a form of facial definition achieved by selecting pencils to draw the brows whose color will closely approximate the basic head-hair shade and applying them to form a specific shape or brow line.

Mascara   An additional facial definition for the eyelashes to accent their length or size.

False Lashes   Extra facial definition of the hairs of

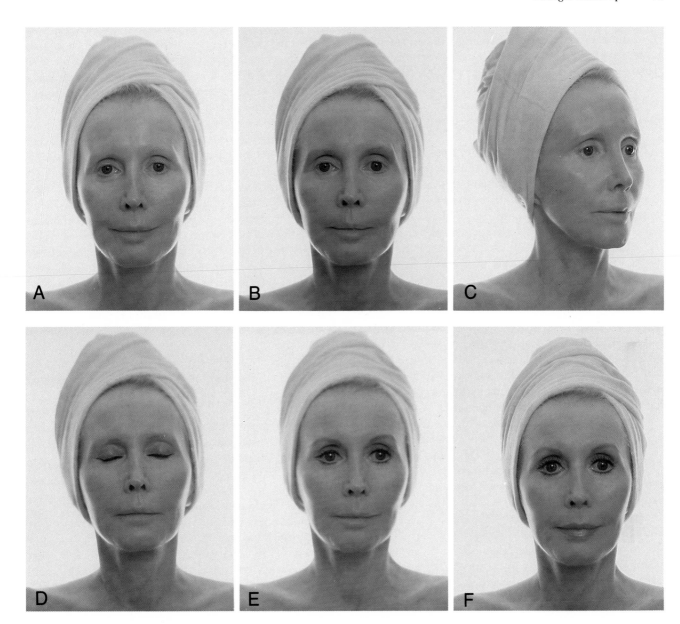

the natural lashes when they are either too sparse or for a particular accentuation of character.

Lipcolor   A color contouring of the lips, which might also include a facial definition if outlined with another color or reflectivity with gloss.

Although the assigned sequence may be varied by some make-up artists, this particular sequence provides a basic progression to a properly applied professional beauty make-up for women with all the elements considered and in an easily followed and progressive manner.

Make-up Styles

Before the application of any make-up, the style of the finished work must be determined. Such style depends greatly upon *current* and/or *lasting* fashion. More often than not, many current styles designed or promulgated by the fashion prognosticators are only partially acceptable to the general public, and the remainder may

FIGURE 7.2   *Beauty production make-up. (A) Before make-up. (B) Foundation (KW-37) applied. (C) Countershading (CS-1); shading (S-13); Cheekcolor (GC Pink). (D) Powder (No-Color); Eyecolor (Lilac); Upper Eyelines (Midnite Brown). (E) Lower Eyelines (Midnite Brown-Pencil); Brows (Silverized Beige Pencil. (F) Mascara (Midnite Brown); Lipcolor (CP-14); False Lashes (Midnite Brown singles added to outer ends of lashes).*

take some time to become generally accepted. As well, what constitutes a current high fashion make-up on a top model may look positively atrocious on someone else who may not have the same facial structure demanded by the world of fashion magazines. Many so-called new trends are experimental probes by the major commercial cosmetic firms to market research the sales possibilities of a new color or cosmetic specialty.

It should be noted that the editorial pages of many of the fashion magazines are filled with such flights of fancy and often amateur-appearing overuse of such attention catchers as bizarre eye, lip, cheek, facial, or even body make-up. Also, manufacturer credits on such editorial pages often have a direct relationship to the volume of purchased advertising space in the magazine rather than reflecting an unsolicited, truly editorial endorsement of a product or an idea. Too many women read or see some of these outlandish displays and immediately rush out looking for the "newest thing"—often a product or shade that has not even been manufactured!

For *production* make-ups for a film or television show with a story and for a woman portraying a regular person in the script, a middle-of-the-road approach must be used in make-up style. Such a make-up should be done with taste and knowledge of the fashions current for the period but, unless the character calls for it, without displaying any product to an excess; that is, we might term such make-up *beauty make-up,* one that simply enhances the more attractive parts of the face and minimizes the portions that might detract from an overall pleasant and natural look. A *high fashion make-up,* in contrast, is one in which products such as bright or glossy eyecolor, outlined or shiny lipcolor, odd shades of cheekcolor, or such are shown in unusual shapes or styles, and one's attention is drawn to the products first. This type of make-up might well delineate a person's character in the script.

Oftentimes, the fashion predictors of street make-up blurt out the news that "all skins will be porcelain" or "foundations will all be on the pink side" for a coming season. This is a misleading and unknowledgeable bit of advice for the general consumer. The more natural a foundation shade is to the skin color and undertone for street make-up, the more attractive the final aspect of the overall make-up will normally be. Weird or unmatching foundation shades are for clowns or fantasy make-ups, not for beauty make-ups.

It should be noted, however, that when applying make-up for a current interview or talk show or a quickly shown series, the make-up artist can often employ many more high fashion trends than for a woman who might portray a housewife in a filmed breakfast food commercial. One good rule to follow is that if the production has a release date of at least six months to a year away and the possibility of a re-release in the future, the use of any extremely high fashion or current make-up, wardrobe, and hairstyle might date the actress's appearance and thereby harm the audience reaction if the selected styles did not last until the film was released. One must remember that just as clothes styles may change drastically from one year to the next, so may eyelines, liplines, and such.

Producing a make-up for a period piece also requires some style adjustment. For example, a 1930s setting of a show produced in the 1980s—that is, a fifty-year difference—might, for the strict purposes of authenticity, show clothes, hair, and make-up styles of the 1930s, but what of an 1830s or 1860s setting? Here there is no living recall of the make-ups of that period,

and in fact, very little or no make-up was employed by women. However, for a 1980s production, to make the players attractive, a modern beauty style make-up can be applied. Such might be termed *make-up license,* but most actresses certainly would not like to appear without make-up, and most assuredly the public would not entirely accept the idea. Make-up for men doesn't run this risk, however, since facial hairstyles and hair dress can define a period (also see pages 129–140).

We shall, therefore, discuss a basic, production, beauty make-up as the first premise and then show how various high fashion touches can be added if required or desired. As today's recommendations for make-up may be employed on the screen, small stages, or even the street, only the small differences between mediums will be noted. A special section on stage make-up will cover larger theaters, ballet, and other stylized forms of make-up and will show the emphasis necessary or the differences required for these mediums (see pages 74–76).

Make-up Application

After the face has been properly cleansed and prepared for make-up (also see pages 34–35), a plastic or cloth covering should be put around the subject, and if the hairdresser has not already done so, any hair strands that may fall on the face should be held back with metal clips to allow full access to all facial areas.

Foundation

For street make-ups, still photos, or even screen make-up (which may be shot on a set that is not heavily lit) or for lighter all-around skin tones, it is best to *color match* the woman's natural facial coloration, taking into consideration her undertone value and natural coloration. To test a foundation shade properly, apply a small patch of the selected foundation on the cheek, the forehead, and the neck below the chinline. If it is the correct shade, the foundation will blend off and closely match the neck area, cover the cheek area, and blend well with the skin value on the forehead. With a bit of experience with this method, a make-up artist can readily prejudge the selection of the foundation in many cases. Learning to judge the correct foundation shades for the medium is one of the most important steps in learning make-up artistry. An improper selection can spoil the entire concept, and all the other elements will be unable to overcome the effect of a skin mismatch or improper shade of skin tone.

If the make-up is to be one for use on a normally lit stage, film, or television set, a bit more *warmth* of coloration is necessary in the foundation due to the intensity of such lighting that will have a propensity to minimize the color saturation of the foundation. In these instances, the correct shade is applied for the medium and the lighting rather than exact color matching to the skin. Nevertheless, the make-up artist must take into consideration the skin's undertone value and employ foundations that will appear natural to that person in the finished make-up. For example, the warm pink tones of KW-37 (the normal production make-up color for a woman with blonde to reddish natural hair) sometimes will look overly pinkish in tone on a brunette with definite olive skin tones. In the case of the latter, a KW-4 might be a better choice of foundation color.

The *undertone values* are best observed when the woman has no make-up on the face so one can see the natural colorations in the cheek, forehead, and neck areas. Of course, combinations are quite common, so the make-up artist must select the foundation shade closest to the skin color and approximate undertone in such cases. Simplifying the color of the skin to *light, medium,* and *deep,* plus *tanned* (for Caucasoids), Table 7.1 gives most women's make-up recommendations: The shades referred to are RCMA colors with equivalents in Appendix A.

The shades of foundation for Negroid skins from KN-1 through KN-9 increase in deepness and vary in undertone. Normally, one foundation number *lighter* than the actual skin color (by the usual forehead, cheek, and neck test) will be found best under most circumstances for dark-skinned people.

William Tuttle recommends the following chart:

NORMAL CAUCASIAN	TAN EFFECT
Medium Beige	Suntone
Deep Olive	Natural Tan
Medium-Dark Beige	Tawny Tan
	Tan-del-Ann

For dark skins, both women and men, all colors between Toasted Honey through Ebony can be used.

Ben Nye shades for screen make-up include N-3, Deep Olive, N-6, and T-1. Although Max Factor does not produce any more special screen make-up shades, the current commercially available Pan-Stik shades are:

Olive	Golden Tan
Deep Olive	Fair
Natural Tan	Medium

The Pan-Cake shades are:

Cream 1	Tan 1
Cream 2	Tan 2
Amber 1	Tan Rose
Amber 2	Natural 1
Amber Rose	Natural 2
	Natural Rose

Some lighter shades of foundation for Caucasoids or Orientals might include (in the RCMA line):

Ivory   A very light cream,
F-1   Very light pinked shade,
F-2   Pale cream.
Gena Beige   A light beige.

Darker shades are:

KW-34   More warmth than KW-24;
KM-2, KM-23, and KM-3   These three are deeper, rising tones in the men's series that have more warmth than the KT colors.

The foundation is a base coat for all other colors and should be applied *very* thinly and evenly with a dry polyurethane sponge. Since the sponge does not tug, massage, or distort the skin, it makes little or no difference whether one starts at the top of the face and works down or vice-versa. Apply carefully, with one stroke overlapping the other to produce a smooth, even coloration over the entire face and neck. Professional make-ups today do not require the same heavy coat of greasy stick make-up that was formerly called "theatrical" or the soft paint of early screen make-up.

Table 7.1   Foundation Recommendations for Women

RACIAL TYPE	SKIN UNDERTONE	LIGHT ON SET	SKIN COLOR		
			LIGHT	MEDIUM	DEEP
Caucasoid	Pink	Under normal	F-3	KW-13	KW-23
		Normal production	KW-36	KW-37	KW-38
	Olive	Under normal	F-4	KW-14	KW-24
		Normal production	KW-14	KW-4	KW-4M2
	Tanned	Under normal	KW-4M2	KW-4M3	KT-1
		Normal production	KT-1	KT-2	KT-3
Oriental	Yellowed	Under normal	Shinto I KW-14	Shinto II KW-24	Shinto III KT-1
		Normal production	Shinto II KW-4	Shinto III KT-2	Shinto IV KT-3
Negroid	Grayed	All	KN-1 through KN-9		

Unless for very large casts where only some color on the face is required, make-up artists seldom employ water-applied cake make-up for the screen. Such gives a very flat and unnatural appearance to the skin that can be masklike and unflattering. Creme-based products have almost totally replaced the Pan-Cake varieties just as they replaced the old tube greasepaint.

When the skin tone is being matched for lighter make-ups, the foundation is normally just blended off on the neckline, and only the blemishes are retouched with a bit of foundation on a brush. However, for production make-ups where the foundation color is higher or warmer in value than the skin, it must be carried to slightly below the wardrobe line at the neckline and must include an application to the shoulders, arms, and hands. One caution is not to wet the foamed poly sponges because this disturbs the oil-wax balance of the foundation and may streak the make-up.

Never apply a thick coating of a lighter color foundation to achieve the desired shade, but use a thin, blended coating of a deeper shade (next on the chart). The end result must appear as a color of skin rather than a coat of make-up—not greasy or spotty—and require little or, in some cases of dry skin, no powdering.

Take care to always blend the foundation up to the hairline and to get a coating just under the chin and nose. Folding the sponge will allow easier application in the eyesocket and around the eyes, and don't forget the ears if they are not to be covered by the hairdo. Before going to the next step of application, recheck the foundation and cover any blemishes with a little extra color on a flat #7 brush. Sometimes a mole or skin blemish requires a color correction as well as covering, so RCMA CS-1, BC-3, or even the 1624 series may be mixed with a bit of foundation to get a complete smoothness and match of foundation color.

Countershading and Shading

The natural shadows on the face, whether of actual skin tone or created by the angle of lights, all have a *grayed* tonality rather than just a deeper tone of the same coloration as the rest of the face. As such, all shading colors employed for natural contours have a degree of gray in their formulation, while the countershading colors are made to have an anti-gray effect.

The best place to judge the natural shading color of the face is to look at the tonal value of the color in the eyesocket area (Figure 7.3). Here the color can vary from a pale blue-gray in light, pink undertone skins through gray-tans, tan-grays and browned gray up to deep grayed browns for deep, olive skin tones for Caucasoids. Orientals will more likely than not have grayed brown tones, while Negroids will have more blackness in the gray of the shadows.

In addition, the shading caused by the direction of light falling on the face from natural or artificial sources

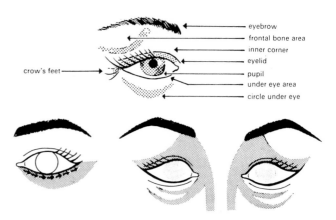

FIGURE 7.3   *The eyes.*

creates a natural grayed shading effect. For example, if one raises his or her head up to a light, the natural shadow area above and below the eye is lessened because it is, as cameraman say, "washed out with light." However, if one lowers the head, this eye area deepens in shadow value due to the lack of light on it. As such, there are only limited areas of the face to which shading color can be added or countershading applied, which will appear natural in make-up for color mediums.

Countershading or *countergraying,* employing the rules of natural contouring, can serve a very useful and corrective purpose in the eye socket area (Figure 7.4). As we have seen, the natural shading colors in this facial area can vary from a pale gray to a grayed tan to a browned gray, a grayed-brown, or a red-brown or a black-brown depending upon the race, undertone value, and age of the woman. Therefore, the purpose of countershading is to color correct these natural skin colors and also to protect the eyesocket area from light-created shadows as well. In the early days of black-and-white television, with the old iconoscope cameras and oftentimes banks of overhead lights only, deep and uncorrectable light-created shadows were formed around the eyes (Figures 7.5, 7.6, 7.7). Make-up artists tried white, but this had a habit of blooming (an overbrightening of the picture area) or worse if there was any bluish cast in the white, even darkening the area, so make-up artists experimented with pale shades of foundations to correct what they could. Even painting out (with white or the light foundation colors)

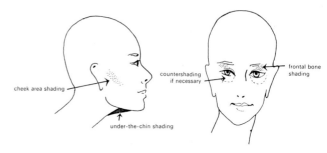

FIGURE 7.4   **Natural contouring areas.**

FIGURE 7.5 *A photo taken of the monitor during an early (1949) television show. It defines the shadow emphasis of the lighting system and television camera work of the day.*

FIGURE 7.6 *A still photo taken of a 1948 TV show on set highlights the inherent contrast of black-and-white televisison of the day. Note that the eye, nasal, and chin shadows are far less in a still photo than on the television screen.*

every wrinkle and fold in the face was tried, but more likely than not, it was only effective if the face stayed in one position frontally (like the old way of doing newscasts) in one specific lighting. Once the subject turned from side to side or moved from that lighting, the overemphasis of the highlighting looked worse as it drew attention to the lightening. When color mediums came into use, it was found that the eye area not only often required lightening but also color correction of the *shade* or *tone* of the natural color around the eye. As such, countershading or corrective colors were found to be best. The most useful RCMA colors are:

CS-1   A very highly-pigmented light color for use on women with pink undertones or light olive skin;

CS-2   The same as above but for those with medium to deep Caucasoid skins or for most Orientals;

CS-3   For tanned skins and for the lighter shades of dark (Negroid or others) skins.

For those with some redness in the eyesocket area, the corrective colors are as follows:

1667   A minus-red,
1624   A warmer ochre for deep color skins,
1624-W   Same but for lighter skins.

Countershading is blended carefully under the eye, right up to the lashline and off the cheek portion of the face so that no line of demarcation shows. As the foundation will cover much of the natural coloration under the eye, the countershading is mainly used as a color corrective so should not be heavily applied or it will become more obvious than useful.

Blend off the countershading in an outward and

FIGURE 7.7 *Make-up for black-and-white mediums.* LEFT: *Without make-up.* MIDDLE: *With a straight make-up in the same lighting.* RIGHT: *Studio portrait with the same straight make-up.*

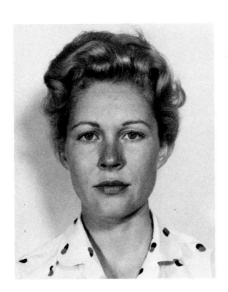  

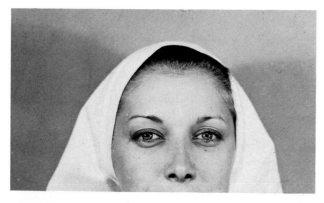

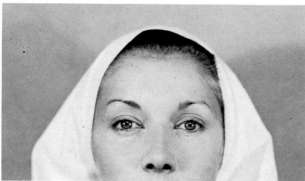

FIGURE 7.8   *Countershading.* TOP: *The area under the eyes has natural shadows.* BOTTOM: *This area is lightened and color corrected to compensate for additional shadows cast by overhead lighting systems as well as any natural discoloration.*

upward direction with a #7F brush to lift the corner of the eyesocket area as well as to accent the eyecolor and eyelining that will follow. Avoid using one manufacturer's foundation and another's countershading to serve this purpose as the incompatibility of the formulations may not be miscible in this soft tissue area. Don't use the blood albumin type of skin tightener before foundation or countershading because they are not compatible.

Again, don't attempt to paint out every line and shadow on the face with countershading. If the foundation is properly pigmented and blended, the smoothness of the overall effect should properly answer the optimum coverage required and allowable for the skin tone. Countershading only aids in the color correction of the eye area and should never appear as a puffy, highlighted under-eye area (Figure 7.8).

Another use of countershading along with shading colors is to correct the effect of a wide or a narrow nasal bridge that makes the eyes appear to be set far apart or close together respectively. For eyes appearing to be far apart, the nasal bridge and inner corners of the eye socket can be shaded. This gives the effect of lessening the space between the brows, thereby making the eyes appear to be closer together. For eyes set closely together, countershade the inner corners of the eye socket and extend the countershading slightly over the nasal bridge to accomplish the opposite of the

shading, making the eyes appear spaced wider apart (Figure 7.9).

One final note on the technique of lightening under the eyes: Take care not to overemphasize the eyebag, or flesh between a dark circle under the eye and the lower lashline, or one is present. As a person ages, this eyebag may become more pronounced, so the make-up artist should countershade only the dark portion of the circle and not the whole area (which includes the raised eyebag). In the same vein, do not attempt to shade the eyebag as this will simply draw more attention to it or shadow under the eye.

When the foundation, countershading, and shading colors cease to be effective in the eye area to such an extent that they are unable to combat the appearance of the fleshy area, cosmetic surgery should be sincerely recommended to performers. Surgery can take the problem far beyond the ken of any make-up miracle, and many performers today take advantage of the cosmetic surgeon's skill (see Figure 8.4).

In addition to eye spacing, there are three basic areas in which shading colors can be applied for color mediums without appearing to be incorrect or overdone (if properly blended with the correct intensity). The first is on the frontal bones above the eyes (Figure 7.10). Here, by matching the shadow tonality found in the eyesocket, shading can be applied with a #7 brush or a folded sponge to minimize the apparent overhang of the bone and flesh structure without drawing attention to it as a bright color might do. Next, to lift or raise the aspect of the cheekbones, most women, even those with rather fat faces, have a slight depression *under* the cheekbone. This small area can be carefully shaded. Since the cheekcolor is to be applied later to the cheekbones, the concert of these two applications accomplishes these purposes together. Finally, the fleshy area that some people have *under* the chin can be minimized by applying a triangle of shadowing color with the edges well blended. This shading should never extend over the jawlines but should be

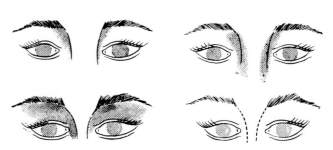

FIGURE 7.9   *Correcting and adjusting eye spacing.* TOP LEFT: *If the area between the brows is too wide, it can be made to appear narrower by shading the inside area of the eye sockets near the corners and extending the shadow slightly over the edges,* TOP RIGHT. BOTTOM LEFT: *If the area between the brows is too narrow, making the eyes appear to be close-set, it can be made to appear wider by countershading the eye sockets near the corners,* BOTTOM RIGHT.

FIGURE 7.10 *Shading on the frontal bones.* LEFT: *Frontal bone area (before).* RIGHT: *Frontal bone shading (after).*

restricted to solely under the chin where a light shadow is often created under normal overhead lighting, natural or artificial, set or stage conditions. As such, the shading colors, selected to be compatible in tonal value with the natural shading colors of the face, can create an appearance of lessening the emphasis created by natural bone and flesh with natural contouring (Figure 7.4).

Frontal bone shading depends upon the degree of prominence of this bone (which can show up more apparently if the brows are plucked too high). Also the proper selection of the color value of the shading depends on the skin and its undertone. Normally, any foundation with a 3 in it (such as F-3, KW-13, KW-23, etc.) will take an S-13 or S-3, while foundations with a 4 (such as F-4, KW-14, KW-24, etc.) will take an S-14 or S-4 (Figure 7.11). Careful selection can be made by viewing the natural shadow tonality found in the inner corners of the eyes and matching

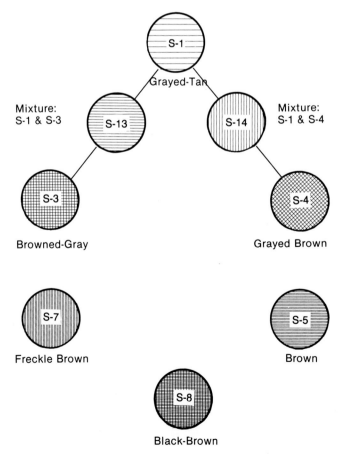

FIGURE 7.11 *Chart for natural contouring for shading colors.*

the gray-brown or brown-gray quality of such. With fashion models, or others with younger faces, this shading area is sometimes replaced with a color highlight or even a reflective one to draw attention to this part of the face. Depending upon the fashion, the structure of a woman's face, and her age level, great care should be taken with the overemphasis of this frontal bone highlighting as this principle is also used to produce an age character make-up (see page 102). Also remember that shading color *may* be used on the frontal bones, but it is not necessary to use it there unless required or desired for the effect.

The application of the shading colors under the cheekbone and under the chin are best done with the foamed poly sponge because the edges must be carefully blended. Especially in the under-chin area, the shading should be confined to the fatty tissue only and not extended too far down the neck or over the jawline.

In black-and-white mediums, jawline shading can be employed to minimize a frontal effect of a wide jaw, but again, care should be taken as to the amount of gray tonality that will be correct for the job. It should be noted that proper lighting can aid such black-and-white corrections (Figure 7.12).

Also in black-and-white mediums, and often on stage, some shading can be done on the nose (Figure 7.13). However, in color film and television, such shading is apt to show as a dark color when the light hits it during a turn of the head. Sometimes S-1 can be used for subtle shading in the nose area, but it should be fully tested on camera before use.

Cheekcolor
The application of color on the cheeks to simulate the natural glow of youth has been employed in professional make-up for many years. Early pot rouge was a thick mixture of red colors in petrolatum and waxes and was used extensively on the stage for both cheek and lip make-up. Later, a pressed cake of dry color made with talc and other fillers and binders was introduced and was called *dry rouge,* or as the Miner Company referred to it, *dry blush.*

In the late 1930s, commercial cosmetic companies coordinated lip and cheek make-up by introducing matching shades of lipstick, dry rouge, and *wet,* or *cream,* rouge. Although the early sales techniques involved the selling of these matched shades, with the suggestion that they be applied to the cheekbone in a circular, an oval, or triangular design, either near to or away from the nose, or high or low on the cheek area so as to define a facial shape, this technique gave way to the very subtle orange-pink Light Technicolor (Max Factor) dry rouge for use with their Pan-Cake make-up and a cream variety for use with their Pan-Stiks. As this pale shade did not register at all on any black-and-white medium, it only served to make the subject appear more attractive and natural in life (rather

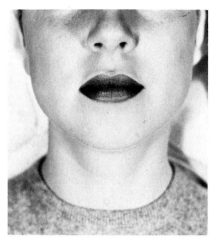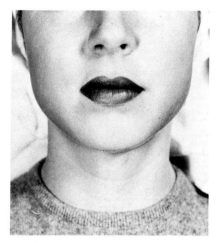

FIGURE 7.12   *Jawline shading for black-and-white mediums.* LEFT: *Before shading.* MIDDLE: *Jawline shaded, frontal view.* RIGHT: *Jawline shaded, profile view.*

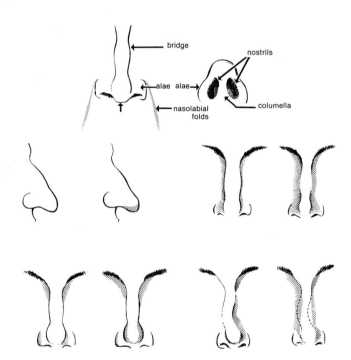

FIGURE 7.13   *Nose shading for black-and white or the stage.* TOP: *The parts of the nose.* CENTER LEFT: *The end or tip of the nose is shaded if the nose is too long or droooping.* CENTER RIGHT: *A light tone down the bridge and shadow on the sides to form a straight wall will make a wide nose appear narrower.* BOTTOM LEFT: *If the end of the nose is too wide or heavy, it should be shaded, keeping the shadow away from the sides of the nose.* BOTTOM RIGHT: *Shadow and highlight can be used to straighten the appearance of noses that are broken or misshapen.*

than on film or television). It was a psychological approach to make-up application rather than an effect required for the screen.

However, for early Technicolor films and color television, this pale orange-pink tone picturized quite naturally due to the red sensitivity in both mediums during the 1940s for film, and later with the introduction of color television. Of course, fashion took this up, and so for some years we had only a faint glow to the cheeks for street make-up.

Fashion, however, thrives on change, so commercial cosmetic manufacturers rediscovered *blush* and started to produce dry cheek make-up in many tonalities during the late 1950s. It came back as a product that could be applied with a large camel hair brush or a wet, natural sponge for use with Pan-Cake make-ups or for the commercial water-based, liquid make-ups.

When RCMA Color Process make-up was designed in the early 1960s, their product was given an original name of *cheekcolor* to describe the new concepts of facial contouring that came about with the introduction of color film and television. Made in the same basic vegetable oil and wax base as the foundation shades, it blended with them perfectly and accepted powdering at the same level and with the same degree (which was not always true of the stickier mineral oil or petrolatum-based wet rouge types). In essence, cheekcolor provides a color contour to the cheekbone area of the face, giving in effect, the appearance of higher cheekbones by the color highlight or accent principle.

Application can be made with any shade of cheekcolor with a dry polyurethane sponge, blending the material in an upward, outward direction on the height of the cheekbones. Avoid carrying the color down onto the side of the face (where the shading or natural contouring is placed to support the cheekcolor). At one period in the 1970s, a rather brick-red shade of dry

cake cheek make-up was often seen on models and was placed *under* the cheekbone. However, as it was placed where the natural contouring *should* have been applied, this make-up often appeared as a smudge of incorrect color or as if the person had rubbed too closely to a red brick wall! Although it did, due to its deep color, have a tendency to appear as shading, its actual color was reddish and, therefore, *bright,* so it highlighted rather than shadowed by gaining too much attention in the wrong place.

Always blend the edges of a cheekcolor application

to a soft margin with the sponge, and if you wish to eliminate some of the color applied to soften further the effect, go over the cheekcolor lightly with a little foundation. After application, and when the entire make-up is completed, rejudge the cheekcolor to see if it is bright enough and of the correct tonality to fit the rest of the make-up. Then, if necessary, add to it or make adjustments. To be sure, fashions in cheekcolor will change as this is a prime product area for the commercial companies. However, a basic professional beauty make-up application of cheekcolor should be subtle and not too high fashion in concept or appearance.

In some fashion directions, a gloss on the cheekbone is also very effective and, if desired, can be done with RCMA No-Color Lipgloss by applying it over the finished make-up with the poly sponge in dabbing motions to control its application. This will produce a very reflective contour in addition to the color contour of the cheeks. Pearlized or golden effects of reflective contouring can also be produced with RCMA Pure Pearl or Gold Lipcolor in the same manner. This produces a softer glow than the hard reflectivity of a lipgloss.

Other effects of color contouring may also be seen in fashion make-ups where cheekcolor is lightly applied to the chin, forehead, earlobes, nose, and so forth to break up the overall flatness of the foundation shade. Women with oily skins should avoid this latter effect as well as the glossing because they have an excess of natural shine on their faces. Everyone should always take care not to overdo any such effect since sometimes too much attention destroys the overall concept of beauty. In all cases, these high fashion ideas should not be put into effect until the basic sequence of make-up application has been done and evaluated.

A note of caution: Don't use a dry cheek make-up material over a cream-base foundation unless it has been thoroughly powdered. Even then, color additions or corrections are best done with the creme variety of cheekcolor. The basic natural colors are RCMA Genacolor Pink and Genacolor Rose, while Genacolor Plum and Flame are also widely used for films and television.

Powder

Due to the relatively greaseless surface of professional creme foundations, a heavy powder is not required. RCMA No-Color Powder with no fillers or pigments will normally be sufficient. Apply with a puff by sprinkling the powder out of the sifter jar on to the puff and then work the powder into the puff. Application can then be made by *pressing* the loaded puff onto the face. In this manner just sufficient powder to dull slightly the small degree of excess halation of the foundation will perform the job required. Don't overpowder or overload the puff, and never rub the powder on the face as this is not necessary.

With the puff folded in half, powder side out, start applying the powder under the eyes, then on the upper eyelids, smoothing out the foundation or countershading with a #7F or #12F brush if necessary, and then continue around the eyesocket and nasal area. Then open the puff and continue, finishing the forehead, chin, cheeks, and neck. Just one loading of the puff with No-Color Powder should be sufficient to do the entire face and neck. When retouching the foundation or eye area during the day, again just *press* the powder into any shiny area.

Disposable cotton-type balls should not be employed as applicators for powder because they often leave fine threads of fuzz on the surface of the make-up. Dusting brushes should not be used either as they afford little or no control, flicking the powder everywhere. A 2-ounce sifter-type container is best for the make-up kit. It can be refilled from a supply-size container when required.

Eyecolor

The most effective and long-lasting variety of eyecolor for professional use is the cake form that is applied with a moistened #3 flat sable brush. With more or less water, a thin to a thicker application can be made and blended on the eyelid. It can also be used to line the eyes when a #1 round sable brush is used. Black, Midnite Brown, Midnite Blue, and Charcoal shades are particularly effective for drawing the upper lashline (or lower for high fashion effects) as they have less tendency to melt or smear under lights and the eye make-up will stay fresher appearing for extended periods of time.

For eyecolor on the eyelid, coordinate the shade with the main color of the wardrobe or the secondary or accessory colors. It can also be used as a separate fashion accent when the wardrobe colors are black, white, beiges to browns, grays, yellows, pinks, or shades of red.

To apply eyecolor, blend it with the brush on the eyelid area only, with the strongest concentration of color near the lashline (Figure 7.14). Never blend it over the frontal bone where the shading may have been placed (except for high fashion effects), and carry the color in a slightly outward direction at the corners of the eyes to accent the lashlines that will follow. Also,

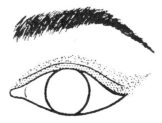

FIGURE 7.14 *Eyecolor applied on the lid area and extended slightly at the corner.*

don't blend the eyecolor down over the area where the countershading was extended at the outer corners of the eyes, but follow the upward swing created by the countershading.

RCMA calls flat-effect, water-applied eyecolors, Eyecolor #1, while the pearlized or gold effect eyecolors are named Eyecolor #3 (applied wet or dry). The latter are more useful when more reflectivity is desired, and they can be applied either with a wet or a dry brush. The wet brush will create a deeper tone than the dry brush when used with RCMA Eyecolor #3.

Another variety of eyecolor that is also a cake variety is RCMA Eyecolor #2, a dry-applied material. Some artists use a #3F sable brush while others prefer Q-Tips for application. The larger the brush, the more the color is apt to flick off to unwanted areas. This type of eyecolor can be applied alone for a delicate filmy coating of color or in combination *over* RCMA Eyecolor #1, used to heighten the density or depth of a color. For example, if the blue shade of Eyecolor #1 is first applied and then a coating of blue Eyecolor #2, a denser, more intense blue than just a heavy coating of one or the other is obtainable. Ben Nye designates these as Cake Eyeliner and Pressed Eye Shadow, and the listing of shades can be found in Appendix A along with William Tuttle's colors.

Some artists also use colored wooden pencils for the eyecolor effect. However, depending upon the hardness of the material, they are either difficult to blend (if hard) or they smudge and run under the lights (if soft). There are some liquid varieties of eyecolor (in highly pearlized colors usually) that are seen from time to time, but they normally will give a definite or hard color to the eyelid and are often difficult to blend (see page 271 for various eyecolor shades.)

Lashlines

In present day make-up, outlining the eyes comes to us from the theater where lashlines have been employed for many years to accent and delineate the eyes due to the special problems created by stage lighting. Historically, lining the eyes goes back to the ancient Egyptians and Mesopotamians (or even beyond), and eyes have gone through the years with more or less accentuation and change of shape. More natural looking eyelining procedures were perfected for screen make-ups, and finally, about 1950, fashion took them under their wing with so-called "doe eyes" for street make-up. Since then, eyelining make-up has varied from little to more and probably will continue to do so.

The lashline or eyeline for professional application serves to emphasize the base of the lashes by outlining them. Therefore, colors of eyelining that are natural to the normal dark colors of the eyelashes are best. The most acceptable for all mediums is a midnight or very dark brown shade or, if the hair is very dark, a

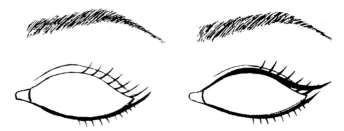

FIGURE 7.15 *Lashlining.* LEFT: *The lower lashline.* RIGHT: *The upper lashline added.*

black for the upper lashline and a softer medium to dark brown for the lower lashline. There are also some creme-style Eyecolors (similar to the old liners) which have highly pigmented bases and which come in a variety of colors that some make-up artists prefer to use. Generally they have to be powdered with a No-Color type powder after application to better set them on the soft over-eye area.

In order to create as little shadow under the eye as possible, many make-up artists don't use mascara on the lower lashes and only use a soft, blended lining with a dark brown pencil near the lower lash-lines. The upper lashlines can be accented a bit more, and so the line will remain in place for a long period of time, an Eyecolor #1 is used and applied with a wet brush. The upper line can be very thin or slightly heavier without it being at all apparent for all mediums (Figure 7.15).

To line the lower lashes with this method, hold a well-sharpened pencil (the 7-inch professional type) to the side, almost parallel to the front of the face, and lay on a thin coat of color close to the lower lashes. Never press the point of the pencil directly on the lid, but sketch the line on. Start the line where the lashes begin to grow (which is often not as far as the inner corner of the eye), and work outward toward the outer corner of the eye. Extend this line only slightly beyond the outer corner in an upward, outward sweep, following the direction of the curve of the upper lashes. This extension is actually a tangent of the curve of the lower lashline and serves as a guide line to draw the extension of the upper lashline. Cake Eyecolor #1 can also be used if a more definite eyelining is desired under the eye.

The upper lashline should be started, again, at the inner corner of the eye where the growth of the lashes starts and should continue to the outer corner where the guide line of the lower lashline is almost met. The brush can then be turned around to draw the extension line back to meet the other line. In this way the curve of the lower lashline is not dropped at the outer corners, and a separation of about 1/16 inch between the upper and the lower lashlines is kept. The ends of the upper lashline should not be abrupt or squared but finished in a needle-point effect. Of course, the degree of extension and the separation, amount, and direction of lashlines will vary with fashion over the years, but this is just a basic method of application that works well for most production make-ups.

Bright colors in pencils or eyecolors are used for eyelining only when fantasy, special, or some form of fashion effects are required. Otherwise, shades of lining that are natural to the lash colors (except of course if the lashes are very blonde) are best.

Eyebrows

The eyebrows are a very definitive part of the face, and actresses often have strong personal preferences about their eyebrows. Fashion magazines are also apt to play with the shape, size, density, and methods of application to a great degree. However, the best brow line is one that is unobtrusive to the overall make-up and that complements the face. Select hair color pencils that are lighter than the color of the head hair for the eyebrows. For example, blondes should use ash, deep ash, blonde (silverized beige), or light brown pencils. Gold can also be used for some hair colors of blonde. For brownettes with reddish highlights, a medium brown is a good color, while brunettes require a dark or a midnight brown. Use black only for those people who have jet black hair. Dark silver gray, gray, and silver gray can be used for women with gray to white hair.

Often, two colors of pencils will be found better to draw the brows. Start coloring the brow hair with a pencil that has the cast and color of the head hair, then finish the outer corners of the brow line (the part where the pencil is more apt to be needed on the skin because of lack of hair in this area) with fine, hairlike strokes of a darker shade pencil. For example, if a woman has a reddish auburn color of hair, use the auburn pencil to color her natural brow hair, and finish the outer ends with a light, medium, or dark brown pencil, according to the tonal value needed to make the brows appear naturally colored on this portion.

Following is a list of appropriate pencil colors to correspond with hair colors:

Black   Black hair only;
Midnite Brown   Brown with black overtone;
Dark Brown   Brown with warmer tonality;
Medium Brown   Brown with more redness in its color;
Light Brown   Brown with a yellowish cast;
Frosted Brown   A good shade for brunettes with no red tones;
Auburn   Orangy tone of brownness;
Silver Beige   Good tan tone for most blondes;
Ash and Deep Ash   Drabber shades for blondes;
Gray   A light gray with no blue cast;
Dark Silver Gray   Deep gray with some highlights of silver;
Silver   To lighten brows with a silver touch;
Gold   Good pencil to lighten darker hair.

Before drawing the brow with pencil, brush it with an eyebrow brush. This action cleans the excess foundation from the hairs, guides then into the desired line and stimulates their growth. Also, after drawing the brows, go over them with light strokes of the eyebrow brush to work the color in and to lose any hard line effect created with the pencil.

For a natural shape, start sketching the brow from a line that is directly over the inner corner of the eye, and continue with upward, outward strokes, coloring the natural hair where it grows or drawing fine hairlike lines on the skin to simulate hair growth if the brow is thin or overplucked in any area. When a point is reached that is in line with the outer curve of the iris of the eye, start drawing the brow with outward strokes—never downward—and end the brows on a line about one-half inch from the outer corners of the eyes. The height or apex of the brow can be carried out slightly further on some eyes to be in line with the outer corner of the eye rather than in line with the outer curve of the iris (Figure 7.16).

The heaviest portion of the brows should be on the inner part, with this width continued to the apex or

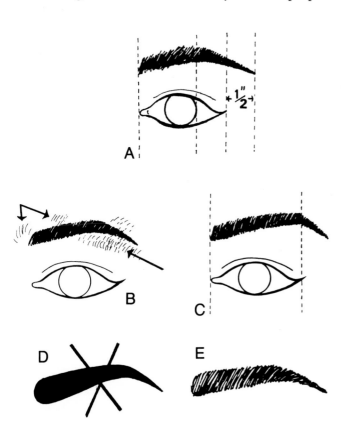

FIGURE 7.16   *Eyebrows. (A) How to measure the perfect browline. (B) Where the brows most often need plucking. (C) For a shorter eye or for a more extreme or striking look, the apex of the brow can be extended to be in line with the outer corner of the eye. (D) The blacked-in or heavily pencilled brow. (E) A correctly drawn brow pencil sketched with hairlike strokes to appear as real hair.*

change point of direction and growth angle of the hair. From here the brow commences to become finer and thinner until it ends in only a few hairline strokes. Never draw a hard outline and fill it in, because the brow should always look like natural hair growth and not a drawn line.

To arrive at a correct brow, it may be necessary to pluck some of the excess hairs. The most satisfactory method is to draw the brow desired, then pluck out the hairs that do not conform to the wanted shape. Don't pluck the brow first, or some hairs that are needed may be removed inadvertently.

Always remember that the brow is the picture frame of the eye (Figure 7.17). If it is too broad, heavy, or low, it will seem to enlarge the eye out of proportion. If it is drawn too high, thin, or narrow, the eye can appear smaller and expose too much frontal bone. Heavy brows are rather unfeminine, while thin brows are remnants of the 1930s.

Mascara

In most cases, black mascara is best for dark brunettes through brownettes, while brown shades serve better for redheads and blondes. Since the purpose of mascara is to deepen the color of the lashes and make them more apparent by this accent, mascara in bright shades of blue, green, violet, and so forth serve more as a color highlight and are not suitable for most production make-ups. Tipping the lashes with silver, gold, or white for certain high fashion effects also falls into the latter category. Some companies make a charcoal gray shade of mascara that can be substituted as a

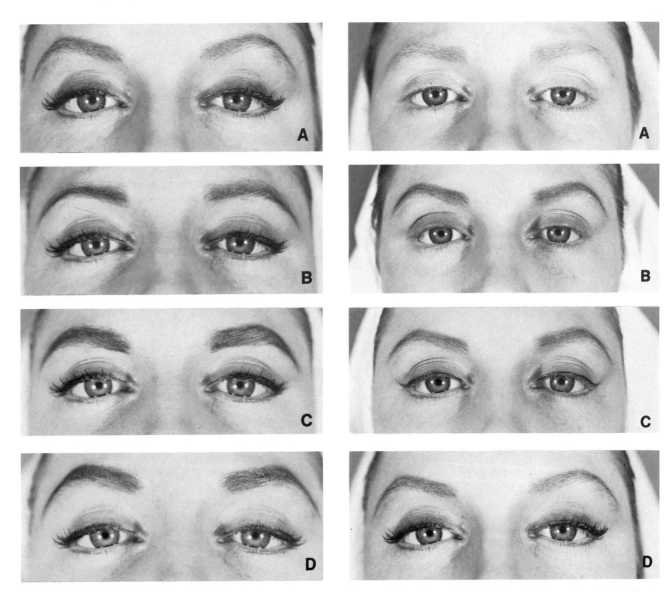

FIGURE 7.17   Eyebrows. Various eyebrows, drawn on the same face to show the effect of the brow shape on the eye area. (A) The natural arch. (B) A straighter brow. (C) Lower, wider brow. (D) Wide brow but with a higher angled arch.

FIGURE 7.18   Eye make-up for black-and-white mediums (A) Without make-up. (B) After shading, countershading, and eyebrow. (C) Eyelines added and eyeshadow. (D) Mascara completes make-up.

milder black or a stronger shade for those who might use browns. (See Figure 7.18.)

In the main, mascara is applied with a lash brush. Brushes with 4-inch handles are best, and a good trick is to cut off the first three rows of bristles nearest the handle to provide an applicator that is easier to control and will hold less mascara.

To use the cake variety, wet the brush in water and rub the bristles on the cake to work up a coating. Don't drop water on the cake and work the brush into it as the excess water will often make a mix that runs too freely.

Creme mascara in a tube can be squeezed out onto a small plastic or glass plate and the brush rubbed into the material to coat it. Avoid saturating the brush with too much mascara, and don't squeeze the mascara out of the tube directly onto the bristles because too much will be picked up, creating a problem in application.

Always load the mascara brush carefully, and apply two *thin* coats to the lashes rather than one heavy one. Take care that you don't bead the tips of the lashes together with blobs of mascara as this appears very unnatural in close shots.

To apply mascara to the lashes, place a finger at the outer corner of the eye, and slightly stretch the upper lashline using a light pulling motion. The hairs of the lashes will fan out and separate, making it easier to coat the individual lashes. With short strokes, first coat the upper surfaces of lashes, and then by using upward curling strokes, apply the mascara to the under surfaces of the lashes. In this manner, the hairs are thoroughly coated and given a natural, slightly upward curve.

Ordinarily, don't mascara the lower lashes unless they are very blonde. In that case, use brown mascara in a very light coat. Mascara on the lower lashes sometimes casts an unwanted shadow on the under-eye area, destroying the effect of the countershading under the eye, and can also smear more readily. However, for some high fashion or special effects, the use of mascara on the lower lashes is called for. Always clean off the mascara brush after each use in RCMA Studio Brush Cleaner and wipe dry.

For personal use, the brush-on type of mascara is good. However, this form of mascara poses the problem to the make-up artist of having to clean the brush thoroughly between uses if used from one person to the next. Avoid the lash-building types of mascaras as they contain small fibers that eventually fall off and spot the cheeks with black specks.

False Eyelashes

For certain make-up purposes, false eyelashes can be more natural appearing than heavily applied mascara (Figure 7.19). However, they should be correctly and expertly trimmed so they appear natural and attached

so they are undetectable. The best varieties of *strip lashes* are made of real hair tied on a fine, hairlike filament with a wigmaker's knot. These are then rolled on a 7/16-inch roller to be curled to give a long, natural sweep to the curve of the lashes. Although some false lashes of this type come precut, they are also available in uncut styles.

To prepare the uncut type, remove the two strips from the box they are stored in, and trim the hairs on an angle so that there is a left and a right lash with the hairs on the outer ends about 1/2 inch long and the inner hairs about 1/4 inch in length. Then clip every other hair about 1/16 to 1/8 inch shorter than the next so that the finished eyelash will not have a straight cut to it but will be feathered naturally.

To apply these, place the heavier curved end on the outer part of the subject's upper lashline, adhering the lash to the lid with RCMA Eyelash Adhesive or Duo

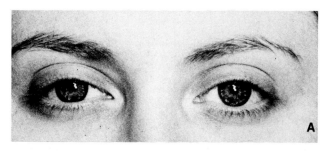

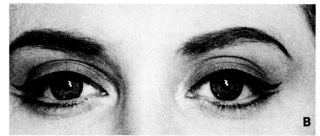

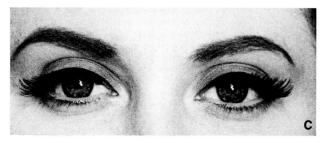

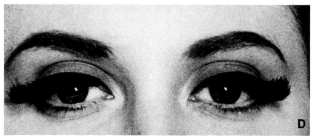

FIGURE 7.19 *False eyelashes. (A) No make-up. (B) Eye make-up but no false lashes. (C) With #2 lashes. (D) With #4 lashes.*

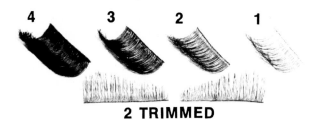

2 TRIMMED

FIGURE 7.20 *Strip false eyelashes. Shows densities of the lashes numbered from #1 to #4 in uncut styles and a #2 style trimmed and feathered for a natural look.*

Surgical Adhesive that has been previously coated lightly on the strip. Dental college pliers are best for all lash work because they have long handles and a curved tip for ease in lash placement on the eyelid. Gum the strip as close to the natural lashline as possible. The length of the strip should not extend beyond the natural lash growth.

It is often necessary to retouch the upper lashline with eyecolor after the adhesive has dried to disguise the strip further. Incidentally, using a pencil for the upper lashline instead of water-applied eyecolor will prevent the best adhesion possible for false lashes. The grease in the pencil material will loosen the strip.

False lashes seldom require mascara since they are dense enough to create the proper amount of definition. However, it is sometimes necessary to retouch the subject's own lashes with a bit of mascara to blend their curve into the false lashes.

There are varying densities of false lashes. The fine types have less hairs and the heavier ones have more closely tied hairs for a more accented look. Colors range from black to brown, but for special effects one can find lashes in shades from blonde to strong colors. Styles vary from #1 through to #4 in density, with the #1 being the most natural appearing (Figure 7.20).

When removing these strip lashes, grasp the outer end firmly in the fingers and peel them off gently. When the adhesive is dry on the strip, it will peel off readily for future use. It is a good practice to replace the used lashes in the box they came in to preserve their shape.

Some false lashes are attached to a small strip with two or three hairs only. These are affixed to the base of the natural lashes in the same manner with the same eyelash adhesive. Although the process is slow and painstaking—each group of lashes must be dipped in adhesive that has been dropped on a plate of glass or plastic and then attached separately in the correct direction of the curve of the lash—they can be controlled for attachment just on the outer ends of the lashline or to replace a space in a natural lashline. After the adhesive has set, the hairs are then carefully trimmed to length and can be coated with mascara if necessary. The results are almost undetectable even in tight close-ups.

Tinting Brows and Lashes

If a natural blonde hair color is deepened to auburn or brunette, the eyelashes and brow hairs can be tinted with a type of haircolor made especially for this purpose, or mascara and pencils can be employed to match the new hairshade. However, when a woman changes from brunette or brownette to blonde or light red hair, there is seldom any need to change the eyelash color as dark lashes are becoming and desirable, but the eyebrows should be matched to the new hair color. Nothing is more of a giveaway to a change in hair color than to leave the eyebrows as dark as they were prior to the change. Generally, a blonde or gold pencil is quite insufficient to lighten the hairs of the brow when they are really dark, so tinting is best.

One professional hair colorist recommends the use of two capsful of Miss Clairol Creme Formula Flaxen (or any similar light shade) mixed with four capsful of 20 volume hydrogen peroxide. This mixture can be applied to the brows for five to ten minutes, depending upon the degree of lightening desired. This mixture is firm enough not to run down into the eyes when the subject's head is tilted back and leaning on the headrest of the make-up chair.

For auburn or red hair, one can use one capful of peroxide mixed with one capful of Lady Clairol Whipped Creme Hair Lightener. One capful of white henna can also be added to form a paste if a firmer working mass is desired. This is applied to the brows for five to ten minutes and will normally lighten the hair sufficiently to match the hair color. Remove all bleaches with extreme care so none will run into the eyes.

Lipcolor

Lipcolor not only is a subject of fashion but also provides a protection for the woman's lips. As a woman ages, lips can require more emollience to keep them fresh appearing, so lip protectant should always be used.

The shape of the lips also relies on fashion, and the natural lipline may not be the most becoming or correct simply because it belongs to the individual since it may not be symmetrical with the rest of the face. Changing the contour of the lips with lipcolor also depends upon personal preference as well as style and even a fraction of an inch over- or under-paint can produce a great difference. The lip style and color should be an asset and not a liability to the overall aspect of the make-up.

To select the best possible lipline, we must first divide the finished lipline shapes into three basic types: the one-third, the one-half, and the two-thirds shapes (Figure 7.21). Using the height of the lipline as a guide, the one-third type has its highest point one-third the distance from the center of the lip to the

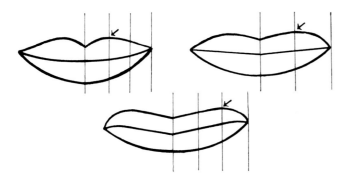

FIGURE 7.21 *Lipline measurements.* TOP LEFT: *The one-third lipline with its highest point one-third of the distance from the center to the outer end.* TOP RIGHT: *The one-half lipline with its highest point at the midpoint.* BOTTOM: *The two-thirds lipline with its highest point two-thirds of the distance from the center to the outer end.*

outer end on one side. The one-half type is half the distance, and the two-thirds is at two-thirds the distance. The first two types are employed (or combinations of them) for most professional make-ups, while the two-thirds variety is seldom seen except for stage singers who require a generous amount of curve on the outer upper lipline to compensate for the general flattening effect on the lips due to lights and distance from the audience as well as the fact that their mouths are often opened to talk or sing. Opera singers, especially in large theaters, may use this style of lipline. On the street, it is an exaggerated style and dates back to certain movie personalities of the 1930s and 1940s.

Most liplines that are becoming will fall between the one-third and the one-half styles, depending upon the conformation of the face in the mouth area and the natural lipline. The accent of the height and curve of these two liplines governs the size of the lipline more than the length of the mouth. However, if a person has a wide jawline, never use a short or narrow mouth, but extend the color all the way to the corners. Conversely, a small, narrow jawline should not be accented by heavily overpainting the lipline at the outer corners (Figure 7.22).

To apply Lipcolor, make certain that the lips are perfectly clean with no trace of old color remaining. If the corners of the mouth are stained with the use of regular staining lip make-up, remove the coloration with RCMA Studio Make-up Remover on a Q-Tip before applying the foundation.

Load the lip brush (#4 flat red sable) with a generous but even coating of the selected lipcolor without distorting the general shape of the brush. Holding the brush between the thumb and first finger (as you would a pencil), place the small finger on the chin to anchor or steady the hand. The subject's lips should be in repose while outlining the lips so the make-up artist can observe the shape of the mouth and the lipline being drawn. Using the sharp chisel edge of the brush, outline the upper lip on one side, then the other,

sweeping from the center to the outer corners of the lip on each side. Then fill in the upper lipline with the flat of the brush. Reload the brush with lipcolor and proceed to outline the lower lipline from the outer corners to the center, on one side and then the other, meeting in a smooth curve in the center. Fill in the lower lipline with the flat of the brush, and then ask the subject to open her mouth slightly so that the inner portion of the lips and the outer corners can be covered, leaving no line of demarcation when the subject smiles or talks. There is no necessity to blot the nonstaining lipcolors as the amount of material to be placed on the lips can be governed by the application with the lip brush.

Finally, examine the teeth for any traces of lipcolor and remove with tissue. Incidentally, if an error is made in drawing the lipline, carefully wipe the unwanted portion away with a tissue or a Q-Tip, correct the area with a bit of foundation on a #7 flat brush, and then redraw the lipline with the lipcolor.

In some cases, to prevent the loss of color separation between the foundation and a pale shade of lipcolor, the outer line of the lips can be sketched in over the light lipcolor with a deeper shade of the same tonality. If the outer lower corners of the mouth have a slight shadow or drop to them, the corners of the lipline can

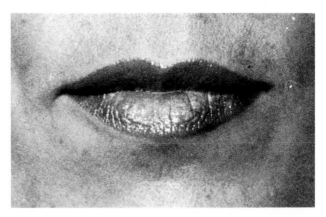

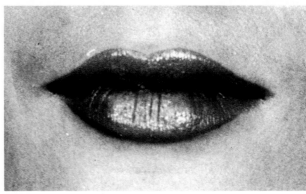

FIGURE 7.22 *Lipline styles.* TOP: *Normal painting for a narrow jawline.* BOTTOM: *Upper lipline painted to the corners of the mouth for a wider lipline effect.*

be firmed and the shadow minimized with a bit of countershading mixed with the foundation color (Figure 7.23).

Don't apply lipcolor to the lips directly out of the tube. No lipstick can give the fine edge control of a brush regardless of the shape in which the stick is molded. For sanitary purposes, the lip brushes must be cleaned with RCMA Studio Brush Cleaner or a similar product after every use. For a completely sanitary application, a bit of lipcolor should be taken from the tube or jar with a stainless steel spatula, placed on a glass or plastic plate (or disposable butter chip), and the lip brush touched to this for application.

Since the lips are one of the most expressive parts of the face (after the eyes), always maintain lips as an important part of the make-up design, and even when all the face is made up, the face appears as blank without the proper lipcolor and design. Any fashion that promotes heavy eye make-up and little or no lip make-up destroys a great expressive part of the face in doing so.

It is quite common practice to do the entire make-up except the lip make-up and then have the hairdresser complete the hairstyle and the wardrobe dress the subject. Then, to be fresh looking, the lipcolor is applied just before going on set.

Smoking cigarettes, drinking coffee, eating lunch, and so on remove much of the lipcolor and may affect the line, so retouching is necessary in these cases. Certainly after lunchtime is a prime retouch period for the lips, but care should be taken all day to watch the lipline.

In scenes where kissing takes place, retouching may be necessary after each take so a make-up artist should have a lip brush and a brush with the foundation ready at all times. Lipglosses and outlining techniques are discussed in the next section on fashion make-ups.

High Fashion Make-up

As stated previously, high fashion make-up is more a display of products than an overall complementary make-up appearance. As such, the amount and placement, accentuation or creation, depth of color or gloss, and any other attention-getting principle produces an unusual make-up display on the face or other parts of the body to delineate what a high-fashion make-up is supposed to be.

FIGURE 7.23   *Countershading applied at the corners of the lower lipline to minimize any shadows.*

To define exactly how the eyes, lips, cheeks, and so forth should be made up for this type of make-up is quite impossible as the direction and style changes too frequently. One can only say that anything is liable to be seen and used. Items such as fabrics, flowers, sequins, glitter, new color mixtures, lighting techniques, and other factors can all contribute to a novel and fresh idea. Many make-up artists prefer to do just that sort of work, and some are extremely clever and innovative in their concepts. Actually, nothing is sacrosanct in this part of the field, and everything from foundation through lipcolor is varied to make a particular point or assume a new direction. However, eye make-up generally leads the list and its varieties are infinite. The lips, cheeks, complexion, and hair can be revamped in countless ways as well. Only a few suggestions are noted here that are rather basic approaches to the art of high fashion make-up. First, eye make-up can be applied with more colors employed, the effects of shading colors reversed from the norm (that is, light colors applied where dark ones usually are placed for a normal beauty or production make-up), or any other out-of-the-ordinary effect that can be created with make-up materials (Figure 7.24). Next, lip make-up can be emphasized with outlining in deeper reds or even brown-tone pencils or lipcolors, or gold, pearl, bronze, or copper nonmetallics added to the lipcolor for extra shine effect. As well, bright, reflective lipglosses can add a slick or hi-shine look to the lips. The possibilities for high fashion make-ups are unlimited and one's imagination can run rampant. Many seemingly outlandish high fashion make-ups in the pages of the fashion magazines start new trends, albeit in many cases less drastic overall, which are reflected in the items brought out each season by the commercial cosmetic companies.

The effect of the eyecolorings shown in Figure 7.25 in color mediums would be the same values and demonstrates the difference between a dark and a light eyecolor on the eyelid. As well, in the second picture, false lashes have been added to accent the strength and depth of the natural lashes further.

After the lipcolor has been applied, lipgloss can be added over it to produce a hard, slick shine. Pearl produces a soft, cold reflectivity, and gold gives a warm tone to the lipcolor. Outlining the lips with a soft edge is done with a lipcolor of the same tonality some shades deeper or a lipliner pencil in the shade closest in tonality but also deeper than the lipcolor employed (Figure 7.26). A medium brown pencil worked into the outline of the lips produces a hard edge to define the lip shape further.

Women with Dark Skins

The approach to Negroid skins and others such as people from India with deep skin colors is best made with the premise in mind that light pastel colors show up weakly on those with paler Caucasoid skins but provide a stronger contrast against dark skins. What might be a bright and dark red lipcolor on a Caucasoid skin might not produce this separation on dark skins.

As such, foundation shades should be applied according to the chart on page 55, depending upon the depth of brown color in the skin and countershading and shading done with the darker shades as noted. Whether or not cheekcolor is called for depends greatly upon the person. Pink cheeks are not indigenous to most dark skins, and the darker the skin the less apt it is to have any pink tonality. However, as a fashion touch, warm reds are sometimes seen in make-ups on people with dark skins.

Eyelining, brows, and lashes require no new directions, but with eyecolors, the make-up artist must

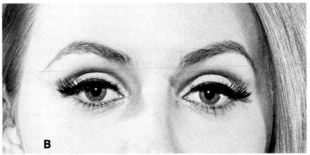

FIGURE 7.25 *Two different effects. (A) frontal bone shaded and darker colored eyecolor (acting as an eyeshadow on the eyelid). (B) Frontal bone highlighted with a pearlescent pale eyecolor, the socket shaded with a deep brown eyecolor, and the eyelid with a very light pearlescent eyecolor.*

bear in mind the principle that light shows stronger on dark skin than dark. For example, a light blue will show up on an eyelid with more effect than an extremely dark blue eyecolor due to the surrounding and basic dark skin color. In addition, many people with very dark skins may have differences in basic coloration in the upper and lower lips so two distinct shades of lipcolors may be required to even out the overall shade.

One note: Foundation *will* even out the differences in yellow to gray areas on dark faces and is very important for dark skin make-up. It should not be omit-

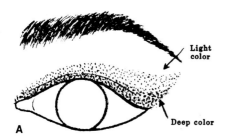

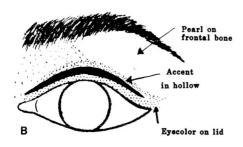

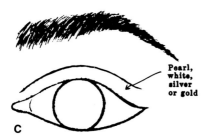

FIGURE 7.24 *High fashion eye make-up. (A) Two colors of eyecolor can be drawn on the eyelid with the deepest shade nearest the lash and a lighter or even contrasting shade blended above it. (B) Pearl, gold, silver, etc. eyecolor can be blended on the frontal bones rather than shading and the hollow of the lid accented with a midnight brown color. (C) Silver or gold eyecolor can be used on the lid as well as pearl or white eyecolor for a very striking effect.*

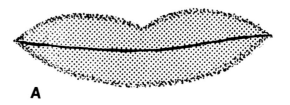

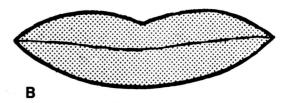

FIGURE 7.26 *Lip make-up tricks. (A) Pale lipcolor outlined with a deeper lipcolor or lipliner pencil in a soft lining. (B) Lips outlined with a brown pencil for a stronger effect.*

ted just because the dark skin *seems* to be deep enough not to require its use. In fact, the yellowed or grayed areas on dark skins can be equated to the rosy or pinked areas on some Caucasoid faces whose differences in coloration are evened out with a properly covering foundation.

Oriental or Mongoloid skin shades are seldom (unless mixed by intermarriage with negroid strains) very dark in coloration and so can be treated much as Caucasoid skins relative to color and reflective contours and definitions.

Readers should also refer to Chapter 10 for certain national and racial differences relative to the judgment of what *is* beauty since in some societies the standards vary, and differences are noted. Research is especially recommended for period make-ups as variances can occur here as well (see Chapter 12 and Appendix D).

MEN

In the main, insufficient emphasis and importance has been placed on make-up for men due to the fact that more effects are possible (and allowable) through colors and shapes, as well as fashion changes, for women than men. Nevertheless, for all mediums, a man's appearance can be facially strengthened and visibly improved through the proper use of correct make-up. Facial areas can be significantly changed, readjusted, and emphasized (or de-emphasized) with judicious make-up application.

Make-up for men takes far less time to complete than for women. However, it is just as important to the actor as it is to the actress and requires its own proper attention.

Historically, although the facial shape theory was applied to women's make-up for black-and-white mediums, it was not employed in the same manner for men, and the basic procedure was as follows:

1. After the man had shaved, and if the beard line area appeared dark—most dark-haired men's beards show a grayish beard outline while fair-haired or gray-haired men don't—the area was covered with a beardcover like BC-1 (a light pink tone to combat the bluishness of the beardline) and powdered.
2. The face was then covered with a foundation at least two shades darker in gray scale than for the female performers.
3. Corrections were made with a foundation three shades lighter than the base color in the under-eye or eye socket areas.
4. Shading could be done with a foundation three shades deeper than the base color in the underchin area if it contained excess flesh (most jawlines were left square rather than shaded). The sides of the nose could be shaded to narrow it, any heaviness of the frontal bones could be toned down a bit,

and if the man had a high forehead, the area near the hairline could be deepened and blended carefully. Prominent ears were also to be covered with the deeper foundation.
5. Powder was applied to reduce shine as a *flatter* effect was generally desirable.
6. The eyelid could be deepened with a dark brown pencil if it was too prominent, but in general, no eyelining was done.
7. Any spaces in the brows were pencilled in, and if the hair was very blonde, light brown pencil was lightly applied to tint the hairs for better definition.
8. The eyelashes were deepened a touch with brown mascara only if they were blonde.
9. Lip make-up was not necessary in black and white, but if they appeared too dark in tone, oftentimes the foundation was carried over the lips.
10. If the front of the hairline was sparse, the area could be pencilled in lightly so that the studio lights would not wash out the edge of the hairline and make it appear more receded than it was. Bald spots on the top of the head could be darkened to match the hair. Hairpieces and toupees were employed when a man was bald enough that pencilling and deepening would not correct it.
11. Most often, necks (that normally fell into shadow with the overhead lights), arms, and hands were not made up unless they appeared too light in tone or were prominently lit for a close-up.

Although basically the same procedure was followed for stage make-up (with ruddier Foundation required for naturalness), color film and television processes did make considerable changes necessary. Although in the later years of black and white many actors with smooth faces and little or no apparent beardline seldom, if ever, wore make-up and simply got a deep tan, color mediums were different. Early color film stocks and television electronics had a propensity for not accepting color as the eye saw it, so everyone was required to wear facial and body make-up. As color mediums have improved greatly, some male performers do now appear without make-up, but make-up artists have always held the theory that *everybody* could be improved with the correct use of make-up—which is quite true in almost all instances.

Naturally, the tonal value of the beardline shows a rather bluish cast on men with dark hair. This must be *color corrected* with an orange tone (the opposite to blue on the color scale) rather than simply lightened. (The pink-toned beardcover used for black-and-white mediums is not correct for color.)

Although countershading for color can be done with either a lighter tone of foundation or a CS-3, shading on men has to be very carefully applied in the correct gray-brown tonality so as not to appear as a dirty face

Table 7.2 Foundation Recommendations for Men

RACIAL TYPE	SKIN UNDERTONE	LIGHTING ON SET	SKIN COLOR		
			LIGHT	MEDIUM	DEEP
Caucasoid	Pink to Ruddy	Under normal	KM-4	KM-5	KM-6
		Normal production	KM-36	RJ-2 KM-37*	KM-38
	Olive	Under normal	KM-1	KM-2	KM-3
		Normal production	KT-2	KT-3*	KT-4
	Tanned	All	KM-37	KM-38 KT-3	KM-8 KT-4
Oriental	Yellowed	All	Shinto II KT-1	Shinto III KT-2*	Shinto IV KT-3 KT-34
Negroid	Grayed	All	KN-1 through KN-9		

*Indicates the most basic shades for testing procedures. Where other numbers are shown, a choice is allowed.

area. Just as for women, the old black-and-white method of a deeper foundation for shading does not picturize naturally in color mediums for men.

Pencils and hair goods have to be exactly matched to a man's hair color and care taken that the lace edges on toupees (where the adhesive is applied) are not touched with the foundation so that no darkening of the lace will take place (as well as prevent proper adhesion).

Skin shades for all races must be tonally correct for any color medium, and again, Caucasoid skins can range from medium pink to deep ruddy, or olive to very deep olive in undertone value as natural coloration, with a more *browned* appearance for tanned skin. Orientals with their yellow undertones and Negroids or other dark-skinned peoples with the gray undertones, range from light to deep colors.

Table 7.2 provides basic test choices of RCMA Color Process Foundations for the three races. Shades for racially mixed individuals can be adjusted according to the undertone value and skin coloration. While the amount of lighting on set determines the necessity of the depth of color value in men's make-up, Table 7.2 is based on the application of the thinnest possible coating of foundation to provide naturalness to the skin. The table's recommendations can be applied to stage, screen, or even on-the-street politician's make-up and for general personal appearance make-up of any kind where the man might appear on television for an interview followed by a still photo session or even a street appearance.

The deeper color recommendations may appear better for sunlit outdoor sequences on beaches or westerns where a great amount of reflectivity is present from the surroundings. The progressive deepening of RCMA Foundations KN-1 to KN-9 for Negroid skins is to allow better skin-matching determinations.

For men, William Tuttle suggests:

NORMAL CAUCASOID	TANNED EFFECT
Dark Beige	Desert Tan
Natural Tan	Xtra Dark Tan
Tan	K-1
	Jan-Tan

The Negroid shades are the same as those recommended for women (see page 55).

Ben Nye's shades for men are M-1, M-2, M-3, M-5, and Y-3, while the Negroid shades are 27, 28, and Dark Coco. As previously stated, there will be slightly different tonalities of recommended shades depending upon the medium and the area where a film or television show is produced, with the more saturated or tanned shades being employed in California and lesser ones in the East or in the United Kingdom.

The basic sequence and procedure for make-up application for men for all color mediums is as follows:

1. Make certain that the beard is closely shaven, and although most men don't require skin preparation prior to make-up, if the skin is oily, a light application of RCMA Liquifresh on a polyurethane sponge is advisable. If the skin is dry, use RCMA Prefoundation Moisture Lotion.

2.. Apply with a dry polyurethane sponge, the *thinnest* well-blended coat of foundation possible, employing a thinner coat of a deeper shade rather than a heavier coat of a lighter one so that the foundation appears as a skin coloration rather than a coat of make-up. This is most important in applying foundation to men for a natural appearance.

3. Color correct any beardline that shows through the foundation with RCMA BC-3 to the desired tone. Most men with dark hair and beards will appear more natural if the beardline is not completely obliterated (as it was in black-and-white mediums) and if some indication of a beardline

shows through. This can be determined by the individual as well as the character being played.

4. Make any corrections necessary to minimize any shadows under the eyes or to reduce the deepness of the eye socket area with countershading CS-3 or with foundation KW-24. The deeper the skin, the less any corrections with countershading or shading will show up.

5. In most cases where shading is required, S-4 is best. The under-chin area (if fleshy) is the main area, and take care not to carry any shading over the jawline. Frontal bones can also be shaded, but most men's brows compensate here. Ears and high foreheads should only have a coat of foundation and not be darkened as with black-and-white mediums.

6. Men will appear more natural with the proper degree of halation if the foundation is not powdered in most cases. However, for some close shots, it may be necessary to apply some RCMA No-Color Powder to reduce some of the natural shine.

7. Cheekcolor is normally not employed for most men's make-up except to provide an excess ruddiness for some characters. RCMA Stipple Sunburn or Cheekcolor Plum can be applied with a stipple sponge if this is desired. For westerns and other outdoor productions, some make-up artists use a light stipple of RCMA Cheekcolor Dark (a deep orangy tone) or Stipple Sunburn on the cheeks, forehead, and nose areas to break up the evenness of the skin.

8. Upper eyelids can be shaded with a dark brown pencil if they are prominent, but the eyes should not be lined (unless the lashes and the brows are so blonde that a bit of emphasis is required). Do any eyelining carefully and blend well.

9. Brows can be brushed with an eyebrow brush to remove any foundation and any spaces filled with matching pencils. Otherwise, do not deepen the brows unless they are very blonde, and do this carefully and so they do not appear artificial or painted.

10. Mascara the lashes only if they are very blonde with a light application of brown mascara.

11. If the lips are pale, a thin coat of RCMA LS-17 can be applied with a brush, taking care that no shape is made—only some color added.

12. The hairline in the front can be lightly stroked with sharp pencils to simulate hair growth and blended in with an eyebrow brush. On the top of the head where there may be a bald spot, deepen with RCMA Eyecolor #1 in the brown shades to match the hair color (Figure 7.27).

Men's hairpieces or toupees can be adhered with RCMA Matte Adhesive and pressed firmly into the scalp with a dry piece of silk or nylon. Some make-up artists prefer a slightly dampened lintless huck towel for this. Most hairpieces are best attached *before* any foundation is applied and the foundation only blended up to the lace area and not over it. Also see Part V on hair goods for new types of hairpieces and other adhesive methods.

13. The hands (excluding the palms) can be made up with a foundation a shade or two deeper than the facial foundation (as most hands will be lighter than the face in coloration on Caucasoids). If the arms are bare, they might also require foundation to match.

14. Don't omit make-up from the neck area if the skin shows up as a lighter color than the face, and if a man must appear bald or with bald spots, foundation should be blended here also. A light powdering will hold the foundation better.

15. Men's make-up can be removed with RCMA Actor's Special or Hydrocleanser and any adhesive removed with RCMA Adhesive Remover.

Hairline Adjustments

For men who are completely bald in front and have only side hair, the hairpiece or toupee can be more extensive. Never apply a toupee too low on the forehead, and watch the blend carefully with the natural hair, filling in with pencilling if necessary (also see Part V on hair goods).

Men who have varying degrees of grayness to whiteness in their hair will often experience lighting problems. The hairline might appear to recede as the light washes out the sparse growth along the front edges of the hairline. Careful pencilling with a dark silver gray pencil well brushed into the scalp will often aid to strengthen the front area, and using the pencil flat to the hair on the sides will allow the hairs to be colored deeper to frame the sideburns better. Most of this extra pencilling will work when there is sufficient hair present to form an outline, but little should be attempted where no hair grows because the pencilling might appear as pencilling—and not hair. Moustaches can be slightly changed in shape with pencilling or to fill a lip scar. In all cases, facial hair such as beards or moustaches do cover some areas of otherwise necessary facial correction (such as beardlines, under-chin flesh, and so on).

Perspiration is one of the great enemies of men's foundations, and the skin coloration must be maintained and retouched if necessary. Also watch that the men do not remove the make-up from their hands by washing them or rubbing them on the wardrobe. Above all, the final criterion for a man's make-up must be that it does not appear that he is wearing any make-up. It must be that effective and that natural looking.

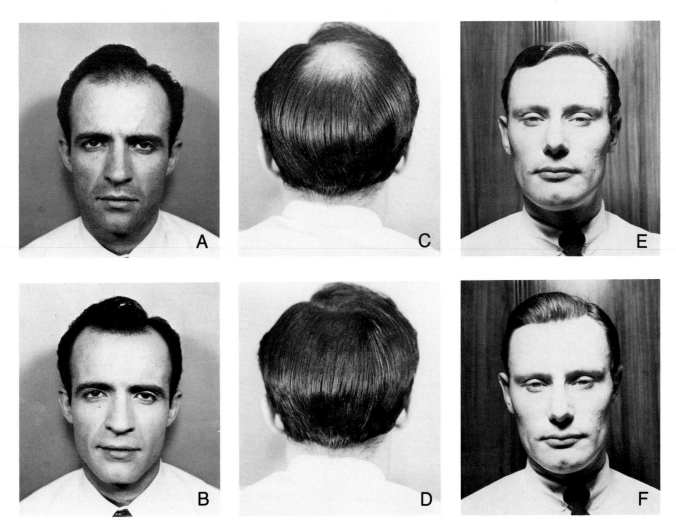

FIGURE 7.27   *Men's hairline adjustments. (A) Before pencilling the front of the hairline. (B) After pencilling. (C) Before scalp coloring with brown and/or black eyecolor. (D) After coloring. (E) High hairline on one side. (F) Hairline straightened and added height given with a small lace hairpiece.*

CHILDREN

Overly made-up children do not appear natural so the make-up artist must judge carefully the skin tone balance between mature performers and child performers. In most cases, a light and thin application of a foundation such as F-3, F-4, KW-1, or KW-13 will produce the right tonal values for most Caucasoid boys and girls under the age of 12 (unless they are supposed to have tanned skin). Oriental and Negroid children's make-up shades will depend upon the density of color value in their natural skins and should be approximately color matched with the light colors on Tables 7.1 and 7.2.

As children's skins normally have a soft smoothness and a gentle reflectivity, powder may not be necessary. Under-eye deepness may be corrected with CS-1 in most cases, and shading is seldom, if ever, required.

Avoid excessive cheekcolor, lipcolor, eyecolor, mascara, or pencilling because such a display of make-up is not attractive on children. Their lips normally have sufficient color, but a touch of RCMA LS-12 can be employed if added tone is desired. Cheekcolor Natural or Genacolor Pink or Rose can be used on the cheeks to restore natural blush, although the foundation, if properly thin, will allow their natural color to come through and generally prove to be sufficient.

Mainly, try to keep parents out of the make-up room, and warn them not to add to the completed make-up or to retouch their child during the shooting session. Unless a child is a star performer, most make-up artists will leave them to last on the schedule not only because they require less work but also so their make-up will be fresh—and will stay like that until the first shot—hopefully.

BODY MAKE-UP

Body make-up on the shoulders, arms, or legs can be done with the regular RCMA Color Process Foundations. Select a tonal value one step deeper than the facial color, and put a few drops of RCMA Foundation Thinner on the cake of foundation or on the sponge to spread it faster and more thinly. Ordinarily, no powder will be required (unless too much foundation thinner has been used).

The ideal body make-up would not come off the skin onto the wardrobe, but it would therefore be just as difficult to remove from the skin. Some body stains are made for tanned effects or deeper skin tones, but they are sometimes hard to clean off as well.

Retouching and constant care must be taken with body make-up, and it should not be neglected. Most major productions have a body make-up woman who will apply any body make-up on the female performers when there is a male make-up artist for the facial make-ups.

For shooting in the sun in bathing suits when the performers have light skin, one trick is to apply a thin coating of a sunscreen with a high sun protection factor on the body prior to foundation application (usually no foundation thinner will be necessary). This will give the proper color to the skin and at the same time protect it from the possibility of sunburn.

If a performer has a tan and there are residual strap marks on the shoulders, the lighter areas can be retouched to match the tanned skin. For those with freckles, a tone of foundation between the skin tone and the freckles will work best to minimize the deepness of the freckles and the lightness of the surrounding skin. See Tables 7.1 and 7.2 for tanned skin colors for Caucasoids.

Moles and other discolorations can be touched with CS-1 on a brush and then foundation carefully stippled over the area.

Shading colors such as S-13 or S-14 can be used to emphasize the cleft when low-cut gowns are worn.

Pancake types of make-up applied with a large-size natural silk sponge are also useful for body make-up. Shades and makes vary, but a tonal value close to, or again, one shade deeper than the facial make-up, should be chosen for use.

Some body make-up, such as for unusual or impressionistic characters, can be done with hair spray type materials (Nestle Tips n' Streaks come in many colors. See pages 39–58). They last very well on the skin and are washed off with soap and water. Experiments are constantly being done to find a good, lasting hand make-up that does not come off on the wardrobe, but can be removed without too much effort. No doubt the make-up labs will produce such an item in the near future.

SPECIAL NOTES ON STAGE MAKE-UP

Although the oldest use of make-up was for adornment, it is difficult to separate this from characterization as it is certain that acting out the exploits that people performed encompassed the use of forms of make-up. When finally the play form took shape with repetitions of words and actions, the delineations of specific characters were depicted by the use of masks or paint in various forms. Early stage make-up was made by the performers themselves until the emergence in the nineteenth century of a few firms that began to manufacture standardized forms and shades of make-up for the stage as we know it today.

The general purpose of make-up on the stage is to counteract the effects of the distance of the audience from the players in terms of facial definition and to compensate for the intensity of the lights on the stage that wash out the natural facial coloring and flatten the features of performers. Make-up for the stage should be applied in such a manner that it does not look artificial from the first few rows of the audience and so it does not lose its character in the middle to the last rows of the auditorium. However, a balance must be achieved that considers that only a certain proportion of the audience in a large auditorium will actually be able to delineate the make-up, while the rest will view it as part of the overall picture of performance, wardrobe and lighting. The smaller the house, the closer the audience is to the stage and generally, the less lighting will be necessary, so the make-up must be compensatory and commensurately applied with less emphasis.

In most theater-in-the-round houses where the audience is small and quite close to the performers, screen make-up procedures must be followed regarding emphasis, style, and circumscription. In larger theaters or those with larger audience capacity, the stress and distinctiveness of both color and form must be enlarged. Nevertheless, it must be borne in mind that only approximately the front half of the audience will *see* the make-up so the application should not be for the last few rows; that is, all the make-up should be viewable as proper and correct from the front row as well as the fifteenth row (approximately) in any theater.

Of course, we must consider the use of make-up as an expressionistic or an impressionistic tool. *Expressionism* in make-up refers to the use of products to portray accurately the character as it would be or appear in real life. *Impressionistic* make-up reveals our fantasies and imagination. For example, if we made someone up to appear as George Washington or Winston Churchill, as close to the original as possible, that would be expressionistic make-up. However, if we made up someone to look like a clown, an animal, or a toy, that would be impressionistic make-up. In addition, there are theatrical forms such as opera, ballet, the circus, mimes, and such that have both expressionistic and impressionistic performers.

All in all, a good stage make-up is not necessarily heavy or overdone today but the right colors and shades applied in the correct densities and areas. A poorly blended or applied make-up for the stage will appear as such to much of the front row, or expensive, seats of the audience.

Reference should be made to Chapter 3 of this book regarding the effects of paint and light colors on the face, and every make-up artist should study stage lighting and craft as part of the required adjunctive knowledge necessary to the profession. Stage lighting is often different from that for the screen, especially when the stage is enclosed by a proscenium. However, the use of footlights has declined considerably in professional theater, and the lighting has become both frontal (from pipe stands and the balconies) and overhead or side from the stage. Thus, many of the old theories of heavy stage make-up must be eliminated, and over-emphasis must be kept to a minimum.

Unfortunately, many of the stage make-up courses given to students at the college level are not kept up to date by the instructors, and much misinformation is passed on to neophytes who take it as proper. Such instructors should constantly renew their own education on make-up periodically to keep up with the advances made by professional make-up artists in the entertainment field.

Today's stage lighting is generally a combination of—or possibly for effect, the single use of—cold and/or warm neutral lighting effects in the playing area of the stage, except when the time of day (early morning, dusk, or night) requires changes in the lighting scheme to create the desired effect. Pale amber gelatins, pinks, and pale lavenders provide neutral effects, while deep blues and strong yellows will aid night and day lighting patterns. In general, due to the use of incandescent lighting, which has a lower color temperature than screen lighting (which must be 3,200° K), most stage make-up should have a pink complement in Caucasoid make-ups to compensate for the added yellow of the light.

Straight Make-up

When stage make-up for performers was first manufactured, the two standard shades were a #5 (yellow ochre) and a #9 (red russet). These were combined in the palm of the hand with more #5 for women and more #9 for men and applied before each performance as the base or foundation for other products. As such, skin color could vary by the addition of more or less of one or the other color, but daily matching was not really necessary as one make-up was not seen next to another as is true for the screen and its editing procedures. Next came the mixtures of colors known as Hero, Juvenile and Sallow to commence standardization of shades (also see Chapter 5 for further information).

Today, Tables 7.1 and 7.2 can be followed quite closely for most foundation shades for the stage as long as the foundations with some pink tone are employed in most cases for Caucasoids. Oriental and Negroid make-ups can stay close to the tables since the pink or ruddy factor is not as important in their skin appearance overall (see page 30 for pink or ruddy tone foundations).

The same sequence of application of make-up can be followed as recommended previously. However, some difference in emphasis in color and shape can be made with reference to the eyes and lips of both men and women of all races.

One of the oldest tricks, but one that is still effective, is to use dark blue eyelines for women (and sparingly on some men) rather than brown or black ones to achieve more brightness in the eye area. As well, bright blue eyecolor for women is still best for most make-ups (except where a green dress or wardrobe is employed).

Women

Foundations should not be applied any more heavily for the stage than for the screen; they should only be of the correct color. Shading and countershading can be the same, but on large stages, due to the lighting, a stronger cheekcolor can be applied, although in the same area and blended well to the edges. Emphasis on eye make-up for large stages can be as shown in Figure 7.28. Sometimes performers will put a band of gold eyecolor or gold lipcolor close to the upper lash line to give the eyes a sparkle for musical shows. Generally, heavier mascara application or false lashes will complete the eye make-up. Bright red lipcolors seem to appear best on large stages. One note of caution: Don't overemphasize the brows—just make them natural as usual for any screen make-up.

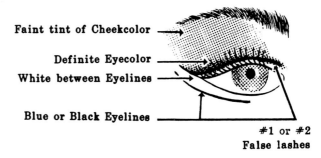

Faint tint of Cheekcolor
Definite Eyecolor
White between Eyelines
Blue or Black Eyelines
#1 or #2
False lashes

FIGURE 7.28 *Eye make-up for large stages.*

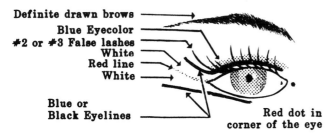

Definite drawn brows
Blue Eyecolor
#2 or #3 False lashes
White
Red line
White

Blue or
Black Eyelines

Red dot in
corner of the eye

FIGURE 7.29   *Eye make-up for ballerinas.*

Special eye make-up for classical ballet is traditional, and Figure 7.29 is an example of this style. Cheekcolor is also often added to the frontal bone by some performers.

Make-up for showgirls is accented by generously painted lips with much lipgloss added, extra heavy false eyelashes, very highly colored eye make-up with much eyelining, eyecolor, and gold, silver, or pearl highlights. Such is in the realm of a fantasy make-up to glamorize the statuesque beauties in that profession. These make-ups also change with fashions, but overemphasis is paramount.

Men

Straight make-up for men on the stage should be as natural as for screen make-up, avoiding lipcolor and bright shades of cheekcolor, overlining the eyes, and emphasizing the brows. Fine brown or blue upper lashlines can be drawn on for large stages, but all the other recommendations for screen make-up should be followed.

For period plays such as Restoration or for operettas of Gilbert and Sullivan or the like, a man can look considerably more painted in appearance. Ballet make-up for men often closely resembles that used for women.

Hair goods such as beards and wigs should be just as finely made as those for screen mediums. For large stages, however, a coarser lace is sometimes employed due to the fact that many performers do their own make-up and do not maintain the hair goods or clean them as well as a make-up artist might.

Character Make-up

The same rules apply as for screen make-ups (see Part III). However, character lines and shadows can contain a bit more redness, and RCMA S-6, and S-10 are useful as well as RCMA Lake Red and Maroon pencils for accentuation for Caucasoids.

One method for creating age is not to employ any foundation on a performer and simply add the lines, shadows, and highlights directly on the skin. This works especially well when young performers are to be aged because the younger characters (who are in straight make-up) will have a healthy appearance, while those who must appear older will look rather sallow and pale in comparison under stage lighting. For quick stage make-up changes from youth to old age, an application of a shading color such as S-6, S-9, or S-10 placed in the corners of the eye socket area and a light coating of S-2 on the lips will aid greatly and appear more natural in many cases than a series of hen-scratching lines around the eyes and on the forehead, which always look false.

For character make-ups on Orientals and Negroids, the shading colors should be kept on the gray-brown to gray-black colors of S-14, S-4, and S-8, avoiding the redder ones (which would be unnatural).

Graying the brows produces immediately an aspect of age, and then one might proceed to the sideburn area and the rest of the hair. Otherwise, the same principles for screen make-up can be followed.

In closing, avoid overpainting children and follow the recommendations on page 73 for screen makeup.

CHAPTER 8

Personal Revitalization Techniques

BEYOND THE SCOPE OF COSMETICS PER SE, AND WITHIN the realm of the secrets of youth seemingly possessed by some performers who appear to have escaped the ravages of time and age in an almost unbelievable manner, is both a medical and a mental attitude or direction sometimes based on either new or even questionable techniques that seems to recreate or revitalize a person's body, mind, and attitude to produce unusual longevity—not of numerical age but of youthful appearance.

To understand further the basis of these techniques, we will need some introspection and open-mindedness that some might find as a new door to the personal revitalization methodology. In this aspect, the human mind is a complicated and interactive organ that controls the activity of our life in many ways. One such course is the desire for escape from the situations or persons that are commonplace to that individual. Such an escape valve is the use of entertainment to relieve the stresses that have become too prevalent in our society.

Entertainment, to an individual, can take many forms, but one of these is to watch the performance of others in creating incidents that occur in life or even fantasy. Such are plays, shows, and films, expressed in the mediums of the stage, television, and film. They are escapism from the personal norm or watching how others cope with their lives and problems. Therefore, here is created the world of those who *do*—the actors and actresses; those who *aid*—the production crews; and finally, those who *observe*—the audience.

In such a context, the audience usually considers itself to be separate from the aura created by the charisma of the performers and their peripheral personnel, and a reverence and awe of such performers becomes a natural direction of thought. Whatever it is that creates the chemistry of this attraction is inexplicable, but it is the ingredient that makes a certain person a star, producing in the audience an almost abnormal dedication and excitement in seeing, hearing, and enjoying the visual and audible performance of such a person. Star quality is attributed to those whose every action becomes unusually interesting to other people, both on and off stage or screen, and special qualities such as beauty of face, body, expression, or other appeal

to the inner mind and imagination are inherent in this state.

Some performers are instinctive in this aspect, while others require assiduous training and/or publicity to bring it forth. The degree of appeal that reaches the audience is often controlled by the amount of exposure such a performer receives—that is, how often he or she is seen or heard. Also, to be able to recreate life in performance, actors and actresses must possess certain qualities that enable them to paint the picture mentally for the viewers. The performer must *play* the doctor, the nurse, the secretary, the boss, ad infinitum and not only encompass the physical strain required to play the part but also enter into the mental and emotional areas of the person he or she is portraying. As such, performers are often subjected to unreal and sometimes overly emotional strain that other people (the viewers or audience) may never have to face. Oftentimes, it is difficult for some performers to separate their personal lives from those they play, and overlapping and enveloping attitudes may result.

Avocations to relieve such strains are often hard to find for performers because, since they can be so many people on stage or screen, little release from their work is achieved. Temperament, loneliness, and a desperate desire to be unrecognizable at times in a crowd may pervade. However, there is always the return to the reality of their hard-won laurels and position, and the maintenance of their face and form becomes important to a serious degree. Performers, it seems, more than anyone else, want to retain their youth or at least their status quo appearance. Make-up does aid this illusion, but there are also other avenues open today for routes unknown or unexplored heretofore.

PHYSICAL AND COSMETIC SKIN CARE

Cosmetic skin care, along with the proper selection of make-up materials, will aid in keeping the necessary moisture level in the facial tissues. Beyond this, there are means of irrigating the pores of the face with steam, sauna heat, ozone machines, and other similar and new directions. Some are effective and others pure puffery, so one must carefully research the methods and equipment beyond the salesperson's viewpoint and pitch. Many systems just soothe out ruffled feathers, so to speak, and are more psychological than truly rejuvenating or effective.

Skin peeling is another method of attack, with both chemical and vegetal peels being employed in addition to dermabrasion techniques. Peels and dermabrasions are often best left to the medical profession although some of the vegetal or enzymatic peels are harmless enough and do slough off some of the dead skin cells on the surface. However, surgical deep peels are another item, and as they contain phenol (carbolic acid), which can dissolve human skin, they must be applied and handled with great care and precision. Never take the advice of a cosmetician or an aesthetician who claims that he or she can do a face peel with this method, and only trust to a cosmetic surgeon this possibly dangerous operative technique. A medically controlled skin peel can give a person a really fresh and new-appearing skin texture, but one must remember that such a peel leaves the skin without its normal protective layer of dead cells so that exposure to sunlight can severely scar or discolor the face. Exposure to the sun and even daylight is not recommended for some days after application and only for limited periods for some weeks later. Amazing results can be obtained in skin coloration and texture after a properly applied and cared-for deep peel, and many Hollywood beauties have resorted to this method when they begin to show signs of aging. A word about sunlight and suntan is necessary here. According to U.S. Food and Drug Administration studies, ultraviolet radiation from the sun is the leading cause of cancer in the human body's largest organ—the skin. More than 300,000 cases just in the United States are reported annually. While skin cancer is estimated to be 95 percent curable if treated in time, experts feel that it is still a matter of grave concern. Premature aging as well as the incidence of skin cancer are accelerated if persons expose themselves to the bronzed look so revered by some. Even those highly touted sunscreens often do not possess sufficient sun exposure rating material to be fully effective, and the drying out of the skin tissue during suntanning will certainly serve to age a person's skin. Light-skinned Caucasoids are especially vulnerable and susceptible to sunburn, which is painful and often unnecessary, to say the least.

For a man to look rugged and suntanned is acceptable (although the exposure is skin aging nevertheless), but for a woman, it spells early loss of suppleness and with it, youthful skin structure. The old film studios sometimes had clauses in the contracts with their important female stars who were making films in Technicolor that prohibited them from exposing themselves to the sun to get a suntan. Some of these now-older female stars have better facial skin than many of the younger ones who delight in the suntanned look. It has always been my theory that if a woman wanted to get a suntan every year until she was 20, it couldn't hurt too much. But after that, every year she got a tan between the ages of 20 and 30 could add another year to her skin age! That simply means that if a performer was 30 and had gotten a suntan every year since the age of 20, her skin would appear as at least a 40-year-old. Between 30 and 40, the incidence increases again by a year, so that her skin could age very rapidly, irreversibly, forming small eye wrinkles at a rather early numerical age. Again, performers have to give up certain life-styles to attain and retain their youthful appearance.

Physical exercise certainly plays an important part in keeping the body vital and supple, and a program of exercise is essential to performers. It is not necessary to achieve the status or stature of a heavy weight lifter, but simple upper-body isometrics along with leg exercises will aid to maintain the body beautiful.

Many performers have a stringent regimen of body care along with facial skin care, while others may resort to periodic concentrated efforts in this direction, attending expensive and highly regimented spas or clinics that specialize in mental, physical, and facial regimes. Also, some beauty salons, men's styling salons, and so-called beauty clinics are giving considerably more attention to cosmetic skin care than heretofore. Certainly the pages of the leading beauty magazines are constantly full of such recommendations, so there is a new awareness pervading the beauty industry.

One beauty clinic or spa recommends the following four-day program of skin care and treatment along with many of the other programs of medical, physical, and mental techniques for their clients.

FIRST DAY
Moisture lotion application;
Witch hazel application;
Analysis with a strong light source and magnifier;
Steam application with steam-producing machine;
Removal of surface impurities (blackheads and so forth);
Light astringent and witch hazel application;
Massage with avocado oil;
Application of a deep-cleansing vegetal masque;
Removal with warm damp towels;
Witch hazel and moisture lotion application.

SECOND DAY
Moisture lotion and witch hazel application;
Enzymatic peel-off masque (35-to-40-minute set time);
Witch hazel and moisture lotion application.

THIRD AND FOURTH DAYS
Moisture lotion and witch hazel application;
Embryo masque (employing fetal chicken egg embryo)
     with a 20-minute set time;
Witch hazel and moisturizer application.

SKIN TREATMENTS

Skin *treatments* differ from skin *care* in that they should be in the province of the medical profession rather than the cosmetic one. Those that have skin problems such as active acne or allergy irritations and rashes should not rely on a cosmetologist-cosmetician-aesthetician for prescribing and treating such but should seek out a physician of dermatology or a cosmetic surgeon. Many highly advertised self-treatments have been found, by investigation of medical authorities, to be based on pure quackery, and the misapplication of some substances to the face or body can cause serious injury when left in the hands of well-meaning but nonmedical personnel.

It is always sound advice to performers who suffer from breakouts like cold sores to avoid putting any form of cosmetic or make-up over them as this can spread the sores or even cause further infection. Make-up artists who are asked by performers to cover problems such as open cuts, cold or acne sores, and such should not place themselves in a position of being liable to lawsuit if the application of any product carried in their make-up kits does cause any infection. It has always been my procedure to report any such medical problem to the assistant director and let management make the decision. If they decide that the sore or cut must be covered, absolve yourself of any blame by taking out a small portion of foundation, lipcolor, or such with a stainless steel spatula and placing it on a disposable surface. Tell the performer that you do not recommend its use and that he or she is on his or her own relative to the make-up or cosmetic material contemplated for use. A witness is a help in these cases for security.

MEDICAL MULTITHERAPY APPROACHES

A new approach to the maintenance, revitalization, or renaissance of both internal and external youth is the use of various multitherapy medical directions. Clinics that promote this type of therapy have sprung up around the world, with some of the major ones being in the Bahamas and in Germany. They employ what they term a *holistic* approach to health and include thalassotherapy (seawater massage and inhalation, embryotherapy (partially incubated fertile eggs), phytotherapy (treatment by plants and extracts), cell therapy (implantation of fetal cells), and a psychological approach to treatments of this kind. Devotees of these clinics, some of whom are willing ambassadors of their techniques while others do not want to divulge what they feel is their particular secret of youthful vitality, include a host of stage and screen personalities, political leaders, and many others who are subject to emotional stress and exertion.

Cell therapy is the implantation by inoculation of homologous or heterologous fetal or juvenile cells or tissue substances for the purpose of compensating for or correcting deficiencies of organs and functions. In brief, such cell implantation is believed by its proponents to work on human body areas that need a stimulus to renew normal functions. The father of cell therapy was Dr. Paul Niehans, whose clinic in Vevey, Switzerland, was a mecca for those who had heard about his researches and results and wished to experience his ideas in this field. Although he commenced by employing cells from live animals, the newest approach is the use of lypophilized (freeze-dried) cells—some 60 different varieties are available—taken from the tissues of the fetus of a lamb or calf and prepared under the strict control of the medical authorities at the University of Heidelberg and the German government, in a clinic and with a method instituted by Dr. Niehans.

Such cells are single-dose ampules labeled according to tissue origin and accompanied by a 5-cubic-centimeter ampule of modified Ringer's Solution as a dilutant. The combination is prepared under controlled medical conditions and administered only by a trained physician. Such treatments may include anywhere from four to six implantations by inoculation, depending upon the results of a very thorough and completely evaluated health and physical examination at the health center.

In some such clinics, cell therapy may be accompanied by other therapies such as embryotherapy, thalassotherapy, and others, making the usual visit or session last about 10 days. Most of the German clinics only deal with the cell therapy, however, and claim that in addition to the possible revitalization of the internal body functions, in some cases it also controls serum cholesterol (a factor in weight gain) as well as being an effective geriatric retardation of degenerative diseases. In addition, cell therapy is claimed to produce an increase in energy level and vitality in a sense of well-being.

The incidence of allergic reactions to such cells is claimed to be practically nil, but those with severe alcohol or drug problems are not encouraged to take the therapy until they go through a period of detoxification to place them in better physical condition. It should be said that those who advocate these special therapies combine them with vitamin therapies, nutrition programs, and other related directions to stimulate biological functions and achieve results not obtained by medication alone, which often aims at symptoms and not fundamental causes in the idea of health.

This new method relies on a utilization of an internal direction that many performers have found to be a great new secret of revitalization in their lives. They complete the program with the psychological feeling that their physical and mental capacities have been aided, and therefore their emotional ones also benefit.

One other medical treatment is that of a form of buffered procaine that is administered by intramuscular inoculation or orally by tablets. It is said to be a vitaminic eutropic factor and regenerator for the treatment of specific phenomena and other trophic disorders in elderly persons and side effects like improved skin elasticity have activated its prescription and use by some clinics.

Again, as for all cosmeticomedical approaches or preparations, such should be prescribed or administered only by a physician trained in their use and possibly after a consultation with a confirming opinion (see bibliography in appendices for further references).

COSMETIC LENSES AND EYES
By George V. Gianis, O.D.*

Performers who must wear eyeglasses often find that by having soft or hard contact lenses fitted to their eyes their appearance and facial aspect will be considerably changed and often improved when the glasses are no longer present. In addition, the long wearing of eyeglasses sometimes furrows the nose bridge and leaves contact point marks, especially when a performer does not wear glasses on the set but then puts them on offstage to read or see better. Here the make-up can get smeared in this delicate area and require constant retouching. Contact lenses eliminate this problem.

Today's contact lenses allow much freedom of movement, and even some football players wear them during a game. Contact lenses may be obtained in a number of forms and styles in both the soft and hard varieties, but for cosmetic eye color change purposes, the hard variety is often employed as the iris can be made any color, being the opaque portion of the lenses, with the subject seeing only through the pupillary area of these lenses (Figure 8.1). The soft type is normally designed only for vision correction, but is also available in colors.

All contact lenses must be individually fitted to the subject's eyes, and such requires the services of an *ophthalmologist* (eye surgeon) or an *optometrist* (eye doctor). *Opticians* are shop or lab personnel who work on the lenses or may fit eyeglasses to a person. Once the cosmetic lenses are fitted to the subject, the subject must be taught how to retain them in position. Perfectly fitted lenses require some practice in applying and retaining the lenses over a period of time, so some

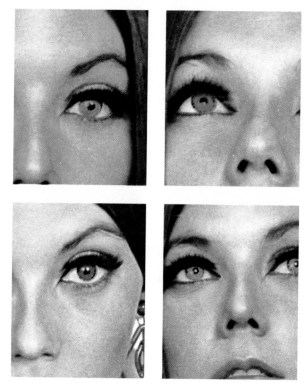

FIGURE 8.1   *Eye color changes with Caprice® colored lenses.*

wearing of the lenses prior to their use on the set is highly recommended.

Lenses are cleaned with a special solution provided by the doctor who fitted them, and the application and removal procedures should be carefully observed by both the subject and the make-up artist. Most physicians today recommend the use of the fingers for this procedure rather than the older method of employing small suction cups. Due to this method of placing and removing the lenses, mainly by feel as it were, it is far easier for the subject to perform this than the make-up artist.

Any make-up that must be applied to the face should be done *after* inserting the lenses, but hair spraying should be done *prior* to insertion unless the face is fully covered. Although most hard lenses can be carefully cleaned of such sprays, soft lenses are often damaged beyond use.

When a completely false prosthetic eye is required for a project, the same doctors can provide these, or one may contact the local offices of either the American Optical Company or Bausch and Lomb for a variety of these. Such firms have offices in many major city areas. These artificial eyes are best selected from the plastic variety rather than the blown glass variety because plastic is considerably safer and less breakable in studio use (also see pages 214 and 117 for character use of false eyes and eyeglasses).

*I am indebted to George V. Gianis, O.D., of Lowell, Massachusetts, for his patience and knowledge in researching this section as well as to Caprice® Cosmetic Lenses for photographs.

HAIR REMOVAL AND HAIR TRANSPLANTS

Excess body or facial hair can be removed by electrolysis—the surest method when permanence is desired. Other methods include the use of depilatory creams, hot and cold stripping waxes, or plucking the individual hairs with tweezers around the eye area or for scattered facial hair.

Depilatory creams are satisfactory for the removal of men's chest or arm hair when appliances are to be adhered to those areas of the body. Shaving only produces a stubble that prevents adhesion, while the depilatory will remove the hair completely, albeit temporarily. Most good brands are quite safe to use and only take about 10 minutes to denude the skin of hair. Hot and cold stripping waxes are widely used for upper lip hair removal from women, and any good beauty salon or aesthetician's studio can provide this service. Electrolysis studios can be found in the Yellow Pages of the phone book in most areas.

Conversely, replacing hair, especially on men's heads, has become a big business today, and many cosmetic surgeons specialize in this procedure. Hairs are normally taken from the man's head at the rear of the neck area for transplanting in the front. Much success in some cases has spurred many male performers to seek this beauty aid, and although the process is a long and often expensive one, it precludes the use of an attached hairpiece for many men. There are, however, some bogus firms that advertise enormous success with this method, but for safety's sake, seek a qualified physician who specializes in this service.

COSMETIC SURGERY

In recent years, the art of *plastic,* or *cosmetic,* surgery has not only improved in fantastic strides but also has become an accepted form of retaining a youthful aspect of the face and body through the adjustment or reconstruction of facial features; the removal of excess tissue that occurs because of aging processes or even obesity; the rejuvenation of the external aspect of the skin to delete facial aberrations such as acne scars, wrinkles, and such; and special injectables that are seeing increased usage to fill out certain areas of the face to minimize age lines. Performers on screen, who are always under close scrutiny by the unforgiving camera, have found that eyelids can be lifted, eyebags removed, wrinkles eliminated, facial folds decreased, sags taken out of necklines, chins firmed and reshaped, foreheads smoothed, breasts augmented or reduced, buttocks lifted, legs and arms reduced in fat content, and many other visually desirable changes made by cosmetic surgery that a make-up artist could not possibly achieve with cosmetics.

The American Academy of Facial Plastic and Reconstructive Surgery publishes a pamphlet, "How to Select a Cosmetic Facial Surgeon" (see bibliography), which has some excellent suggestions and information on cosmetic surgery and its various specialist surgeons. Although certification by a board of specialists in their area of surgery is important, all doctors must be considered on the basis of individual qualifications. Selecting a surgeon who is a specialist in certain aspects of this work is important, and the person who elects to have the surgery must have full and complete confidence in the surgeon. It is also important that a person is psychologically prepared for the changes that will occur as well as healing time.

General Plastic Surgeons

These are the general practitioners of plastic surgery in that they operate on all parts of the body. Not all plastic surgeons do cosmetic surgery. Some perform reconstructive surgery only; many do both. Others may limit themselves to cosmetic surgery on the face alone or on the entire body.

Head and Neck Surgeons (Otolaryngologists)

Otolaryngologists specialize in the head and neck area. Some do cosmetic surgery only. Other do both cosmetic and reconstructive surgery in this region of the body.

Eye Surgeons (Ophthalmologists)

Ophthalmologists do cosmetic and reconstructive surgery about the eyes. Very few limit themselves to cosmetic surgery alone.

Skin Surgeons (Dermatologists)

Dermatologists do mostly skin sanding, hair transplantation, and removal of small new growths. Some do skin surgery of the eyelids and other types of cosmetic surgery of the face.

COSMETICS AND SURGERY
By Robert O. Ruder, M.D.*

Every make-up artist is called upon to create a better illusion by color, appliances, and hair design. However, these techniques are temporary alterations, comprising the initial phase of achieving a better facial aspect. When one's features require greater modification and improvement, the reconstructive and cosmetic surgeon should be consulted.

In this case, make-up artists must assume the position of both a specialist and a liaison because they

---

*To get a medical opinion on this work, we have asked Dr. Robert O. Ruder, assistant clinical professor in the Department of Head and Neck Surgery and Facial Plastic Surgery at the UCLA Medical Center, to comment on the various aspects of his professional work.

should have realistic information about basic and common cosmetic surgical procedures, their advantages, and their limitations as well as a list of surgeons who perform these operations. As well, clients who expect minor miracles from make-up artists often expect major ones from surgery and as such, risk disappointment. *Improvement* is a more realistic goal than perfection in both cases, thus blending the features in a better proportion. In some cases, performers, familiar with their own particular features, often self-diagnose their problems with surprising accuracy, while others merely are aware that they don't look good. Surgical alteration should be considered when make-up techniques can no longer achieve an acceptable or reasonable improvement. These changes can include skin resection (brow lift or blepharoplasty), skin dissection and repositioning (facelift), feature reduction (rhinoplasty or otoplasty), and feature augmentation (chin implant, collagen injection, and so forth).

Nasal Surgery

Aesthetically, the face is divided into thirds, with the nose occupying the middle area. Attracting the greatest attention, this middle third can be further subdivided into an imaginary triangle with the eyes and eyebrows as its base. The eyelids may bulge with age, but the nose drops, distracting one's attention from all other facial features. A well-proportioned nose will serve to highlight one's eyes and cheeks and will de-emphasize a protruding or weak chin, thin or thick lips, and early laxity of the jowls. More important is the vital function that the nose performs. It not only is our organ of smell but also conditions the air entering the lungs through its cleaning, moistening, and warming abilities. Obstructed, inadequate nasal breathing may cause headaches, sinus problems, sore throats, or changes in senses of smell or taste. The procedure for correcting deformities of the nose is called *nasal reconstruction,* or *rhinoplasty.* This operation consists of the removal of excess bone or cartilage and the subsequent rearranging and reshaping of the new nose.

Photographs are usually taken before the operation to assist the surgeon in analyzing problems and in planning the surgery. At this time, the various areas can be defined, such as the profile (e.g., a hump or depression to be corrected), the bridge (too wide or too narrow), and the tip (shape and position). Such planning is necessary to set the parameters of change and to establish the desired goals. However, certain inherent surgical limitations may make it impossible to guarantee that a particular ideal can be accomplished in some cases.

All surgical work is usually performed inside the nose, leaving no scars on the outside. Nevertheless, there are exceptions. If it is necessary to make the nostril smaller, for example, an incision is made where the nostril meets the upper lip. As this is located in

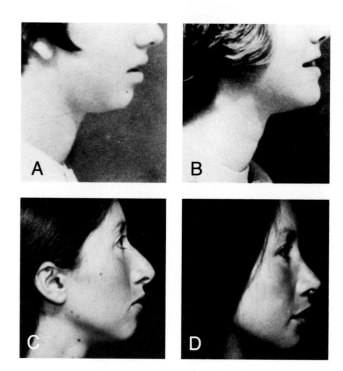

FIGURE 8.2   *Two examples of nasal surgery accompanied by chin implantation.* TOP: *Before and after.* BOTTOM: *Before and after.* (Courtesy Dr. Robert Ruder.)

a natural body skin fold, the scar is practically invisible within a few weeks. Following surgery, tape and splints are applied to the nose for one week. After this time, most of the discoloration and swelling has disappeared. Within the ensuing six weeks, the majority of the swelling has subsided. Generally, it takes up to one year for the last 1 to 2 percent of the final healing to take place. Chin implantation is often performed at the same time as nasal surgery to improve the profile further (Figure 8.2).

Eyebrow and Eyelid Surgery

The eyes complement the middle triangle of the face. However, they compete with the nose for attention. Unlike the nose, the expressive action of the eyes magnifies even the slightest blemish and emotion. The surgical procedure to correct wrinkles, puffiness, and excessive skin folds of the eyelids is called *blepharoplasty.*

The skin of the eyelids is the thinnest and most delicate in the body. This thinness and progressive loss of elasticity and tone of the underlying muscles and tissues allows the aging process to begin cosmetically in the eyes.

The development of eyelid bags, starting in the mid-twenties, is intimately associated with herniated orbital fat that can only be improved by surgery (Figure 8.3). As with every other type of surgery, operations intended to improve sagging skin or wrinkles necessarily leave scars. All external surgical scars are permanent and cannot be erased. The job of the surgeon is to place these scars in natural lines of the eyelids where they are least noticeable and are more easily camouflaged by cosmetics. The operation is usually done on both eyelids and on all four lids (upper and lower), often at the same time as a facelift. This procedure, done in the surgeon's office under local anesthesia, takes one to two hours. Most of the sutures are removed after three days, but some are left in for about 1 week. Make-up may be worn three to four days after the removal of all the sutures to cover any discoloration around the eyes. Usually, most of the noticeable swelling and blueness will disappear within 3 weeks, with a better result, photographically, in 12 weeks.

However, all wrinkles around the eyes cannot be removed by blepharoplasty alone. If any noticeable, disturbing creases persist, chemosurgery (chemical peeling) can be performed two to three months after the surgery. The surgeon's goal is to enhance the eyelid impression and to produce more youthful-looking eyes but not to remove all of the lines of life's experience.

How long a blepharoplasty will last varies with each person and, particularly, with the individual's style of living. Smoking, alcohol, and excessive sun exposure cause fluid accumulation. Due to the fact that the

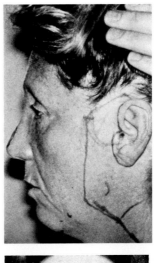
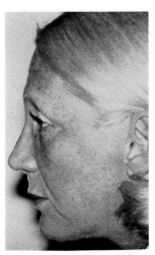
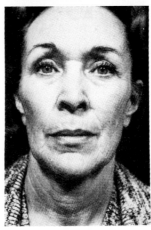

FIGURE 8.4 *Facelifts.* TOP: *Before and after.* BOTTOM: *Before and after. These before and after facelifts demonstrate the general amelioration in facial aspect frontally and in profile. The red pencil markings on the top left photo shows the correction of a nose hump and the incision indications on the ear area. (Courtesy Dr. Robert Ruder.)*

upper one-third of the face stretches and sags less with aging than the lower two-thirds, eyelid and eyebrow corrections are more lasting than facelifts. The person continues to age, but the eyelid skin will be smoother and tighter than it would have been without surgery.

Facelift Procedures

Four major differences set the old face apart from the young:

1. Changes in skin color,
2. Changes in the skull (which becomes thinner through bone absorption and occupies less volume or shrinks),
3. Natural lines of expression of the face that deepen and become more apparent,
4. The sagging of the skin into bags, jowls, pouches, and wattles due to the loss of skin elasticity and atrophy of fat, plus the effects of gravity.

These changes are more pronounced around the eyelids, the corners of the mouth and nose, and the front

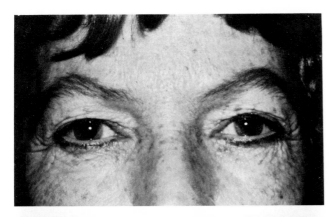
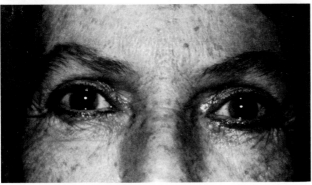

FIGURE 8.3 *Correcting upper eyelids.* TOP: *Pre-operative overhanging upper eyelids accent the age of the person and prevent a good eye make-up.* BOTTOM: *Postoperative photo shows a vast improvement in the upper eyelid area with a successful lessening of the overhanging flesh and allowing the proper placement of eyecolor on the eyelid. (Courtesy Dr. Robert Ruder.)*

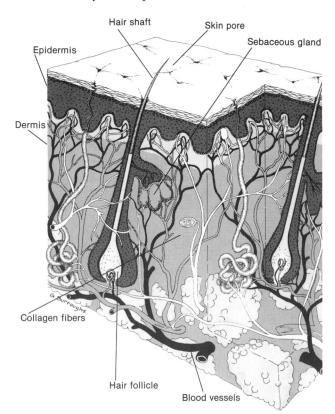

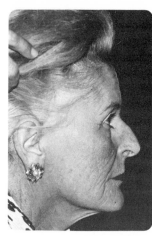

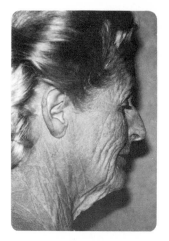

FIGURE 8.5 *A cross section of human skin.*

FIGURE 8.6 *Chemical peel on the face. Before and after photos showing considerable skin texture improvement and smoothness. (Courtesy Dr. Robert Ruder.)*

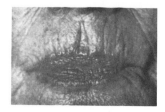

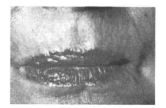

FIGURE 8.7 *Chemical peel around the mouth, before and after. The purse-string lines around the mouth are greatly minimized. (Courtesy Dr. Robert Ruder.)*

of the neck. Only an operation known as *rhytidoplasty* (facelift) has been able to achieve a satisfactory illusion of youth by reducing the wrinkles, jowls, and loose, hanging skin (Figure 8.4).

The facelift can be performed in a hospital or as an outpatient surgical procedure. The surgical method is to make incisions in front of and behind the ear, placed in the natural creases of the skin, and in hair-bearing areas of the scalp in order to camouflage the scars. The skin is separated from the underlying tissues and rotated to remove only the excess tissue. The excess skin is removed and the remaining skin sutured back into place.

Following surgery, dressings are applied for five days. Swelling and discoloration, usually not painful, will generally subside by the second week. After three weeks, one may resume other outside activities. Some may experience occasional twinges of pain and possibly a numbness of the neck, face, and about the earlobes, but this is a normal healing process and should disappear in six to eight weeks.

If performed properly, performers need not worry about loss of facial expression due to facelift. Gentle, conservative surgery will not produce a tight, masklike appearance because facial movements are controlled by the muscles, not the skin. The skin, which is intimately attached to the underlying muscles, is merely a drape and moves in unison with muscle contraction.

The proper time for surgery is based on aesthetic rather than chronological age. The largest percentage of facelifts, however, is performed on those in their forties and fifties.

The Skin

The skin constitutes the largest organ of the body and plays a most significant role in reflecting one's health, protecting and performing basic functions, and acting as a showcase (Figure 8.5). The skin is alive, versatile and variable and is the area where the make-up artist and the surgeon must work together to improve it. We can cream it, bleach it, lift it, and paint it, but we literally cannot shed it.

Skin color is determined at birth and cannot be altered under normal conditions. Five basic pigments admixed between the layers in various proportions achieve assorted blends. Oxyhemoglobin (hemoglobin containing oxygen) imparts a rosy glow; reduced hemoglobin (lacking oxygen) gives a bluish tint; carotene gives a creamy golden hue; melanoid adds shadowy tones; and melanin imparts the subtle light and dark variations of skin color.

Lines and wrinkles are caused by the skin's loss of elasticity as collagen and elastin fibers within become brittle and thin. If the skin further atrophies, it loses its velvety texture and color. Creams, lotions, and gels can lubricate the skin and cosmetically conceal a multiplicity of faults, but surgery has the most long-lasting effects and yields the smoothest skin. Three different surgical techniques are now commonly being performed to improve some aging effects of the skin not corrected by facelifting and blepharoplasty operations.

Peel

Chemosurgery (chemical peeling) mechanically removes the outer layers of the skin (Figure 8.6). Peeling particularly improves shallow and moderately deep facial wrinkles, freckles, pigmentations, and skin blemishes. Eyelid wrinkles and vertical purse-string lines around the mouth are dramatically improved by the peel (Figure 8.7). Face peeling creates a gentler, more natural firmness and smoothness of the skin, and its texture becomes softer while the complexion brightens.

The main chemical agent used in peeling is either phenol (carbolic acid) or trichloroacetic acid. One should be extremely selective in choosing an experienced surgeon for this work as harmful side effects are most likely to occur at the hands of those who have little or no training in this procedure. (Author's note: Chemical peeling should *only* be done by trained medical personnel and *never* entrusted to a cosmetician/aesthetician.)

Facial chemosurgery is performed in the surgeon's office with the patient sedated. Tape is often applied to all areas of the face to render a deep peel for severely weatherbeaten skin. After four days, the tape is removed and powder is applied. Nonallergenic cosmetics can be used ten days postoperatively to cover and protect the smoother, softer, pinker, and more resilient skin.

Dermabrasion

Dermabrasion, also known as *sanding,* is another way of treating facial scars and blemishes. While chemosurgery is now being utilized for finer wrinkles, dermabrasion finds its best use on patients with more deeply pitted acne scars. Although abrasion will improve facial scarring considerably, it will not completely smooth the skin. The goal of treating irregular surfaces of the skin is to sand down the high parts so the lower ones (depressions) appear to be less deep. As the face heals after dermabrasion, a new layer of epidermis forms over the underlying dermis.

The face responds better than other body parts to dermabrasion and peeling because it contains the highest density of regenerating tissue (hair follicles and sweat glands) and has a rich blood supply for optimum healing. The skin of the neck fares less well because it is thinner and less vascular and has less regenerating tissue.

Patients having dermabrasion heal similarly to the peeling procedure, but because both techniques remove the outer layer (epidermis) where the pigment resides, areas will have a slightly lighter look for three months that can be covered with cosmetics. Although one should rigidly avoid direct exposure to sunlight for this period, most patients resume their normal activities after three weeks.

Collagen

The newest, most innovative method to soften severe wrinkles, is the use of injectable collagen. The furrows that appear on the forehead between the eyes (known as glabellar lines) and wrinkles from the nose to the sides of the face (nasolabial folds) seem to respond best to collagen injections (Figure 8.8). Purse-string wrinkles around the mouth and tiny creases in the skin surface respond less favorably.

Collagen is a fundamental protein of the body, and the human body accepts this material as belonging to itself. Therefore, injectable collagen becomes indistinguishable from the surrounding tissue. Some patients may see little improvement after the first injection of collagen since the first session is to stretch and soften skin defects. Most people begin to see good results by the end of their second or third treatment.

Other Procedures

Many other innovative procedures such as lip recontouring, fat suctioning, hair transplantation, and scar revisions are extremely helpful and can be performed at the same time as these other procedures. Before recommending any surgical procedures, make-up artists should consider if the person is exotic, fashion oriented, conservative, subdued, or flamboyant and what *they* hope to see when a surgical procedure is finished. The surgeon must see that all changes should appear natural and well blended to the person concerned. Finally, the main criterion should be that when facial goals cannot be accomplished by make-up alone, surgical procedures should be considered for improvement of the facial contour.

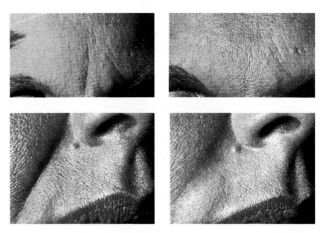

FIGURE 8.8 *Collagen injections.* TOP: *The forehead, before and after.* BOTTOM: *The nasolabial area, before and after. (Courtesy Dr. Robert Ruder.)*

In all cases, it is best for the make-up artist not to attempt to perform services and procedures that should be left strictly up to the medical profession. As well, covering with any form of make-up skin problems such as active acne, broken pimples, cold sores, and any other variety of open cut or abrasion should not be done by make-up personnel as there is always the possibility of an infection. Never take the verbal acquiescence of performers to place make-up over such skin problems, and always inform them that you will furnish the make-up, if they so desire to cover it, in a separate plate or container for them to apply to themselves. It is always a good idea to have a witness to your words—and the performer's—or to get a written release. Only then is the make-up artist not liable if any infection does occur as a result of make-up application on an open skin area.

COSMETIC DENTISTRY
By Jean Marie Doherty, D.M.D., D.Sc.

There are a variety of dental conditions that, in addition to being unhealthy, can create an unappealing appearance. Maintaining good oral health is especially important to performers because teeth are critical to speech as well as aesthetic aspects. With the advances made in dentistry in recent years, it is much easier than before to achieve proper dental health and excellent cosmetic results with less pain and inconvenience than heretofore.

Jaw Relationships

One important function that teeth perform, and perhaps the least understood by non-dental professionals, is the maintenance of jaw relationships. The teeth hold the upper and lower jaws a specific distance apart, providing support for the lips and facial muscles. If this distance is decreased, then the facial muscles and skin are not supported, and deep lines may be seen around the nostrils and wrinkling around the lips. Loss of *vertical dimension*, as this condition is termed, may result from wear of the teeth or dentures or undererupted teeth. This condition may be corrected by any number of techniques including crowns or caps, bridges, or removable dentures.

Orthodontics

The relationships of the teeth within the jaws is also an important factor in an aesthetically pleasing smile. Most people are familiar with the term *orthodontics*, or the use of metal bands and wire braces to move the teeth into a better alignment, correcting overbites, crooked teeth, and such. Although most commonly done on children and teenagers, recent advances in materials and techniques have made orthodontics more popular among adults, especially those in the public eye. Clear plastic brackets bonded to the teeth may be

substituted for metal bands. The use of bonded brackets has also eliminated the pain that formerly accompanied separation of the teeth prior to placing the metal bands. Lingual braces, also called *invisible braces,* use brackets and wires applied to the inside of the teeth near the tongue. These braces (though more difficult for the orthodontist to adjust and, therefore, more costly) are favored by many performers since they pose little worry in close-ups.

Crowns or Caps

Crowns or caps have been employed for a number of years to improve the appearance of the teeth. In this procedure, the natural tooth is reduced by grinding, evenly all over, and a crown is placed over it, restoring it to a natural-appearing color and shape. Crowns may be made of porcelain fused to gold or other metal or porcelain alone. The porcelain and metal crowns are stronger and, for most people, provide an excellent aesthetic result (see Figure 8.9). Occasionally, all-porcelain jacket crowns may be preferred by certain individuals, like singers, who may object to the small margin of metal that is sometimes visible on the tongue side of the crowns when the mouth is opened very wide.

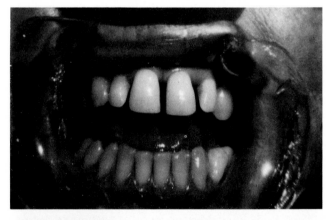

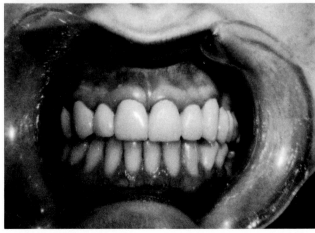

FIGURE 8.9 *Jacket crowns placed for aesthetic purposes over existing sound natural teeth improve the total aspect of the mouth. (Photos and jackets by D. Lawrence Fadjo, D.D.S.)*

Bridges

Dental bridges are used to fill spaces where teeth are missing and can be constructed so they are quite undetectable. The teeth on either side of the space are prepared for crowns. When the crowns are made for these teeth, a false tooth is placed between them, filling in the space, so the three teeth appear quite natural.

Bonding

There are alternatives to the traditional crowns and bridges when spaces need to be closed or teeth replaced. One method is *bonding,* a process utilizing monomer/polymer plastic material that is applied to the tooth surface to change the shape and color of the tooth.

The bonding procedure can improve the appearance with little or no grinding of the teeth in a single visit to the dentist at a much lower cost than jacket crowns. Although bonding will last for a number of years, it is not as satisfactory a procedure as are crowns and may require repairs or repeat applications in time. The plastic material (usually acrylic) also may be stained by smoking or by drinking coffee or tea.

Bonded bridges are a new alternative to traditional bridgework and do not require that the adjacent teeth be capped. Instead, a tooth with small metal wings is made to fill the gap. This is bonded with plastic material to the adjacent teeth on the tongue side, making the attachment invisible. Bonded bridges are most often employed where the surrounding teeth are healthy and without fillings.

Periodontics

Good oral hygiene, which includes daily brushing and flossing, is of utmost importance in maintaining healthy teeth and preventing gum disease. Even the most beautifully made crown will not last if the gums and supporting bone are affected by pyorrhea or periodontal disease. The first signs of such diseases are a redness of the gums as well as bleeding while brushing or flossing. Prompt treatment can prevent or minimize receding gums and mobile or loose teeth.

With all the advances in dentistry, no denture or appliance looks or functions as well as the real thing. Regular dental check-ups and cleaning are essential and a real investment for the performer for health as well as appearance.

Conclusion to Part II:
The *Right* Make-up

THERE IS NO PERFECTLY RIGHT MAKE-UP FOR ANY MAN OR woman in a professional sense, only the visual interpretation of a concept that must be shared with the individual and those that work on the person or judge the make-up and hairstyles. Make-up artists often disagree on semantics as well as placement values of products to achieve what they consider to be the right make-up, but in the end, the client and the viewers must be satisfied.

Many times, performers will consider that only one make-up artist can do their make-up, only to try another and find that they like the second interpretation better. Preferences as well as services may lead to this conclusion, so a make-up artist should never feel slighted that a performer insists upon another to do his or her particular make-up. After all, it is strictly an artistic judgment in the end, and *right* is what they want it to be.

Interlude

THERE MUST ALWAYS BE A SEPARATION BETWEEN BASIC and advanced techniques in any field, but few other areas of endeavor find the line so clearly drawn and the talent for continuing the study so obviously drawn as between straight or beauty make-up and character make-up.

Many people who enter the make-up field learn to do a beauty make-up with certain degree of facility and success, oftentimes with considerable innovative and fashionable results. However, when many continue their studies and efforts in the direction of the more advanced area of character make-up, they either fail in interest or drive, or display little talent for the perfection and patience that is demanded in accomplishing the tasks required. It is a realm neither for the diffident or dabbler nor for those who do not have confidence in their skill. The climate is one of competition with challenges at each turn and of sharing knowledge with their peers. It is the ultimate in being a professional make-up artist.

Today, the old paint-and-powder styles are still important, but tri-dimensional make-up by the employment of appliances or prostheses entails considerably more diverse study than the old stage methods of utilizing nose putty, cotton and collodion, muslin, and so on. Molded plastic and latices, foamed materials, gelatins, waxes, and many other products and means are part of character make-up principles.

Another required skill is proficiency in the laying and cutting of facial hair for men. More applicants to union examinations fail in the correct manner of laying and trimming a beard and moustache than in any other phase of character make-up. Therefore, studying and learning character make-up that entails paint-and-powder techniques to mold making and appliance manufacture, to a sound expertise in hair goods is the purpose of the following parts of the book.

Every make-up artist with full capacities in such character make-ups will always be more in demand for challenging jobs and, therefore, will gain more respect in the profession. In addition, they learn that any created effect is never the last word on the subject but that further research will always show the way to something or some way that is better. All those who excel in character make-up are constantly researching and testing new ways either to do the old or discover the new or more exciting effect. It is insufficient to make one beard, one cast, one appliance, one effect and consider that one then knows all. Constant practice and redoing will aid to ensure proficiency and to learn something new or different each time the exercise is performed. That is the answer to why some make-up artists can lay an excellent beard in five minutes, while others will take a half-hour to do the same only adequately.

Part
III
Character Make-up

Share willingly what you have learned,
and others will share knowledge with you.

CHARACTER MAKE-UP

CHARACTER MAKE-UP IS THE APPLICATION OF MAKE-UP TO change the appearance of a person as to age, race, characteristics, or facial and/or body form. This change may be in any one of those elements or a combination of them. Character make-up is intended to aid the performer in the portrayal of a role by giving a facial and/or body likeness to the character being played. More often than not, additional preparation for the application of character make-up is required since many of the needed items may not be carried by the make-up artist in the everyday work kit.

Typically, the make-up artist will have a script conference with the producer and/or director of the project both before and after reading the script. Performers may be called in if there are any special problems or effects that are dependent upon them. As well, the make-up artist may submit preliminary sketches and ideas at that time. When the job is set, the preparation time then commences and, depending upon the complexity of the make-up, can take a day to possibly up to a year of research, tests, and manufacture of special items. During this time the make-up artist will often refer to stills and other films or tapes for matching a character or designing a certain or special effect. Sometimes motion picture or videotape tests are made to try the make-ups, and many concepts may be explored and discarded before the agreed upon make-up is put into work.

As character make-up often takes more time than ordinary beauty make-up, additional time and make-up personnel must be put on the schedule for the shooting days so production will not be delayed. It is common to have two or more make-up artists working on one subject for a complicated make-up procedure or character as time can be saved by this in the make-up room and eventually on the set (Figures III.1 and III.2).

A filing system of pictures is a great aid in selecting the various facial properties of a character. Illustrated journals and periodicals of general interest are good sources as are travel or geographical magazines. Old history books that can be cut up serve to fill out the period and historical characters, and art and natural history museums often have publications that are of interest. How to arrange the accumulation of material for a filing system can be found in Appendix D.

However, in the design of character make-up certain limitations in face and form must be borne in mind. Although make-up artists can perform seeming miracles of make-up, directors and producers sometimes have no conception of what can and cannot be done. A fat face or body cannot be made thin, and a large nose cannot be made smaller with make-up, but a thin face and small nose can be made larger. Although the director of amateur productions may take many liberties, asking the make-up artist to attempt to make changes or use people for a part who are not suitable, professional filmmakers or television and stage producers should not cast a performer who may be obviously wrong for that part. Unfortunately, however, many do commit this error and expect the make-up artist to bail them out!

Great care thus should be taken with casting, especially in the area of known historical characters, that the performer really appears as the character and not a caricature of him or her. For the most part, old stage techniques must be discarded for films and television make-ups, and although paint and powder can be relied upon for some changes, latex and plastic appliances provide a better tri-dimensionality.

Even these items have to be carefully applied so that the edges blend into the skin and the surface resembles the adjacent area. For close shots, character make-up ought to be well nigh impossible to detect as make-up. The camera's eye is often more critical and difficult to please than our own, so each make-up should be carefully checked and watched closely through the production. If the make-up is not convincing, more than half the battle is lost in maintaining a realistic characterization no matter how clever or talented the performers may be. They will simply not "look" the part.

Charts for individual characters and indications of the make-up to be tested, or after testing make-up that has been approved, can be made employing the information in Table 9.1 for male and female characters. An example of this use can be seen in Figure III.3 for a simple age character make-up.

FACIAL DISTORTION

In years past, actors on stage and early film make-up artists sometimes relied upon the use of silk fabric, chamois, and thread to distort certain areas of the face to achieve changes in structure. One such method was to pull down (or lift) the tip of the nose by running a fine, strong thread around the head to tie in the back. The tip of the nose was protected by a small piece of chamois leather to which this thread was attached. When the thread was pulled tight, it either depressed or lifted, depending upon where the chamois and thread was placed, the tip of the nose. The chamois was generally covered with some nose putty and heavy base make-up, and the thread was invisible to the eye of the audience. A fine example of this was displayed by Cavendish Morton, a British actor in an early make-up book, when he showed the Three Witches from *Macbeth* all done on his face using this method on two of the faces (Figure III.4).

Another way to distort the face is to use RCMA Scar Material to catch a few hairs of the lashes and adhere them to the face. The scar material, upon drying,

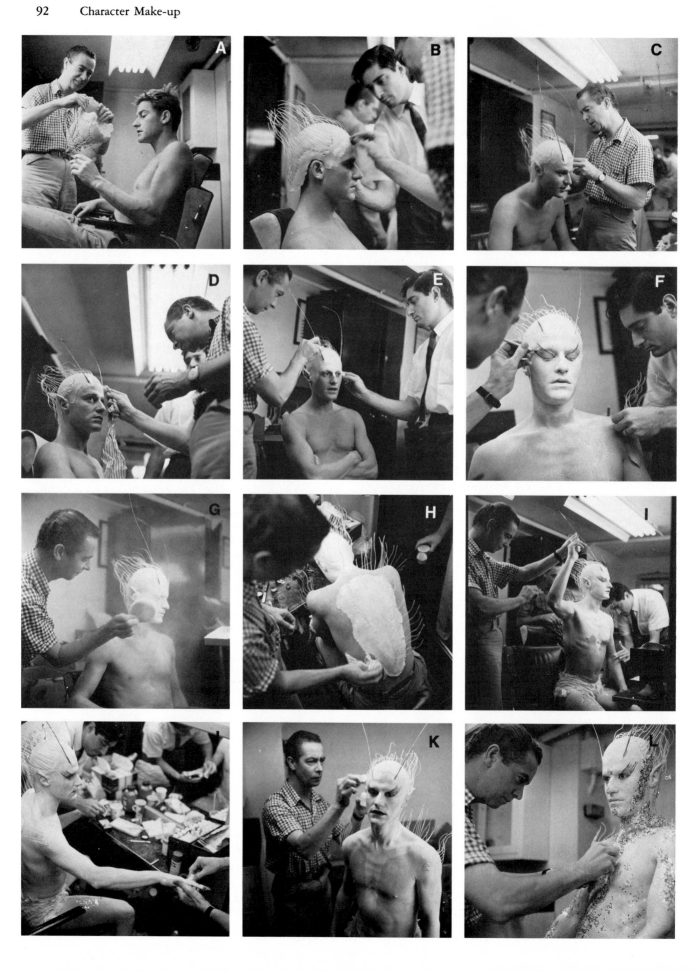

contracts and pulls the hairs away from the eyelid, distorting the eye. A similar use of this can be seen on the first witch in Figure III.4.

Silk muslin can also be employed for facial distortion (see Reversal Techniques, page 110), and heavy flesh under the chin or on the neck can be minimized using such lifts. Eyebrows can also be raised somewhat using lifts, providing that the hairdo allows any excess wrinkles caused by the lifts to be covered. Wigs and costumes can do much to disguise those portions of facial distortions that might otherwise give the illusion away. Lon Chaney, Sr., used facial distortion make-up methods in many of his famous characterizations, and one shows how the costume was cut around the jawline to effect sunken, hollow cheeks (Figure III.5).

MULTIPLE CHARACTERS

It becomes a different and often exciting challenge for a make-up artist when one performer must be made up to play many or varied characters in a series or within a production. Much depends upon the budgetary factors involved, but foamed appliances would normally be the ideal method to create the different faces. When expense is a factor, one must seek less

FIGURE III.1 *Character make-up for actor Roddy MacDowall for his portrayal of Ariel in Shakespeare's* The Tempest *(WNBT-TV).*
(A) The author prepares to attach a foamed latex cap piece set with plastic spines, designed and made by Robert O'Bradovich. (B) After the cap is attached with Matte Plasticized Adhesive, the edge is stippled with eyelash adhesive and dried. Assisting in the make-up is Bob Laden. (C) The antennae (made of long feather spines) are attached. (D) Foam latex ear tips are adhered. (E) The face and cap are covered with a white AF-type base. Speed is accelerated by the use of two make-up artists. Even so, the make-up took 5 hours to complete. (F) The eyes are made up with blue pencil, while the shoulder pieces are fitted and adhered. (G) The make-up base is heavily powdered. (H) The foam latex back with its plastic spines is attached. (I) The body is covered with an initial coat of white PB-type base and then powdered. (J) Long plastic fingernails are adhered, and the hands are made up. (K) An additional coating of PB foundation is heavily applied where the sequins will be sprinkled. (L) The sequins are held to the body and face by the sticky oils in the foundation as no adhesives work well over this type of base.

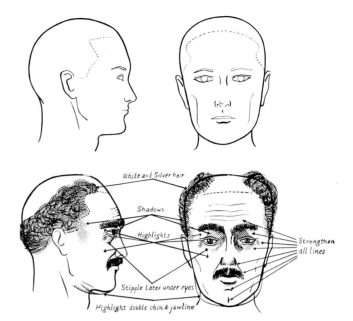

FIGURE III.3 *A men's make-up chart and how it is used to record a character make-up.*

costly ways. When the character is male, the nose becomes a good starting point, then the hair, both on the head and on the face.

In a series of make-ups that were part of a live television show starring Paul Winchell, one segment of which was a short dramatic playlet, he played a different part each week (Figure III.6). In this case, most of the noses were of the paint-in latex variety (due to cost factor), while others were simply made with plastic wax molding material (also see pages 95 and 232). The moustaches and beards were generally on lace, except for the unshaven effect in the lower middle photo. Wigs or dressing his own hair completed most of the make-ups, and where age levels were required, the addition of highlights and shadows created the make-up characterizations.

One of the most outstanding multiple character make-ups was done by Bill Tuttle for the *Seven Faces of Dr. Lao* for MGM. In this film some excellent foamed

FIGURE III.2 *The completed make-up. The palms of the hands were also made up in this case due to the overall make-up effect required. As he entered for each scene, the sprite sprayed a handful of sequins at the camera and then stepped through them as they settled downward to add to the ethereal effect he created. Roddy MacDowall contributed this idea for his entrances, which shows the coordination and cooperation required for a special make-up like this. Both make-up artists worked on set retouching as the part was a very active one throughout.*

FIGURE III.4   *Cavendish Morton's portrayal of the Witches from*
Macbeth in a photo montage that has inspired many a make-up artist in
the creation of these spectral creatures.

FIGURE III.5   *The use of costume as well as facial expression on an early*
Lon Chaney, Sr., make-up. Note how the headgear has been pulled into
the cheek and jaw area to narrow the chin line effect. (Courtesy Metro
Goldwyn Mayer.)

pieces were employed to change Tony Randall into the
various parts he played.

A most facially versatile actor of the New York stage
and in television in the 1950s was Maurice Manson.
Seen in Figure III.7 in a number of make-ups, he never
appeared the same twice (unless asked to repeat a famous make-up—Santa Claus was his most played).
These make-ups were either done by himself or by
some of the make-up personnel in the 1950s period.
The photo with the telephone is Maurice at his most
natural!

MAKE-UP SPECIAL EFFECTS

There is also a burgeoning tangential direction to character make-up that is called *make-up special effects* or
special effects make-up. Here regular make-up methods
are combined with dummies, puppets, special visual
effects, camera tricks, and editing plans to produce
illusions or deceptions of vision for the audience. It
entails the close amalgamation of appliances, mechanical devices, electrical motors and instruments, and
other craft materials to produce the results. Make-up
special effects are closely and generally allied to horror
make-up designs, blood-and-gore concepts, science
fiction creatures, and the like that go beyond the normal character make-up principles of changes in age
levels or racial or national facial structures. Most of
the change effects start with the human and progress
as far as possible, then are cut by editing to a dummy
form to complete the effect. In this way the effect goes
far beyond the norm of the human form and shape.

Not all make-up artists neither are interested in—
or can devise and execute—these make-up special ef-

Figure III.6   *Changes of make-up on Paul Winchell for various roles on his NBC-TV weekly show. (Photos by Albert Freeman.) (Make-up by the author.)*

FIGURE III.7   *Maurice Manson as various characters in history.*
(Manson Collection.)

fects nor have the vast equipment or laboratories (that are well nigh factories), so it has become a limited field for a few talented people. The conceptions, drawings, sculptures as well as the execution of the devices required for some of the make-up special effects are often the children of many parents, with a number of persons contributing to the final effect. Although most make-up artists who specialize in this division of the field become knowledgeable in many phases of it in time, each often becomes a specific technician in one or two particular phases; that is, some become mechanical device experts, others sculpture marvelously well, others dream up the designs or draw the sketches, while some perform the lab chemistry, employing the myriad of products necessary.

Make-up special effects *are* make-up but in a very special area that produces effects that could not oth-

erwise be seen. They stimulate the imagination and appeal often to the macabre in a truly graphic and explicit sense. Make-up special effects artists are constantly researching new ways to "scare the pants off" the avid and demanding audience and to revolt the human senses, and they wish to leave little doubt that the results are real. It is in this illusionary realm that the accomplishment of visual expectancy makes it magic, for here are the make-up magicians in the true sense.

The films made for this special effects genre may not be dramatic successes or have any classic performances, but in addition to being very lucrative, some of the effects and characters created do become classics for those who follow this type of film entertainment. For examples of this art, see Chapter 15.

CHAPTER 9

Age with Make-up

AGE LEVELS ARE SUBJECTIVE AND MORE OFTEN REFER TO ingrained concepts of youth, man- and womanhood, middle age, old age, and extreme old age. However, numerical age can be deceiving in many senses and often has no relationship to appearance on the screen or stage. More often than not, due to the aging procedures, older men are cast to act with younger women in the concept that a few lines in a man's face add maturity, while a few lines in a woman's face place her in a character part category. Of course, today we have great cosmetic surgeons to perform their art, and the working life of the actress can be extended considerably.

At approximately the age of 35, certain facial changes begin to occur in the face to increase the mature look of a person, with some people showing this trend more than others. Here the separation may commence from playing young to somewhat older parts, and this trend will continue to the age of 50 or so. This, then, is a category that is often termed *middle age*.

MIDDLE AGE

During this middle age period the skin undergoes various changes. People who have spent a great amount of time outdoors take on a ruddier or more tanned look from exposure to the elements, while those occupied in a more sedentary or indoor life are apt to appear much less so and most often have smoother complexions. Healthy people maintain a good, normal coloration in undertone as compared to those who may have been ill for some time and whose color may range from a sickly hue to the over-red tone of those with high blood pressure.

The fresh bloom of youth commences to leave the cheeks of women, and the beardline of men with dark hair begins to show up more strongly than when they were in their twenties. The hair often becomes gray, or in some cases with men, baldness creeps in at the hairline in front or at the top of the scalp. Lines can start to appear in the fleshy portions of the face near the eyes, mouth, nose, and neck. Cheekbones either protrude more as the cheeks start to sink in on thinner people, or the cheeks, jowls, chin, and neckline can take on more weight on heavier people.

Women's brows and lashes can thin out somewhat, while many men's brows get a trifle bushier or even grayer. While this is occurring, the face begins to take on a certain set of folds and wrinkles that often remain throughout this period of life and beyond.

Of course, the screen and in-person appearances of people can be much more discerning of the commencement of aging than the stage due to the distance of audience from performer. However, many women are photographed or televised with a diffusion lens or filter to soften the effect created by age lines or more make-up. Sometimes, a slightly out-of-focus effect is employed for softening the picture as well.

Creating a middle age look for a young performer must be a combination of elements and crafts (Figures 9.1 and 9.2). First, the overall wardrobe is important. Next, the use of a severer or not quite up-to-date hairstyle can aid. Certainly, eyeglasses with frames that are appropriate are a good aging means, and a moustache on a man helps at times. One very convincing recourse is not to use a foundation when one wants to age a young face. In this way, the natural blemishes and colorations are not *corrected* as they would be in a straight make-up when the foundation evens out the overall skin coloration. Make-up artists must vary their aging tricks so that all the characters do not have the same glasses, moustaches, hairstyles, wardrobe, and so on and must choose the most appropriate one or two means for each of the characters and then use the others as secondary plans. The figures given here are only basic premises and show only one style of added age to youth, and the study of many faces will give a wider understanding to this critical type of character make-up.

Especially difficult is the convincing aging of a woman who has taken care of her skin, hair, and dress, so many women in the forties look as well (albeit different due to fashion mostly) as they did in their twenties. Here, performance and the script must serve to establish their age level more than make-up in some cases. The less the character being played has taken care of himself or herself, so to speak, the easier it will be to depict age in the face with make-up.

Usually it is easier to show age in a man's face than a woman's as the facial and head hair can be grayed, and gray hair doesn't have the same stigma for men as it does with women! However, some established

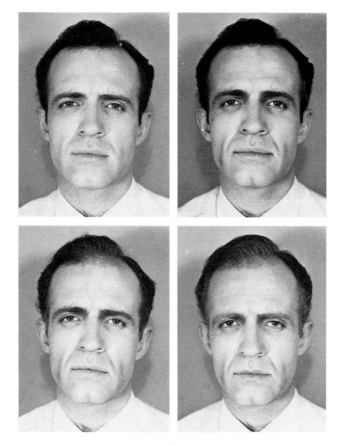

and can become the basis of make-up determination when aging a young person to older age. The old theatrical make-up terms of *Sallow Old Age* and *Robust Old Age* were names of stick foundations in some lines and were supposed to describe the skin coloration best suited for the indoor or sickly and the outdoor or healthy, respectively. With screen make-up, the skin coloration created with make-up must be of a more natural appearance and tone, and the lines and shadows drawn to look as real as possible.

Making a person in his or her twenties look 60 is far more difficult generally than making a person in the forties appear 60 due to the aid that some wrinkles and facial folds gained at 40 that need only be strengthened, while for the 20-year-old, they must be created. The main mistake that most make-up artists make in doing an old age make-up is that they try to do too much with foundations, shadows, and highlights for the screen. Where it may work on large stages, the close shot immediately discerns the work as make-up

FIGURE 9.1   *Middle age make-up for men.* TOP LEFT: *Straight make-up for contrast in progression.* TOP RIGHT: *With no foundation on the face, shadows are blended in the corners of the eyes, nasolabial folds, jowls, and chin. Highlights are added to accent shading.* BOTTOM LEFT: *To increase the effect, hair whitener can be rubbed into the hairline to recede it and the hair slightly grayed in front and in the sideburn area.* BOTTOM RIGHT: *A few more years of age are added by increasing the area and amount of hair whitening to include the eyebrows and by deepening the age lines a bit.*

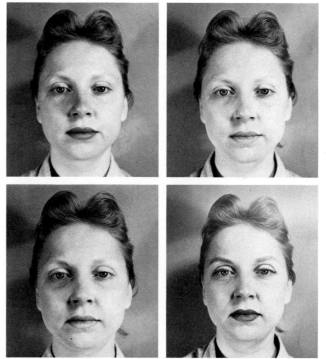

stars will want to hold on to their hair color and facial appearance for a long period, so coloring the hair and appropriate make-up correction will be asked for with these men. Imagery is important and the make-up artist must maintain this direction. To maintain the most youthful appearance possible, cosmetic surgery will answer most needs, and reference to the methods can be found in Chapter 8.

OLDER AGE

Supposedly, once a person is over 50, he or she enters into a state called *older age*, with 65 being considered to be the age of retirement. However, from 50 on often becomes today the most productive period of a person's life based on the experience and knowledge gained in the first half-century. The expression a *strong fifty* is sometimes heard, denoting that such is a prime period, while the ailments of older age sometimes begin at that age. All this is often reflected in one's face

FIGURE 9.2   *Middle age make-up for women.* TOP LEFT: *No make-up, but with hair set in a severe style to date it.* TOP RIGHT: *Highlights are added to the nasolabial folds, jawline, and around the mouth. For this face, it was determined that making it appear puffier would add to the age aspect rather than more shadowed as in the man's make-up to make the appearance more gaunt.* BOTTOM LEFT: *Shadows are added on the temples, corners of the eyes, nasolabial indentations (which were only slight) and under the cheekbone area.* BOTTOM RIGHT: *Eyebrows, some mascara, lipcolor, and a few lines drawn on the neck folds, corners of the mouth, and the flesh on the frontal bones given additional highlight to droop the corners of the eyes are done before powdering. Eyeglasses were used to give an added touch of age. Aging a woman is always a subtler make-up than for a man in most cases.*

FIGURE 9.3 *Joaquin Sorolla. (A) First, we see him in his youth at about 25 years of age with only normal lines and a wispy moustache. The hairline is strong and dark. (B) Next, in his early middle years, a change in hairstyle along with a thickened moustache and a beard. The flesh on the frontal bones has thickened, and the nasolabial folds have deepened slightly. Only some indication of eyebags has crept in, while the brows are full and dark. (C) The next stage of later middle age shows gray speckled in the hair with a receding of the hairline on his right side, along with gray in the brows, moustache, and beard. The nasolabial folds are much stronger and more defined as are the frontal bones and the circles under the eyes. The mole on the nose is more pronounced due to the lighting but is in all the photos. (D) Last, a fine example of old age with receded hairline, light gray to white hair on the head, brows, and face. Although the face has remained essentially of the same dimensions, the lines and wrinkles now show strongly around the eyes, with the overhanging frontal bone flesh and eyebags are quite apparent and defined as are the nasolabial folds. Due to the beard and moustache, the jawline, lips, and neck do not show the aging process, a good trick to remember when doing age make-ups with facial hair. The oldest age photo also shows that the ears have changed along with the rest of the face, but the effect is only really noticeable in this last photo. Ears should not be omitted when planning a make-up if they are seen and not covered by the hairstyle.*

rather than reality. True, the accentuation of the proper areas of the face with highlights and shadows does provide a good black-and-white make-up for stills or motion pictures and television, but in color mediums of the screen or stills, tri dimensionality is necessary.

A good study is to view photographs of progressive old age of persons without make-up. An excellent example is in the four photos of the Spanish painter Sorolla (Figure 9.3).

Hands also follow the aging process and should be made up accordingly. In middle age there is not too much change in the back of the hands (palms see little change even to older age), but as age progresses, and if the hands are to be seen on screen, they must be made up as well. As all character make-ups are a progression of make-up exercises, let us take the advancement of age levels through the various stages of application principles.

MAKE-UP APPLICATION

Whether or not to employ a foundation for an age make-up depends upon the character being played by the performer and the health level, exposure to ele-

ments, and environmental status in the area of the script. However, as a general rule, it is best not to cover the face with any foundation but to commence on highlighting and shadowing first and then adding any skin toning that may be required. Wrinkles made with shading colors and countershadings always go on prior to the powdering. Dark or tinted powders are not recommended for any type of character make-up application as they will darken and sometimes obscure any fine lining or highlights that may be important to the make-up. Hair whitening is most often last, but the application of hair goods to the face for men sometimes precedes the sequence of application if the hair is to block out a good portion of the lower facial area.

Primarily, a basic stage or even a black-and-white screen age make-up often relies heavily on the art of *chiaroscuro,* which is a pictorial representation in terms of light and shade without regard to color and what make-up artists often refer to as age with paint and powder only. Such basic training is necessary and seemingly creates a tri-dimensional look by highlighting the portions that should appear to be raised and shadowing those that appear to sink in.

With color mediums such as screen make-up and with stage make-up where the audience is close to the performer, only a very limited amount of facial areas can be successfully highlighted and shadowed for these effects, and then skin texture methods (such as Old Age Stipple) and tri-dimensional appliances made from latex or plastic are required.

FIGURE 9.4   *An age make-up with highlights and shadows only. At the time of this photo, Paul Muni was 60 years old but was to play 90-year-old senator. To accomplish this appearance, his normally dark hair was bleached white, and after some false brows, only shadows and highlights were added to complete the make-up by the author for a black-and-white television production on the General Electric Theatre on CBS-TV.*

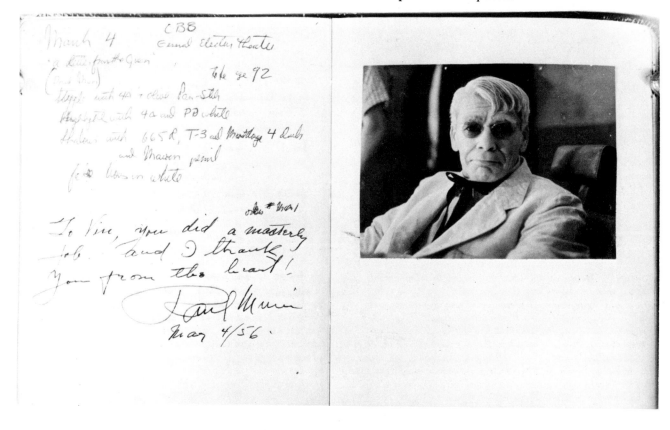

Foundations

Although a basic aging make-up with no prostheses or just some Old Age Stipple can be done as recommended with no foundation, when any appliances are used, foundation will be required. For Caucasoid women a basic shade would be KW-2 or the slightly deeper KW-4. Both lack pink tone and give a good starting point for facial coloration. If more pink is desired, a KW-36 may be used. For Caucasoid men, basic olive tones would be KM-1 or KM-2, and for a pinker or even ruddier complexion, KM-4 or KM-5 (in thin coats) should be tested.

Negroid and Oriental races can follow the normal charts for skin color and foundation shades. The darker the foundation (or natural skin color), the less any age lines, shadows, and so forth will show up.

Depending upon the skin tone required, Ben Nye, William Tuttle, and Stein's make a number of shades that are suitable for age make-ups for both men and women, and their lining colors also include many colors that are useful for creating lines and shadows. A full listing of these is found in Appendix A as well as on pages 31 and 267ff. Nye's #40 Creme Brown Shadow has a purple-brown color that works well both for stage and screen age characters, while the #42, #43, and #44 have a browner complement so as to range through all racial types from Caucasoid to Negroid skins. Highlights can be done with the Extra Lite and Medium Creme Highlights or tinted with a Creme lining color.

Age Spots and Coloration

Age spots often occur in aging faces and vary from a light to a dark brown tone. Many times they are freckles or sunspots, while in other cases they may be other discolorations. Using RCMA Shading Colors S-1, S-14, S-4, S-9, or S-10 in uneven dots of varying sizes will produce this effect. They can be added after the make-up is completed and then powdered down. Also see RCMA's Old Age Spot Kit.

For an outdoor or western film, requiring a ruddy or long sun exposure effect, a general overall stipple of S-6 or S-7 in the forehead, cheeks, nose, and chin areas will produce this look. Also, a bit of Color Wheel Blue rubbed lightly on the lips with a brush will kill the youthful redness. For young men with little or no beardline grayness, a light application of a beard stipple in gray tone will create the effect of a shaven beard area.

Highlights and Shadows

Although the oldest manner of showing age on stage might have been with a carved or formed mask of an elderly character, the use of highlights and shadows was the first (and still effective) means of changing the facial structure with light and dark (Figure 9.4). Blending is the most important point to learn, with both hard and soft blends being required depending on the placement on the face. As black-and-white screen make-up required only shades that televised or photographed light and dark gray, browns or grays were employed along with lighter and darker foundation shades. However, color mediums often require the combination of gray and brown plus, in some cases, blued reds (as was recommended for the stage as *lake* color). As such, it is best to do any make-up as if it were for color as it will more likely than not picturize correctly if a print or telecast is viewed in black and white.

Table 9.1 shows that the depth of the shadow area can vary from a gray to a grayed brown or browned gray to a laked brown or reddened brown to achieve more emphasis. (See also Shadings on page 59). Care must be taken with the highlights, and the use of white or yellow that was formerly recommended for the stage is seldom employed for color screen make-up.

Due to the undertone value of Mongoloid or Oriental skin, any laked or reddened shading should be avoided, and the light pinked countershadings (such as CS-1) are generally too light. RCMA CS-2 and Foundations KW-2 or KW-24 are suitable for highlights from light to deep skin tones, while shadows can be made with S-14 and S-4 in most cases.

Negroid skins are the most difficult to highlight, but Foundations KW-2, KW-24, and KW-4 are best (in a rising scale of deepness), and shading colors S-4 and S-8 can be employed for shadows. Sometimes RCMA 1624 can also be employed for highlighting,

Table 9.1 *Highlight and Shadow Recommendations for Caucasoids*

DEGREE	HIGHLIGHT		SHADOW	
	MEN	WOMEN	MEN	WOMEN
Slight or Soft	KW-24	CS-2	S1	S1
Medium	CS-2	CS-1	S-14 S-11	S-13 S-14
Medium to Strong	CS-1	CS-1	S-14 to S-10	S-14 to S-9
Strong	CS-1	CS-1	S-6 S-4	S-2 S-9-10

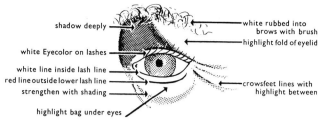

shadow deeply →

white Eyecolor on lashes →

white line inside lash line →
red line outside lower lash line →
strengthen with shading →

highlight bag under eyes →

← white rubbed into brows with brush
← highlight fold of eyelid

← crowsfeet lines with highlight between

FIGURE 9.5 *Basic application areas for eye make-up.*

and some black added to the shading colors will further deepen them.

Wrinkles

The signs of age in the human face generally begin to show in the areas around the eyes, mouth, and neck in the form of wrinkles as these parts have soft tissue. Forehead wrinkles are apt to appear in later age, and care should always be taken with these since they can look painted on if improperly applied. Frown lines between the eyes on the bridge area of the nose are also convincing in creating age level—providing they fall into the natural frown pattern of the subject. Also, the creases that are formed between the cheeks and the nose (the nasolabial folds) are good places to accent if they are present in the face being aged. Highlights and shadows are particularly effective in the nasolabial folds, but the edges should be very well blended. In all cases, following the natural wrinkles in a face provides the best method for aging with highlights and shadows, while tri-dimensional appliances can provide age accents where the natural lines and flesh do not occur.

Eyes

The eye socket can be divided into the upper-eye portion, from the inner corner of the socket to the flesh on the frontal bone, and the under-eye section, including the eyebags and the wrinkles or crow's feet at the outer part of the socket and on to the upper cheekbones. Also refer to Figure 6.4 for a drawing of the human skull and the important bone structure and to Figure 7.3 for illustrations of the eye socket areas.

Figure 9.5 shows the basic application areas for eye make-up. Although some aging in the eye area can be done with shading and countershading to create shadows and highlights respectively, to create eyebags or emphasize the frontal bone areas, better effects are possible with foamed latex or plastic appliances. Examples of aging in the eyes in Figure 9.6 show the variety of shapes and depth of the fleshy portions of the eye socket area (see page 165ff for molds and casts relative to eye appliances).

In all cases, the eyes are the best parts of the face to age with make-up, whether it be with paint and powder or appliances. The simple addition of eyebags will produce a drastic change to the youthful face.

Nasolabial Folds

The next prime area is the nasolabial fold on each side of the nose because here one of the deepest wrinkles in the face can be formed. A study of the photographs in Figure 9.7 will show a hard deep line followed by a blend of shadow up to the point where the upper part of the fold becomes a protuberance, providing a natural highlight on this part of the cheek. Some nasolabial folds continue down to the chinline on each side of the mouth, while at other times a separate mouth wrinkle will be formed at the corners.

Chins and Jawlines

This leads us to the chin and jawline area that can either show added weight and flesh or deep folds and wrinkles formed as muscle tone loss occurs in age (Figure 9.8). Jawline and neckline appliances are called for in almost all cases here as simple highlighting and shadowing cannot provide this effect for color make-up.

Foreheads

Forehead lines and wrinkles are where the most obvious make-up errors are made by beginners who commence to sketch in elaborate wrinkle patterns on young faces to denote age level. Study the truly wrinkled forehead to see the variety of patterns that occurs in the forehead with wrinkling (Figure 9.9). Highlight and shadow seldom, if ever, work unless to accent an already lined forehead—and even then the effect is often ineffective. Here too, a good, thin appliance may do the job in a far more efficient and correct appearing manner, but the forehead should be the last facial part to rely upon for producing an age make-up.

The Head

Baldness that occurs during a man's youthful years can be corrected today with hair transplants or toupees. Many famous actors including Bing Crosby, David Niven, Ray Milland, and Burt Reynolds have worn toupees or hairpieces on screen to cover the inroads of baldness. Other actors, delight (it seems!) in partial baldness as it allows them to portray many character roles by the addition of various styles of hairpieces. Sir Alec Guiness, in an interview, remarked that his partial baldness was an asset to a character actor. However, baldness employed to denote age can be created by various means, some of which are only relegated to amateur theatricals (like the soaping of the hair to achieve a bald effect), but plastic appliances known as *bald caps* provide the best method for a full or partial head baldness (Figure 9.10). (See Figures 13.14 and 13.15 for bald caps and molds.)

In the photos of the bald heads (Figure 9.11), we have chosen an actor (Vince Barnett) who has been partially bald since his youth, and with growing age,

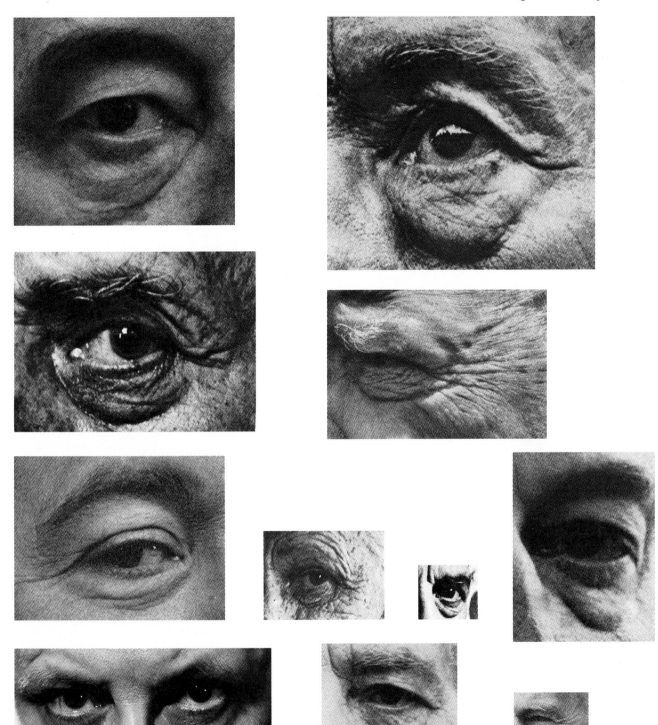

FIGURE 9.6   *Aging eyes.*

the skin has tightened up to a very smooth pate but with the normal freckling that usually occurs. On most bald men one can almost see a clear line of demarcation between the forehead skin texture and the smoothness of the balded area. This is why we do not make the same skin pores on a bald cap mold as for a facial one. Although most bald caps are made by spraying the bald cap material on a head form, an excellent cap can be made that is quite heavier and more durable (and

easier to apply) by brushing the cap material on the form (see make-up for Emperor Franz Joseph in Figure 14.20).

Partial Bald Effects

Men's hairlines can be made to appear to recede somewhat by lightening the hairs on the edge of the hairline with a hair whitener like RCMA HW-4. Although this procedure works well for any black-and-white me-

FIGURE 9.7 *Nasolabial folds.*

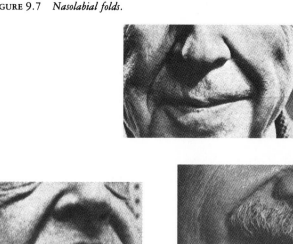

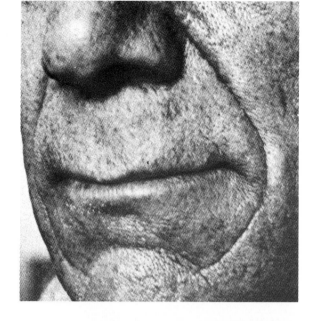

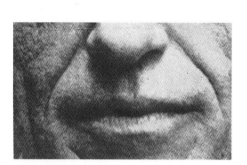

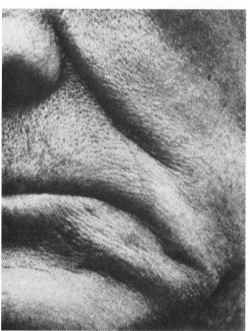

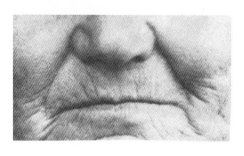

dium, it must be judiciously handled for color where the shade of the hair whitener must be matched to the hair shade (opposite, that is, on the color scale). Another method is to work in a thin coating of plastic wax molding material in a skin-matching shade into the edge of the hairline to minimize the hair growth.

Another more drastic measure is to thin out the hair by cutting off small tufts at the roots of the hair in an uneven pattern and then lightening the deepness of the scalp color with foundation. This was the procedure employed by Dick Smith to age Walter Matthau in *The Sunshine Boys,* and it was very effective. The alternative was to cover Matthau's full head of dark hair with a plastic bald cap each day and then

add a thin hairpiece over this. Rather than this daily and lengthy procedure, Matthau opted to have his hair cut—and to wear various forms of caps for public appearances until the hair grew out! As well, Jack Hawkins had the front portion of his head shaved for his role in *Lawrence of Arabia* to play a specific character in the film. Even though these latter methods seem drastic, the hair does grow back in and the actor is saved much time in the make-up room during the filming, and the bald effect is far more realistic and easier to maintain.

An oldtime method of receding the hairline, used by some amateurs, is to slick the hair down with a heavy soap and water mixture until it dries flat to the head. However, the ingredients of any soap held for a long time in contact with the human hair might have a serious effect on the hair and scalp and is not recommended.

Hair

During the progression of age the hair color can go from a streaked gray to gray to white. Also, the dress or cut of the hair can make a difference in depicting age level, especially with women's hairstyles. Most older women have a tendency to keep a certain style for a longer period than a younger woman who is constantly trying to keep up with the dictates of fashion.

A touch of hair whitener applied at the temples of a man will commence the aging process, as will streaking the eyebrows with hair whitener, white mascara, or such. Another excellent method to gray a partial head or even a full head of hair is to employ white and

FIGURE 9.8   *Chins and jawlines.*

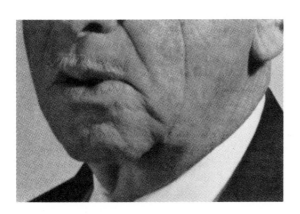

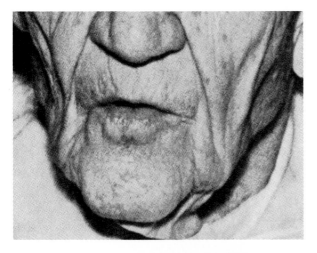

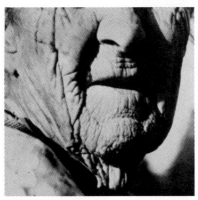

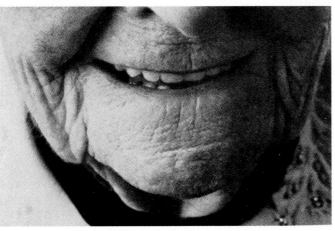

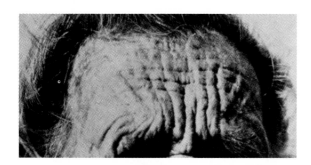

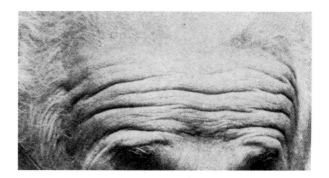

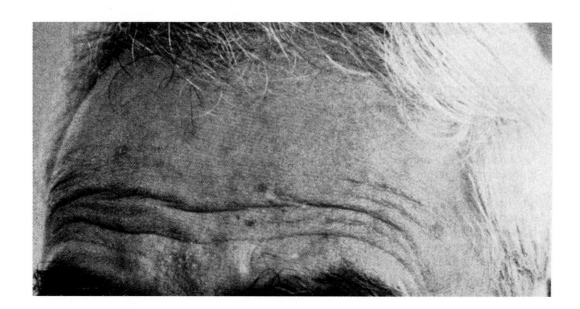

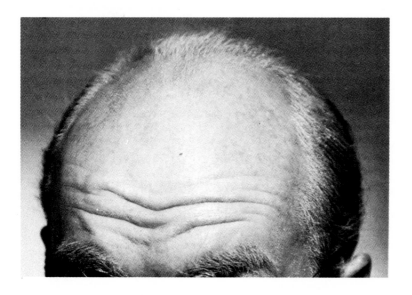

FIGURE 9.9   *Foreheads.*

silver hair sprays. The face should always be carefully covered with a towel when spraying hair so that none of the spray will get into the eyes or on the make-up (as it is difficult to remove without major retouching). Take care in using this type of hair spray because one can easily overuse it and the hair becomes matted down and unnatural looking. Some make-up artists spray the color into a small container and apply the spray with a brush (like a hair-whitening brush) to control the application better. This method also works well for facial hair graying.

It is well to remember that all men do *not* grow moustaches as they age, so save this trick for those who might require it to achieve an age level. Beards added to a face to denote age depend a great deal on the character being played, the period of the production, or the current style of fashion.

It is, therefore, more difficult to depict age in a woman's face than a man's using the hair as a tool. Most women will continue to color their hair in youthful shades, while to the majority of men a bit of gray or even a full head of gray hair is quite acceptable.

Bleaching the color out of hair is another way of whitening the entire head of hair, but it is seldom done except for performers willing to go through this procedure (Figure 9.4). Also see Part V for additional information on hair and hair goods as well as Chapter 11 for period and historical hairstyles and facial hair.

The Hands

If the hands are to be seen they must also be made up to the age level of the character being portrayed (Figures 9.12 and 9.13). However, only in old age do the hands change that much, and only in a close up will they be apparent.

Old Age Stipple

One of the major advances in producing an age make-up was invented by George Bau of Warner Brothers during the early 1950s. Prior to that, age make-up consisted of highlights and shadows plus the use of any facial appliances required, but the *skin texture* of old age was always a problem where the appliances did not cover the skin. Working with a mixture of pure gum latex, powder, gelatin, and a ground up pancake make-up, he produced a mixture that, when applied to the face with a coarse sponge, stretched, dried, then restretched, produced excellent surface wrinkles. This formulation was improved upon by the author to produce RCMA's Old Age Stipple (Figure 9.14).

It should be noted that just latex—of any variety— is insufficient to produce this effect even though some theatrical make-up books still carry this information. The addition of a single layer of pink or peach-colored Kleenex tissue at the time of the stipple application

FIGURE 9.10 *Bald head. Walter Slezak in R.K.O. Pictures* Sinbad the Sailor *with a partial bald cap and wig in addition to brows and moustache.*

will accent the wrinkle formation greatly and is of use when a performer's skin has little natural stretch for the application. Usually, Old Age Stipple works best with the middle age face to produce older age, so this tissue method does help with younger faces. Heavier paper products used with this stipple will also produce heavier skin textures for fantasy characters (Figure 9.15).

Generally, the application of Old Age Stipple works best when one make-up artist applies the material while the other stretches and holds the skin taunt during the process. RCMA Color Process Foundations can be applied over Old Age Stipple to accent wrinkles and so forth. Foundations that contain mineral oil or petrolatum are not recommended for use over Old Age Stipple as they either tend to break it down or appear chalky. Old Age Stipple is made in skin shades so little or no foundation is required over the application.

The stretch on the skin area must be maintained during the application of the stipple, the drying (with a hand hair dryer), and the powdering (with no-color powder). Only then can the skin be relaxed. Also, restretching the skin in a number of directions will

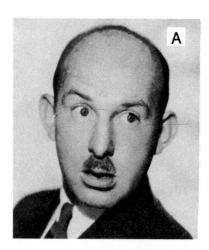

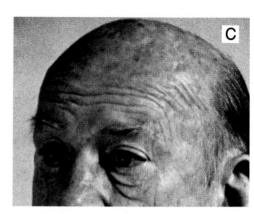

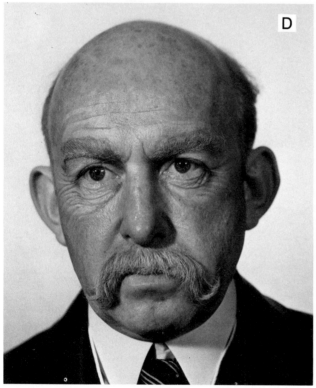

FIGURE 9.11   *Bald head. (A) Character actor Vince Barnett in a straight make-up in one of his varied film roles. Early baldness aided him to portray many types with the addition of hairpieces. (B) With the addition of a wig and a change in facial hair, a new character is formed. (C) In later years, sunspots appear on men with bald heads. (D) Another character with the eyebags increased by pencilling, heavier brows, and a moustache on lace added.*

cient stretch to the neck area. The upper then lower areas of the eyes are next and then the cheekbones and the chin. A paint-in molded eyebag is added with some RCMA Eyelash Adhesive and colored with KM-2 to match the surrounding skin. Before adding the eyebag, the skin is restretched to produce full wrinkling. Age spots have been added with S-1 and S-9. The actor's hair has been whitened with HW-3 and the brows augmented with gray hair. Glasses and special tooth caps have been added for the final effect. No extra foundation has been applied over the medium shade of RCMA Old Age Stipple. Some age stipples on the market are made from plastics, but they do not work as well and often have a tendency to shine excessively as they are not matted.

The original formula, as quoted in the author's 1958 *Technique of Make-up for Film and Television,* uses a neutral face powder (1½ tablespoons) ground with a pancake make-up (1 tablespoon) to which is added 1 teaspoon of powdered gelatin. To this is added 3 tablespoons of hot water to form a paste. Then add 4 ounces of pure gum latex, and mix until combined. The mixture can then be poured into wide-mouth jars. To use, it was recommended to place the jar in hot water to liquefy the gelled mixture. RCMA Old Age Stipple usually requires this hot water bath before use and needs to be well and completely stirred before application. The latter has a shelf life of over two years if kept tightly capped, while the original formula seldom lasted a few months before it hardened.

produce better wrinkling. Overlapping each coat is necessary so as not to leave any gaps in a full-face application of the stipple. As well, Old Age Stipple can be applied only in the eye area for a limited wrinkle effect for less age. If the effect must be repeated for a number of days of shooting, it is important that the same stretching and application procedure be followed each day so that the wrinkles will fall in approximately the same places.

Full-face stipple by make-up artist Gary Boham on John Neville-Andrews for *The Glorious Romantics* shows how this material is best applied in small areas at a time to produce the best effect (Figure 9.16). Note that the position of the head pulled back gives suffi-

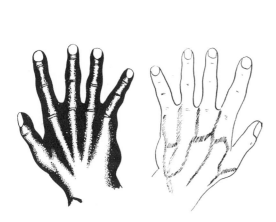

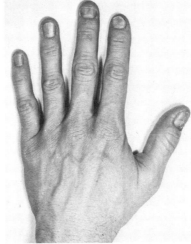

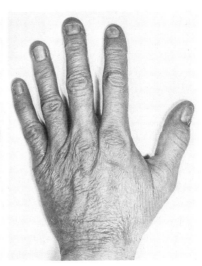

FIGURE 9.12 *Hand.* LEFT: *Basic bone structure can be emphasized with highlights and shadows.* RIGHT: *The vein structure can be delineated with a blue color.*

FIGURE 9.13 *Hand.* LEFT: *Hand without make-up.* RIGHT: *Old Age Stipple applied, dried, and powdered. The veins have been accented slightly.*

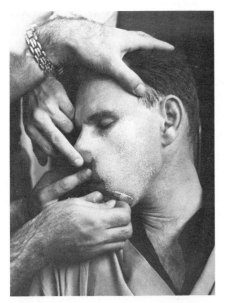

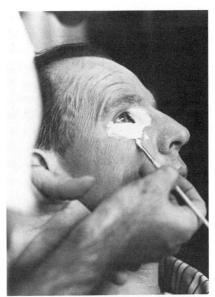

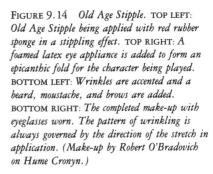

FIGURE 9.14 *Old Age Stipple.* TOP LEFT: *Old Age Stipple being applied with red rubber sponge in a stippling effect.* TOP RIGHT: *A foamed latex eye appliance is added to form an epicanthic fold for the character being played.* BOTTOM LEFT: *Wrinkles are accented and a beard, moustache, and brows are added.* BOTTOM RIGHT: *The completed make-up with eyeglasses worn. The pattern of wrinkling is always governed by the direction of the stretch in application. (Make-up by Robert O'Bradovich on Hume Cronyn.)*

FIGURE 9.15 *An old age fantasy make-up done by Robert Philippe for an NBC-TV production, using paper towelling with Old Age Stipple to produce a heavily wrinkled skin effect.*

Prostheses
Partial Appliances

If the make-up only requires some aging, eyebag appliances are usually sufficient. However, jowls or an under-chin prosthesis may also be called for in a basic use of applicances (Figures 9.17 and 9.18).

Some make-up artists do an entire face in partial appliances, and this is how Stan Winston and Rick Baker did their Emmy-winning make-up on Cicely Tyson in "The Autobiography of Miss Jane Pittman" (Figure 9.19). The application of this make-up took some six hours, and the photo progressions show the separate appliances. Although a basic clay sculpture over the life mask of the actress gave them some idea what the completed make-up would look like, the final result was far more spectacular on the human face, where it was transformed into lifelikeness by the performance as well as the make-up. This make-up will stand as a classic in its field and an inspiration for all make-up artists who aspire to do fine age character make-up.

Full Faces

Whenever the full face (and probably neck) must be covered with an appliance, the make-up artist has the choice of making the prosthesis in one full piece or in several adjoining ones. There are times when the full

facial appliance is considered to be best, and some examples are shown in Figure 9.20.

Another example of Bud Westmore's work can be seen on Yvonne DeCarlo as she ages 50 years to play a 75-year-old grandmother in another Universal production (Figure 9.21). The same facial technique is used, but the hands are just covered with an old age stipple material. The neck cords are particularly effective in the profile shot of the completed make-up.

Progressive Age

One of the most important principles in a progressive age make-up is that the lines, shadows, and highlights as well as any prosthetic appliances follow the same pattern of aging at the various levels. One of the ways that make-up artist Stan Winston accomplishes this is to make sketches on overlays on the face to be aged, and then he follows these sketches carefully when sculpturing the appliances. Figure 9.22 shows a series of these drawings of Jason Robards to plan a progressive age make-up. Note how the facial formations follow the pattern of the face to age in approximately 20-year increments.

Figure 9.23 shows a progressive age make-up done by the author. The model's body positioning as well as the slump in the chair aided the effect throughout.

REVERSAL TECHNIQUES

Taking years off to reverse the age of a performer often involves more than just the shading of a double chin or heavy flesh on the frontal bones (Figure 9.24). Facial lifts have been employed to lift sagging skin from the neck area, but they take some time to apply properly and must be watched carefully so they don't lose their adhesion. Most performers today who have a chinline problem seek out a good cosmetic surgeon and have a surgical lift that looks considerably better and is certainly easier all around than wearing the uncomfortable strip lifts.

To make the lifts, cut strips of mousseline de soie (silk muslin) with pinking shears about 3 to 4 inches long and ½ to ¾ inch wide (Figure 9.25). Cut a notch about ¾ inch down from the top of the strip, and fold it over. Place a piece of white No. 40 linen thread or one of the newer clear thread types in the notch, and cut it so that it extends about 1 foot on each side. Place the lift on a plate of glass, and spread a model maker's type of glue (good brand is Sig-ment) over the folded portion. Dry thoroughly. Then, with a razor blade, separate the lift from the glass and trim any excess cement. Make two of these and tie in a dressmaker's hook on one lift and the eye on the other. An elastic band can be employed on the eye side along with a series of knots (instead of the eye).

These lifts are attached to the face just below the

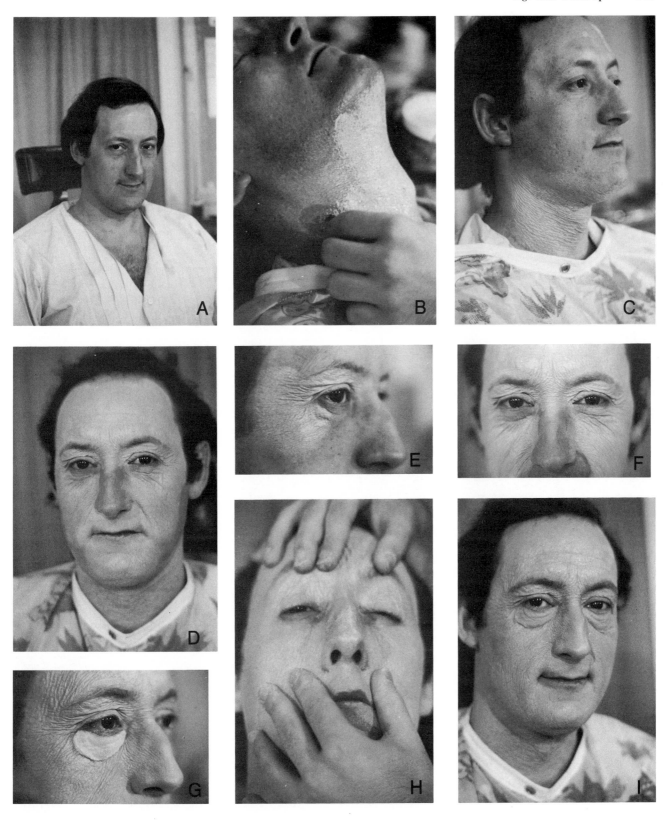

FIGURE 9.16   *A full-face Old Age Stipple by make-up artist Gary Boham on John Neville-Andrews for* The Glorious Romantics *shows how this material is best applied in small areas at a time to produce the best effect. (Photos courtesy Gary Boham.) (A) Before make-up. (B) The head is pulled back and with the neck stretched, the Old Age Stipple is applied, dried, and stretched. (C) Showing the neck wrinkles when head is in relaxed position. (D, E) The eyes are done, upper and lower lids. (F) Continued stipple on the face. (G) A set of latex eyebags is added with eyelash adhesive. (H) The entire facial skin is stretched to fully expand the wrinkling effect of the Old Age Stipple. (I) The eyebags are blended into the make-up with an application of KM-2 foundation.*

(J) Age spots are added with brown colors and the hair whitened with HW-3 (ochre-white), with some hair added to the brows to make them appear shaggier. (K, L) Glasses and special teeth caps are added to

complete the make-up. No extra foundation was applied over the medium shade of Old Age Stipple on the rest of the face.

FIGURE 9.17   Prosthetic appliances for aging the neck. LEFT: The natural neck of Vaughn Taylor in a straight make-up. RIGHT: After the addition of a prosthetic appliance, giving sagging folds and a double chin (make-up by Dick Smith).

FIGURE 9.18   Prosthetic appliances for aging the neck. A set of foamed latex appliances made by Stan Winston on Robert Reiner in an excellent old age make-up. Note extended neck prosthesis. (Photos courtesy Stan Winston.)

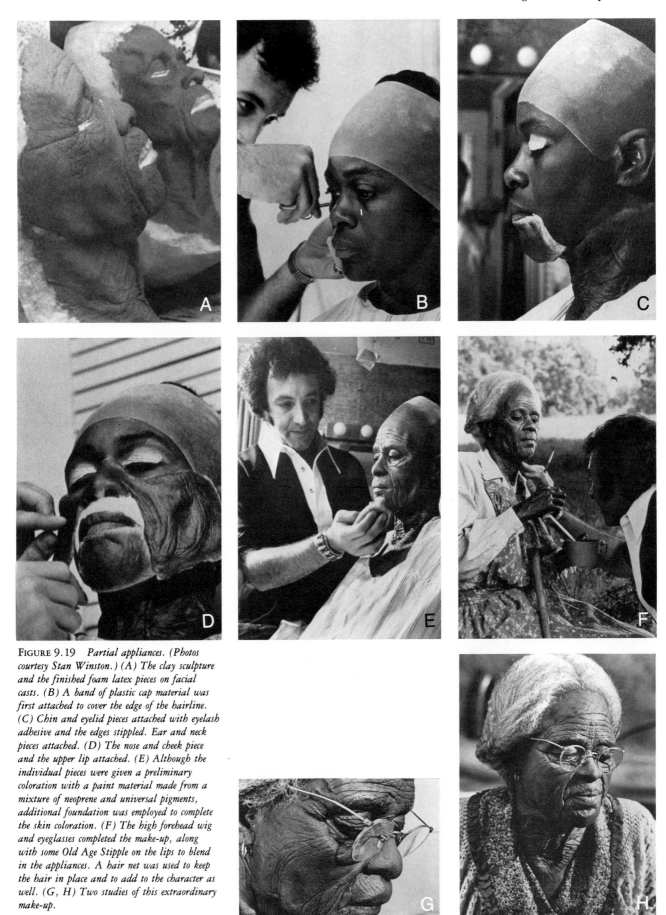

FIGURE 9.19   *Partial appliances. (Photos courtesy Stan Winston.) (A) The clay sculpture and the finished foam latex pieces on facial casts. (B) A band of plastic cap material was first attached to cover the edge of the hairline. (C) Chin and eyelid pieces attached with eyelash adhesive and the edges stippled. Ear and neck pieces attached. (D) The nose and cheek piece and the upper lip attached. (E) Although the individual pieces were given a preliminary coloration with a paint material made from a mixture of neoprene and universal pigments, additional foundation was employed to complete the skin coloration. (F) The high forehead wig and eyeglasses completed the make-up, along with some Old Age Stipple on the lips to blend in the appliances. A hair net was used to keep the hair in place and to add to the character as well. (G, H) Two studies of this extraordinary make-up.*

sideburn area with RCMA Matte Plastic Sealer and dried. When the sealer is fully dried, pull the two ends of the cord up over the head and secure at the proper tension with the hook-and-eye arrangement. The use of the elastic band will give some leeway to the fit, but each subject must have the lifts adjusted for proper tension. Too much pull will form an unwanted large crease near the ear, so be careful to avoid this.

Any RCMA foundation can be applied over the lift and dried sealer. It is always more difficult to attach lifts on those with short hair, but a wig normally covers most of the lift and pull creases. Lifts can also be used on the neck, pulling them to the back of the neck (providing the hair covers them) or even on the forehead to pull up the brow area (again with the aid of the hairstyle). Normally, a heavier coating of any foundation is required to reverse age to cover the defects in the skin and even out the complexion.

Brows that have become too thick must be thinned out, and any gray hair corrected with color. Consult a good hair colorist for a permanent change or use mascara or hairsprays for a temporary coloration. If glasses are normally worn, contact lenses will provide a more youthful appearance.

Removing men's facial hair such as moustaches and

beards will generally make a man appear more youthful. Hairstyle is also a consideration, and the hairstylist should be consulted. Men with heavy or dark beardlines should have them minimized by a rather heavy application of beardcover before the application of a foundation.

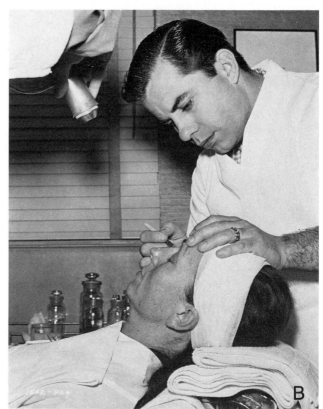

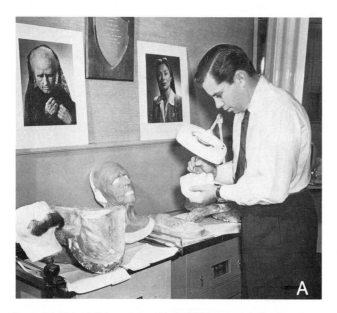

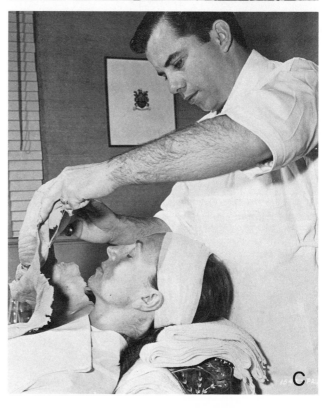

FIGURE 9.20 *Full latex faces. (A) Bud Westmore with the photos before and after as well as the facial casts and mold for the full face foamed latex appliance used on Agnes Morehead for the Universal Studios film* The Lost Moment *(1947). (B) A dilute solution of eyelash adhesive is applied to the eyelids. (C) The mask placed over the face. (D) The appliance adhered around the mouth and ears. To hold the appliance to the face better, a reject piece of the forehead portion of the appliance was attached. (E) Careful attachment in the eye area is done. (F) The entire face is coated with a latex prosthetic foundation. (G) The hands are rested upon specially built holders as the knuckles are built up with small foam pieces. (H) A latex skin is then applied over these to produce an old age effect. The wig is added and the brows laid on. (I) With a dark costume and lace mantilla, the effect is complete.*

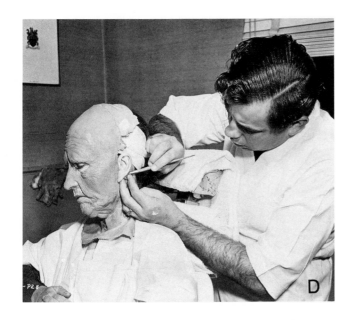

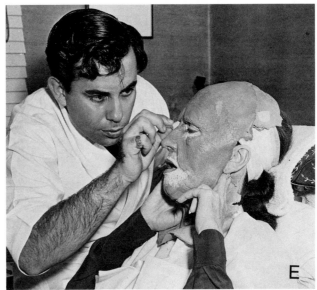

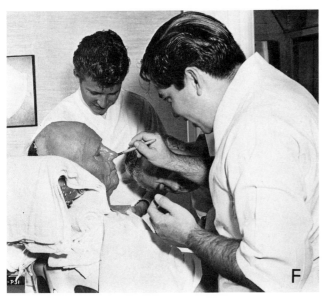

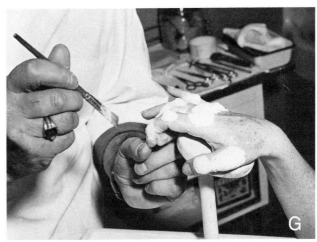

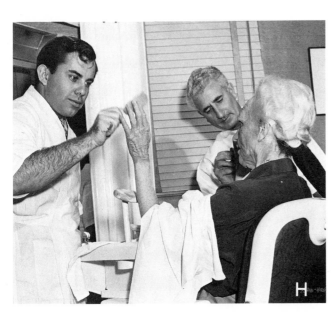

 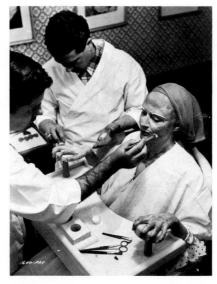 

FIGURE 9.21   LEFT: *Ready to apply the face on Yvonne de Carlo.*
MIDDLE: *The coloration and the hands being made up.* RIGHT: *An on-set
study showing the marvelous neck cords.*

FIGURE 9.22   *Three line drawings of a proposed make-up by Stan
Winston on Jason Robards.*

FIGURE 9.23   *Progressive age.* TOP LEFT: *For the youngest man, a model with a naturally thinning hair front and some loose skin under the chin was given a corrective make-up by pencilling in the hairline to correct it and make it appear fuller. The chinline was shaded and a straight make-up applied for the black-and-white photo.* TOP RIGHT: *In the second stage, the pencilling of the hairline was removed as was the foundation and under-chin shading, and a moustache and glasses were added.* BOTTOM

LEFT: *The third stage took more time as Old Age Stipple was applied along with some lines and shadows. The hair was whitened somewhat, a grayed moustache attached, and the glasses changed.* BOTTOM RIGHT: *In the final make-up, the hair was whitened considerably more, heavy false brows were attached, and more highlighting and shading was done. (Photos courtesy Carl Fischer Studio/make-up by the author.)*

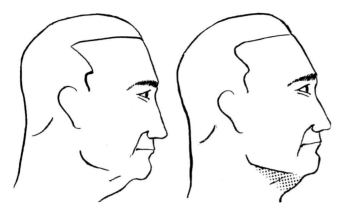

FIGURE 9.24 *Reversal of age techniques.* LEFT: *The natural neck showing loose flesh.* RIGHT: *After shading on the jowls and double chin area.*

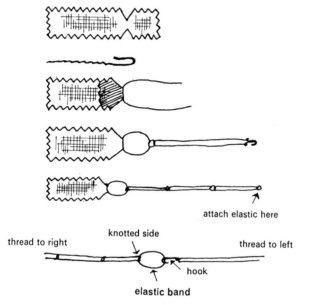

attach elastic here

thread to right   knotted side   thread to left

hook

elastic band

FIGURE 9.25 *Facial lifts.* ABOVE: *Construction of facial lifts using mousseline de soie (silk muslin).* TOP RIGHT: *Another style of lifts.* BOTTOM RIGHT: *Attachment of lifts and their pull-up action when hooked.*

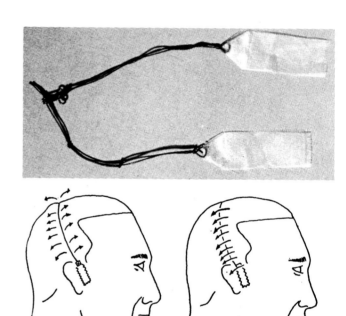

CHAPTER 10

Racial and National Aspects

ETHNOLOGY IS A SCIENCE THAT DEALS WITH THE DIVISION of humankind into races and their origin, distribution, relations, and characteristics. Accordingly, it divides all peoples into three main *racial* types: Caucasoid, Mongoloid, and Negroid (Figure 10.1). Out of these, and combinations of these races by immigration and intermarriage, are made up all the *national* types of the world. There are a great many national groups that are only occasionally encountered as characters in screen and stage scripts, while others appear constantly. We shall, therefore, for purposes of clarity and correctness, refer to these racial types rather than the political categories of white, black, and so on.

BASIC HEAD AND FACIAL SHAPES

Although Caucasoids may have a great variety of head shapes and facial structure differences in various parts of the world, there is more typification in the Negroid and Mongoloid peoples (Figure 10.2). In general, in addition to the skin coloration of the Negroids, an increase in lip size as well as eversion is seen along with larger alae and other nasal structural enlargement as compared to the Caucasoids. The hair structure and shape are also different, and the eyes and hair are usually quite dark. The Mongoloid race is typified by the epicanthic fold in the upper eyelid, straight black hair, and dark eyes, and the nose is usually broader and flatter than the Caucasoid. As they predominate in the Far East, they are often referred to as Asiatics or Orientals, and a yellow undertone to the skin is quite pronounced in most. They also may have a very square jawline along with lips of a deep purple shade.

Interbreeding will, of course, produce some mixtures of racial strains, and one will find Negroids with light eyes, straighter, lighter hair, and feature changes when crossed with Caucasoids, while a mixture of Caucasoid and Oriental strains will produce offspring with the epicanthic eyefold. Examples of such will be seen in the section on national types.

HAIR

While the Caucasoid hair colors range from the lightest of blonde through a variety of tones of red to auburn and light to dark brown, the Mongoloid hair shade is predominantly black and very straight, while the Negroid hair is prevalently black but quite kinky. The actual shape of each hair in the various races changes from quite round in very straight hair to a long, elliptical form for the curliest or kinkiest hair form (Figure 10.3). Straight hair is the heaviest and longest in growth, while kinky or wooly hair is wiry, coarse in texture, and lighter in weight than the straight or curly types. Wavy or curly hair is oval or elliptical in shape and is found on many Caucasoid types.

Facial hair growth is cultivated and trimmed to various shapes by some national groups as adornment and by others as a religious mark. Others may pluck the hair to inhibit its growth. National characteristics

FIGURE 10.1 *Racial types. There are three main racial types: Caucasoid, Mongoloid or Oriental, and Negroid. The normal characteristics of the Caucasoid type have been illustrated in other parts of this book, but common facial characteristics of the Mongoloid and Negroid types are shown here.* TOP RIGHT: *Male Oriental or Mongoloid (Chinese).* TOP LEFT: *Female Oriental or Mongoloid (Chinese).* BOTTOM LEFT: *Negroid male.* BOTTOM RIGHT: *Negroid female.*

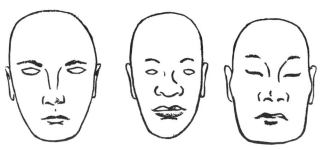

FIGURE 10.2   *Head shapes.* LEFT: *Caucasoid.* MIDDLE: *Negroid.* RIGHT: *Mongoloid. The general differences can be seen in the size of the lips, the shape of the nose and eyes, and in some cases, the jaw structure. These are only prototypically indicated here simply to display the general differences.*

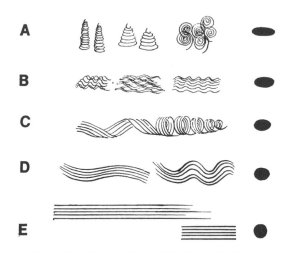

FIGURE 10.3   *Hair shapes and forms. (A) Wooly. (B) Kinky. (C) Curly. (D) Wavy. (E) Straight.*

covered by the Color Process make-up shades of RCMA ranging from Ivory to KN-9. To change from one racial type to another in skin coloration, one need only follow the normal charts for the desired type as listed in Chapter 5 due to the high coverage afforded in that line (also see Appendix A).

It should be noted that the deeper the foundation used, the less the shading will show up, and care should be taken with any lighter colors used as highlights or countershading. When a dark-skinned performer is made lighter, a heavier application of foundation may be necessary, and sometimes an undertoner like an orangy color such as BC-2 or BC-3 must be applied prior to a foundation to color-correct the undertone value as well as the skin coloration.

When skin color changes are made, the make-up artist must take care that the exposed body parts such as hands, neck, shoulders, arms, and so forth also be matched accordingly. Using a bit of RCMA Foundation Thinner on the sponge will aid to spreading the foundation more rapidly. If thinned too much, the resultant coloration may be too thin or light, so take care with the application. At times, one shade deeper foundation (than the facial shade) may be employed for body make-up to get a proper skin match when the face has more natural color than the body.

Eye Shapes

The Mongoloid eyefold, or epicanthic fold, is one of the basic facial differences that must be created with

are often particular to a certain era or style, and these are seen in Chapter 11 and delineated in the section in this chapter on national types regarding facial hair on men. Head hairdos may also take on national characteristics as well as style influences in different eras and ages and can greatly aid in characterization and visual effect.

MAKE-UP TO CHANGE RACIAL TYPE

Since fine performers exist in all racial types, it is seldom necessary to change a person of one race to another. However, some instances do require such, so the basic methods of making these changes can be separated into skin coloration; eye shapes; nasal changes; adjustments of jawlines, cheekbones, and lips; and hair—facial and head—styles.

Skin Coloration

The skin tones of the peoples of the world range from pale pink to deep blackish brown and are generally

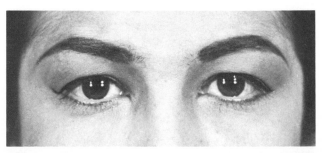

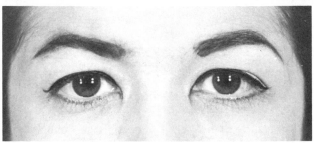

FIGURE 10.4   *Changing eye shapes.* TOP: *The frontal bone and inner eye socket shaded with S-14 to make the eyes appear typically Caucasoid.* BOTTOM: *The frontal bone and the inner eye socket are highlighted with CS-1 or CS-2 and a small upper eyeline drawn to the inner corner of the eye to simulate an epicanthic fold. Not all eye formations will lend themselves to this technique or effect to make the eyes appear more Oriental or Mongoloid.*

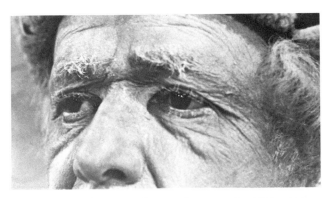

FIGURE 10.5   *Changing eye shapes. A latex epicanthic fold is added to the upper eye socket to turn a Caucasoid actor into a Japanese character.*

make-up when a Caucasoid type is to appear as an Oriental. The appearance of the fold can be created with a highlight on the upper lid, starting in the inner corner of the eye, and extending over the lid where the eyecolor and shading are normally placed. The eyes are then lined in black as shown in Figure 10.4.

Although this technique is effective in long shots on some eyes or on large stages, a prosthesis is necessary for closer shots to create a natural appearance. These appliances can be made of painted-in latex, foamed latex, or plastic and adhered to the periphery of the eye socket with RCMA Matte Plasticized Adhesive or RCMA Prosthetic Adhesive A. The new lids are then covered with foundation and powdered. As various eyelids will require differently curved appliances, if these are made for stock, some variations should be made (Figures 10.5 and 13.13). Most camera angles from a lower point than directly front should be avoided because the appliances can sometimes be seen. It should also be remembered that some eye sockets lend themselves to the application of these appliances and some do not.

An effect to produce a sloe-eyed look can be obtained by gluing the upper lashes on their outer ends to the upper eyelid (Figure 10.6). This eye shape is often desirable for American Indians, Mexicans, and some Asiatic types. Also see information on page 256 for eyebrow blocking.

Nasal Changes

The nostrils can be enlarged by shadowing the inner walls of the outer area of the alae with dark brown pencil. This seemingly creates a wider nose. Actual enlargement can also be done by inserting small pieces (¼ to ⅜ inch) of plastic tubing that is slightly larger in diameter than the nostrils.

The flatness of the bridge of the nose can be accented with a highlight as can the area between the brows just above the nose. Of course, tri-dimensional effects can be created with appliances to remake the nose into

any shape larger than the actual one. However, making a nose smaller is only accomplished by cosmetic surgery.

Jawlines and Cheekbones

Highlighting can be done on the cheekbone or jawline areas with lighter shades than the base color to some small degree. These areas can also be strengthened by shadows on the fleshy portion of the cheek between the bones. All colors must be blended carefully in any such application for color mediums because such could show up in a profile shot. A spot of shine applied with a dabbing motion with a polyurethane sponge with clear lipgloss on the cheekbone will also produce a reflective highlight to accent it.

Prognathism, or a projecting jaw, is best created by employing prosthetic teeth over the performer's teeth while the addition of an appliance over the chin can accent the effect. Prognathism is a facial structure that the make-up artist may be called upon to create for prehistoric types or for horror films but seldom to define a racial type.

Lips

Most dark-skinned peoples have a maroon or blued-red-brown color to their lips—both males and females. This can be created by using deep lipcolors like LS-17 or by adding S-2 to CP-6 lipcolor. An everted effect can be simulated by heavily shading the underside of the lower lip with S-8 or even Black and slightly highlighting the area above the upper lips then overpainting the actual lip shape. Sometimes outlining the lips with a lighter shade of foundation or a very pale beige tone of lipcolor will aid in simulating this effect. Sometimes, those with very dark Negroid skins will have a lip color that is almost the same tonal value as the rest of the face with no apparent redness at all.

Hair Shape and Color

With the aid of a hairdresser-colorist, most hair can be changed in coloration and texture as well as style.

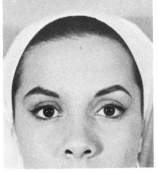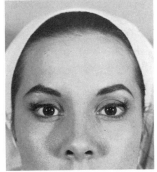

FIGURE 10.6   *Changing eye shapes.* LEFT: *A sloe-eyed effect is created by applying eyelash adhesive to the outer ends of the lashes and to the eyelid.* RIGHT: *When dry, the lashes are pressed to the skin to pull up the lid to make a sloe-eyed appearance.*

Curly and woolly hair can be straightened as well as colored, and straight hair can be curled.

Oftentimes, hair must be stripped of all color with strong bleaching action and then tinted to the desired shade. Changes within a racial type, such as from southern to northern European, can be affected strongly by a hair color lightening, while the opposite change can also do the same.

NATIONAL TYPES

In discussing national types, it is well to divide them first by continent, then by area, then by country or division. In our constantly changing political world, countries change their names as quickly as they change their leaders so we shall try to stay within the bounds of geographical, not political, distinction. An additional consideration is that intermarriages produce fewer culturally *characteristic* national types than in previous generations; national and racial barriers are down in many countries. Thus, it is often difficult to distinguish nationals until they speak.

Even national costumes have largely disappeared today, except within the confines of the home or special feasts and holidays. However, costume is an excellent aid in national characterization for period productions.

Although we should try to avoid stereotypes, we must still *characterize* different nationalities through prototypic distinctions whenever it is necessary for make-up purposes. What follows are general guidelines for approaches to make-up for different national types but is not a replacement for careful and thorough research. Pictorial examples and accurate physical and costume descriptions are extremely valuable. A well-thought-out, individualized approach to the portrayal of a typical national or other class will always be more effective.

Europe

The presentday Europeans can be divided into the following general groups: *Mediterranean* or *southern European, Alpine* or *central European,* and *Nordic* or *northern European.* All are Caucasoid except the Laplanders, who are a Caucasoid-Mongoloid mix.

These European groups have variances from each other in skin tone, hair coloration, and facial features, but in general there are no great racial differences. Mixtures of these basic groups are found in all countries of Europe so these prototypes (southern, central, and northern) are not confined specifically to the areas mentioned but simply typify the basic physical traits of the *cultural* idea of the national type.

Make-up for any of today's European types can be the same as for any of the straight or age make-ups previously discussed, bearing in mind the particular differences noted under the specific undertone values. Men's facial hair depends on fashion and period.

The Mediterranean or southern European prototype has olive skin, dark hair and eyes, oval face, and a long head. They are seen in Spain, Portugal, southern France, Greece, Italy, the western Mediterranean islands, and some parts of Wales and Ireland.

The Alpine or central European group is, prototypically, round headed, fair or medium skinned, with wavy brown hair, medium to dark eyes, and broad faces. They may be found in Austria, central France, Switzerland, Czechoslovakia, the USSR, and the Balkan countries.

The Nordic or northern European type is found in Scandinavia, parts of Holland and Belgium, northern Germany, Poland, and Great Britain. They are typified by light hair, light skin, blue eyes, a long head, and prominent, well-cut features.

North America

North America is comprised of two major countries, the United States and Canada, whose racial stock varies greatly and whose population represents a conglomerate of peoples who exhibit no true national physical or facial characteristics. Their portrayal on stage or screen can only be defined by the period or by speech sectionalisms or dialectical accents. (Mexican peoples will be dealt with under Central America.)

However, as far as we know, the original inhabitants of the American continent were of Mongoloid stock and known as American Indians (a misnomer given to these peoples by early European explorers of the continent). A somewhat detailed account is given here as they are frequent figures in films of American life based on the colonization and settlement of North America.

Related to the American Indians are the Eskimoes in the far north who exhibit the strongest Mongoloid features. They have long heads, but very broad faces; large jaws; moderately narrow noses; coarse black hair; and the eyes very often have the true Mongoloid eyefold. They are usually short and stocky in build, and their skin color is somewhat yellower than the American Indian types further south. Make-up shades can be the same as for the American Indians, but war paint is not employed as facial adornment.

The American Indian population spreads from upper Canada to the furthest south of the United States. American Indians normally have a brown skin, which often has a yellowish to even reddish cast; very dark brown or blackish eyes; straight, coarse black hair with very little natural body or facial hair. Many have a broad face, often with prominent cheekbones; aquiline to flat noses; round heads, with some longheadedness in certain tribal groups; and their height varies from short to tall (Figures 10.7, 10.8, and 10.9).

In their aboriginal state American Indians wore their hair in various fashions. The women generally had long hair, sometimes parted in the middle and braided on each side to hang over the front of the shoulders. Some

FIGURE 10.7 *American Indian types. Early seventeenth century Iroquois chiefs. (Library of Congress.)*

hair ornaments were used, such as decorated headbands or colored material tied to the ends of the braids. The men, as is still the custom with many aboriginal groups around the world, often had far more elaborate hair and headdress styles than the women. In some tribes, they wore their hair with a center parting and braids, while, in other tribes, pompadoured the front of the hair and had two side partings. Some southern tribal groups left their hair long and simply tied a band of cloth around the head to hold the hair from the face. The scalp lock was sometimes worn by the northeastern groups, and it consisted of shaving or burning off the hair on each side of the head, leaving only a long strip about an inch or two wide running down the center of the head. Sometimes just a tuft of hair was left at the top of the head while all the rest was removed. Many wore birds' feathers as adornment in the hair, but some tribes wore eagles' or hawks' feathers in the hair or on headdresses as badges of office or marks of courage.

Face and body tattooing and painting were practiced by almost all North American tribes. The designs were sometimes symbolic, but most of the time they were just used to adorn the face or body, or, in times of war, to strike fear into the hearts of their enemies by the fierceness that their painted faces presented.

In the early days, the main colors were red for war (made from earth colors) and black for death (made from charcoal). Both colors were called *war paint*. The leader often painted his face black and streaked his body with red, hence the misnomer of *redskins*. The brighter colors that were later seen on the Indians of the Plains were often supplied as trade goods by settlers, and bright blues, reds, greens, yellows, and whites were employed.

Considerable research is important before doing make-up for a specific American Indian tribal type as each one may be different and quite particularized. Consult the Bulletins of the Smithsonian Institution and other museums for the distinctive painting patterns and features of the various tribes.

For women, use RCMA KM-3, KT-2, KT-3, K-1, K-2, K-3 depending upon the depth of color required. Line the eyes as usual, accenting their obliqueness slightly. S-4 can be used for shading and cheekcolor rose used sparingly. Do not overdo the cheekcolor as Indian women do not have much natural redness to their cheeks. Lipcolor should be applied only in the natural lip area and of a brownish-red shade, as Indian women did not wear any Lipcolor as adornment. Young women's lips are soft and full, but they aged early due to the outdoor existence. Body make-up can be of the same matching shades. Modern day American Indians can be made up according to the charts for olive skin type Caucasians using the best shade for skin matching.

American Indian men can be made up with KM-8, KT-3, KT-4, K-2, or K-3, again depending upon the depth of skin tone required. It is good practice to vary the skin tones or foundation shades if many Indian characters are to be seen in the same production. Eyebrows should be black and it is often necessary to softly line the upper lid to give the eyes depth and definition with the dark foundation shades. As most Mongolian types have eyebrows that are sparse on the outer ends, the brows can sometimes be drawn in an almost upward-slanting, oblique manner. To achieve this, the outer ends of the brows can be blocked out with RCMA Matte Plastic Sealer prior to foundation application. Some artists use a small piece of plastic cap material over the outer ends of the brows for this effect. Any eyeshading can be done with S-4 or S-5. Lips can be slightly tinted with LS-17 for a deep coloration. Body make-up can also be the same as the foundation or one tone deeper if required. So as to maintain a shine,

FIGURE 10.8 *American Indian types. Five Plains Indians of the nineteenth century. (Signal Corps, National Archives.)*

don't powder the foundation application. If an oiled body effect is required or desired, use glycerin or RCMA Tears and Perspiration (which will not affect the Foundation). War colors can be as previously mentioned.

Wigs may be used for hairdress, while a scalp lock effect can be created with the application of a bald cap to which is attached a scalp lock made of yak hair on

lace. Color the cap with Foundation and then stipple the normal hair growth area with S-3 or beard stipple to give the effect of a shaven head. Without this stipple, the hairline will look unnaturally bald rather than shaven Indian style. Extras often have skull caps made of latex with hair attached (often paint bristle fibres for the hair) and the front of the cap covered with a streak of warpaint to disguise the edge. Modern American Indian men follow current hairstyles although some do hold to their old appearance customs in their tribal areas. As many films have been and will be made of early America, many Caucasian performers are made into American Indian types. This is possibly the widest use of racial change that is prevalent in films today.

Ben Nye makes a series of colors in the American Indian I series from golden copper through dark bronze, which are excellent, while in the Mexican MX series, the olive brown and ruddy brown can also be used for a less ruddy effect. In addition, some of the lighter 20 series can be used.

William Tuttle has a number of Custom Color foundations suitable for American Indians with both an olive tan or a ruddy tan undertone value, and Stein's velvet stick or cake make-up series has suntans and Indian shades that are equally good in color tone value.

Central and South America

The native inhabitants of Central and South America were of Mongoloid stock until the mixture of Spanish blood infused a definitely different basic type. Some will show a predominance of the Indian features while others will be closer to the Spanish southern Mediterranean type.

The Yucatan Maya, ancient Inca, or the presentday Jivaro are all Indian types and bear a general resemblance to the American Indians, but often with a particularly curved nasal bridge that is quite distinctive. Skin colors are quite similar. However, hairstyles are different from those of their northern neighbors as are their decorative ornaments.

The Mexican national is a mixture of other national groups and a combination of races. Many exhibit mixtures of the Indian and the Spanish in their features. Skin colors cover the same range as for American Indian types to the southern European, with a tendency to be on the yellow-tan side. Black hair and dark eyes are prevalent. Facial hair can vary, with often more on men who have more Spanish than Indian blood.

Most of the other Central and South American countries vary from the Mexican type, in those closest to Mexico, to the Spanish influences of the Southern European further south.

Southwestern Asia

The basic population of this area is of the Mediterranean stock type. The Armenians and Asiatic Turks are characterized by a yellowish white skin, medium height, and a prominent, aquiline nose with a depressed tip and large alae. There are some Syrians and Iraqians in this facial group as well as some people along the southern coast of the Arabic area. Few grow beards, but when they do they are dark and often quite curly.

The Jewish people can be divided into two cultural groups—the *Ashkenazim,* who are from the USSR and central and western Europe, and the *Sephardim,* comprising those of Spain, Portugal, Asia Minor, Egypt, and the Arabian area. They often have light skin, medium to dark eyes, and a nose formation with a rather prominent bridge, which is said to have descended from the early Philistines or Hittites (see page 127 for description of other Jewish and Arab peoples who are from the northern coast of Africa). Many beards and moustaches are worn both as adornment and for religious purposes.

In Iran there is often a mixture of Mediterranean and Mongoloid elements with some of the traits of the Arabs and the Armenians. This stock has black, wavy hair, light brown skin, a long head and face, a prominent but rather narrow nose, and dark eyes (for beard types see Figures 17.6 and 17.7).

Southern Asia

India can be divided into three main groups of individuals: those from the Himalayan Mountains, those from the plains of Hindustan, and those from the southern area called the *Deccan.* India contains many peoples whose physical characteristics are a combina-

FIGURE 10.9 · *American Indian types. Geronimo, the Apache chief (nineteenth century). (National Archives.)*

tion of a prehistoric Negroid population, a Mongoloid group from the northeast, and a Dravidian stock from the northwestern frontier area. The main population of the Deccan consists of Dravidian types with brownish-black skins, dark wavy hair, a long head, and medium build. The Tamil people of Ceylon and southern India also resemble this group.

The Veddas of Sri Lanka have coarse, wavy, long black hair; dark brown skin; a long narrow head with a slightly receding forehead; prominent frontal bones; thin lips; a broad face and nose; and short stature. They are survivors of the earliest inhabitants of India and are of Australoid stock, being a Negroid-Caucasoid mixture.

The Kashmiri of India have light brown skin, a long head with a high forehead, a long and narrow face, regular features, and a prominent well-formed nose. They are relatively tall as compared to other people of India. The Benares of north-central India, the Bengalis of northeastern India, the Jaipur of northwest India, and the Rajputana of northwestern India are all of Caucasoid stock. Except for their brown skin color, many of the people of the North Indian group closely resemble Europeans. (Make-up shades KN-8 and KN-9 are useful for dark-skinned Indians.)

The women of this northern area are often very beautiful, with dark sloe eyes and pale brown skin. Beards are plentiful in India where there is a great Moslem population.

The Indian Ocean Islanders of Andaman, the Semang of the Malay Peninsula and eastern Sumatra, the Aëta of the Philippines, and the Tapiro of New Guinea are short people with short black hair, black-brown skin, a small round head, broad face, and full lips that are not everted. Little facial hair is grown by the men.

The Burmese represent the southern Mongoloid type and have black hair (which is very sparse on the face and body), a broad face and nose, a round head, yellow to brown skin, and often slanted eyes with some Mongoloid eyefold. Those in the localities farthest away from China tend to have less yellow in their skin.

Southeastern Asia

The Malay Peninsula and Archipelago have peoples that can be divided into a large southern Mongoloid group and a more primitive combination group. Of this latter, the Semang are a pygmy, negrito group whose height is normally 5 feet or less. They have short, frizzy black hair that is sparse on the body and face, chocolate brown skin, a round head, thin lips, and a short, flat, broad nose. The Sakai, a Negroid-Caucasoid mixture group, are lighter in skin color, with long wavy or curly black hair that has a reddish tinge, and they are tall people. The Jakuns have dark red or copper-brown skin; coarse, dark hair; oblique, dark eyes; and a round head with high cheekbones.

They are of Mongoloid stock and are found on the Malay Peninsula.

The Malayan or Indonesian family can be found all over the Malay Archipelago. The Malay of the Malay Peninsula; the aborigines of the Philippines, Borneo, and Celebes; the Javanese and Sudanese of Java and Bali; and the Bataks of Sumatra are members of this group. There are also branches of this family on Formosa and Madagascar. They are short and have dark wavy hair, tawny-yellow skin, prominent cheekbones, slightly prognathous jawlines, and somewhat oblique eyes with some Mongoloid eyefold. Little facial hair is grown.

Eastern Asia

The Chinese are of medium height with intermediate sized heads. They have yellowish-brown skin, oblique eyes with definite Mongoloid eyefolds, and straight black hair. The cheekbones are often high and the face square in appearance. The nose is sometimes flat and spread at the alae. The brows are seldom heavy on the outer ends. This brow growth is a typical feature of all Mongoloid types and seems to be more apparent as the eyefold increases in size. Today, little facial hair is grown by the men as the growth is always sparse. The women have about the same coloration as the men.

There are two main Japanese types. One is tall and slender with fine features; a prominent, narrow, arched nose; and eyes either straight or oblique with some not having much Mongoloid or epicanthic fold. The other type is short and stocky with coarser features, a broad face, a short rather concave nose with wide alae, oblique eyes with the Mongoloid eyefold, and a darker skin than the other type. Both have black hair and dark brown eyes. Little facial hair is grown today with the exception of some modern moustaches.

Another Japanese group, called the Ainu, is confined to a small area. They possess heavy, black, curly beards and bushy or wavy hair; a tanned color of skin; a short, narrow, concave nose; a broad face; and large, dark brown, straight eyes. They are an ancient prehistoric stock different from the other Japanese, being an Archaic Caucasoid-Mongoloid mixture.

Central and Northern Asia

This area comprises Tibet, Mongolia and Chinese Turkistan, and parts of Siberia. Among the people are the Chukchi of northeastern Siberia, the Koryaks, the Kamchadales, the Giliaks, and the Asiatic Eskimoes. They have straight, black hair; brown or reddish brown beard hair; yellowish to brown skin; prominent cheekbones; a flat face; oblique eyes with Mongoloid fold; and a straight or concave nose. The face often has a square jaw with a pointed chin typical of the Mongoloid type. Some straggly moustaches and beards are grown by the men.

Oceania

This is the area that extends from New Zealand to Hawaii and from Australia to Easter Island. The six main divisions of type are found in Australia, Tasmania, Melanesia, New Guinea, Polynesia, and Micronesia.

In Australia the English stock are similar in characteristics to the comparable European types. The aborigines of Australia have black curly or wavy hair with often heavy beards, dark chocolate brown skin, a long head with a flat receding forehead, prominent frontal bones, and a deeply set, very broad nose. They are of a so-called Australoid stock, being an Archaic Caucasoid-Negroid mixture.

The aboriginal Tasmanians became extinct during the last part of the nineteenth century. In general they had similar physical characteristics to the aboriginal Australians. There are English European types there now.

Melanesia covers the Bismarck Archipelago, northeastern New Guinea, and the Solomon, Santa Cruz, New Hebrides, Loyalty, Louisiade, New Caledonia, and Fiji Islands. The native inhabitants have hair that is usually very curly, a long head (some have a round head), light to very brown skin, a rounded forehead, prominent frontal bones, and a broad nose, and they are medium to short in height.

New Guinea offers a great variety of types in its Negroid group. The three main classes are Negritos, Papuans, and Melanesians (see earlier description of the latter). The Negritos are the Tapiro Pygmies of Dutch New Guinea and are similar in racial type to the Semang Pygmies of the Malay Peninsula. They have short black hair that is plentiful on the face and body, yellowish brown skin, a straight nose that is medium wide, and a deep convex upper lip, and they are about 4 feet, 10 inches tall (full grown male). The Papuans have dark skin, a long head, black hair, a retreating forehead, prominent frontal bones, an often prognathous jaw, and a prominent broad nose that is convex with the tip often turned down, and they are of medium stature. Both the Papuans and the Melanesians are of Negroid stock with some Australoid aboriginal features.

Polynesia consists of a group of islands in the central Pacific area that includes New Zealand, the Hawaiian, Society, and Marquesan groups, as well as Tonga and Samoa. The people here are tall, with straight to wavy hair, olive to brown skin, an oval face with high cheekbones, a prominent and sometimes convex nose, and a round head. Few men grow facial hair as it grows sparsely.

North of Melanesia are the Marianas, Caroline, Marshall, and Gilbert Islands, which are called the Micronesian group. The people here have yellow to brown skin, black straight to kinky hair, black eyes, and prominent cheekbones and are of medium height. They are a Caucasoid-Negroid mixture.

Africa

Living in the northern and northeastern parts of Africa are Caucasoids who have dark brown to black hair that is curly or wavy, reddish brown to dark brown skin, a long head with a long oval face, thin lips, a pointed chin, and a prominent narrow nose. They are slender in build and are medium to tall in height. They include the Berbers, Tuaregs, Hadendoans, and Bisharins. Many grow beards and moustaches (see pages 260 and 261 for facial hair growth).

The Arabs and Jews or Israelis of this area are usually of medium height, dark haired, oval faced, and with a long narrow nose that is straight or sometimes hooked at the end. Their skin ranges from a pale to a deep olive for the women and from deep olive to a brownish tan for the men (depending upon the amount of facial exposure to the sun). Also see Jewish types on page 125. Many grow beards and moustaches.

The Negroid population of Africa consists of the natives of the West Coast, the Upper Nile, and the Northeast Coast, including the Bushmen, Hottentots, Pygmies, and many others. Their color ranges from a light yellow-brown to a very dark chocolate brown, and in general, they have a flat, broad nose; large everted lips; dark brown to black hair that is either very curly or kinky; and dark eyes that are often narrow and slightly oblique. They practice some body painting, tattooing, and scarification for adornment in the less-developed sections. The painting shapes are different from the Native American variety, but the effect and use is similar. Whites, yellows, and blues are common shades. Their height ranges from the 4-foot Pygmy of the deep forest jungle area to the slim 7-foot people of the Watusi. Hairstyles in the aboriginal groups are often very elaborate as are decorative head gear. Few grow any facial hair.

In any of the racial groups (but most often in the Negroid), albinos can occur. They are people with a lack of pigmentation in their bodies, although they possess all the other physical characteristics of their particular racial and national type. The hair is either white or very light blonde, and the eyes are a pinkish shade. They generally always wear glasses (when available to them) because their eyes are quite weak. They must also keep out of the sun as sunburn is one of their most painful (often fatal) illnesses. In addition to the RCMA Foundation recommendations, see Appendix A for equivalent shades by various other manufacturers.

It should be kept in mind that political upheavals constantly occur around the globe. Thus, although the names of the nations may change for political and geographical purposes, the ethnic groups and traits do not.

The best short work done on ethnology is *The Races of Mankind* (Chicago Museum of Natural History) by Dr. Henry Field, formerly curator of physical anthropology at the Chicago Natural History Museum, and revised by Dr. W.D. Hambly, curator of African ethnology, and illustrated with bronze sculptures by Malvinia Hoffman, long recognized as the foremost sculptor in the field of the world's racial and national types. The information for this chapter was drawn from this comprehensive pamphlet.

CHAPTER 11

Period and Historical Characters

THE RECREATION OF THE PEOPLES OF THE PAST IS A MOST fascinating branch of character make-up work, involving not only research into the lives of our ancestors in portraits and pictures but also in the molding and sculpting of likenesses to make appliances to achieve resemblances and the painting of the features to continue the reconstruction of the personality and to provide the performer with a likeness that places him or her in the character being played. Without the skill and technique of the make-up artist, this visual transposition would be lost, and the audience would not believe the performer who says that he or she is playing the character.

PREHISTORIC MAN

We are indebted to the archeologists who have uncovered remains and patiently reconstructed the likenesses of our remote forebears. For make-up purposes it is not necessary to go into a lengthy discussion about fossils or prehistoric man, so three basic types out of many will serve to show some of the evolutionary changes that have occurred in the development of *Homo sapiens*, or today's human (Figure 11.1).

The broad developments can be listed as follows:

Lower Savagery Lower Paleolithic (Old Stone Age), *Pithecanthropus erectus* type; Middle Paleolithic, Neanderthal man.
Higher Savagery Upper Paleolithic Cro-Magnon type; Mesolithic or interim period.

The domestication of animals by humans led to the transformation of Mesolithic savages into Neolithic barbarians, preparing the way for civilized man. *Primitive barbarism* came in the Neolithic or New Stone Age and then followed another transition period of Pre-Dynastic people. Finally, civilization is roughly calculated from the First Dynasty in Egypt, about 3100 B.C.

One of the earliest types of man-like creatures was *Pithecanthropus erectus*. Specimens that have come from Java show that the reconstructed skull of this type was long with a narrow, slanting forehead and enormous frontal bones. He probably had the power of speech, but his brain was only about two-thirds the size of modern man's. He is believed to have been about 5 feet 8 inches tall, weighed about 150 pounds, and walked in an upright posture but not holding his head very erect. He is believed by some authorities to have lived about half a million years ago.

Homo neanderthalensis, or the Neanderthal man, lived at a later period and had a low, narrow, receding forehead but a brain equal in size to modern man's. His face was long and the eye orbits projected. His nose was flat and broad, the upper jaw prognathous, the lower jaw heavy and powerful, the chin small and receding, and the neck extremely heavy, indicating that his posture was stooping and possibly less erect than *Pithecanthropus erectus.* The bones of his arms and legs were heavy and strong, and the tallest among this type must have been about 5 feet, 5 inches in height.

The two preceding types of lower savages were precursors of *Homo sapiens* although some researchers believe in a parallel rather than a direct connection. Certainly they are the prototypes of what we can now call prehistoric man and will serve as a basis for constructing a make-up when this type is called for in a production. Much reconstruction work must be done on the face to make a person resemble these prehistoric types so plastic or latex appliances are necessary for this work. Years ago, make-up people laboriously built up the face to resemble the prehistoric character with spirit gum, cotton, and collodion, but today such methods are as archaic as the characters themselves (relatively so, that is!).

FIGURE 11.1 *Prehistoric man.* LEFT: Pithecanthropus erectus. CENTER: *Neanderthal.* RIGHT: *Cro-Magnon. (Courtesy of the American Museum of Natural History.)*

In the Higher Savagery we find another type that is close in physical appearance to modern man. The Cro-Magnons had a long narrow skull and a brain capacity equal to that of today's humans. The forehead was broad and high, the nose narrow and high bridged, the jaw large but of a modern form, and a strongly made chin. Specimens indicate that Cro-Magnons were about 5 feet, 11 inches tall and robustly built. The women of all eras resembled the men but had smaller features and generally less physical development. The men are also believed to have been bearded in most cases.

Museums of natural history throughout the world have many books and pictures as well as reconstructed forms of prehistoric people. See the displays if possible, and consult books on anthropology for more study material when required to apply make-up for prehistoric types.

A number of films have depicted prehistoric creatures, among them, *One Million B.C.* (Hall Roach, 1940) that employed more facial hair than appliances on the men, with the women being quite modern in features. However, in *Quest For Fire* (20th Century Fox,

1981), facial appliances of various types in addition to hair were employed to create the characters. Although a number of make-up artists worked on various stages of this production, Christopher Tucker designed many of the important characters. For the three leads, he altered the performers' profiles with forehead prostheses and changed their jaw structures with upper and lower dentures, which fit over their natural teeth. Although not an exact reproduction of primitive man—in modern man there is too much head above the brows—the make-ups were effective enough to win the prestigious British Academy Award and the Academy of Science Fiction Award.

ANCIENTS

Civilization as we know it and the preservation of historical records are synonymous. About 3100 B.C. the First Dynasty was founded in Egypt, and it developed a distinctive culture that has come down to us through the ages. For our purpose we shall treat all the ancient civilizations in a prototypic fashion that can be recognized for production purposes. As such

FIGURE 11.2   *Three principals, Ron Perlman, Everett McGill and Nameer El-Kadi in* Quest For Fire *(Courtesy 20th Century Fox).*

FIGURE 11.3   *Ron Perlman before and after make-up (Courtesy 20th Century Fox).*

we have the Egyptians (all dynasties), the Mesopotamians (which include the Chaldeans, Babylonians, Assyrians, Medians, and Persians), the Greeks, the later period Romans, and the Hebrews throughout the biblical period. Additional study may certainly be required if a production is large or covers a wide scope, and R. Turner Wilcox's *The Mode in Hats and Headdress* (1949) is invaluable for references on the facial hair, hairdress, and such on both the ancients and the eras that followed to the present time. Most professional wigmakers have this book and references are often made by page and drawing for a specific hair or beard style by the make-up artist in ordering hair goods.

Egyptians

In general, the head was shaved by both sexes, and wigs were worn made of human and animal hair or vegetable fibers. The priests did not cover their shaven heads. No facial hair was worn and both sexes used facial make-up for adornment (Figure 11.4). Black kohl, powdered malachite green, red clay, and vegetable dyes and white lead were used. The wigs were often dyed brilliant reds, blues, and greens and were elaborately dressed. In later periods, after Rome had conquered Egypt, the women often let their hair grow long and dressed it in the Roman fashions.

The make-up consisted of black lines drawn around the eyes from the inner corner and extending outward

in a straight line to a point an inch or so beyond the outer corner of the eye (Figure 11.5). Eyeshadow in blue or green was often applied to match the wig color. The cheeks and lips were reddened and the eyelashes and brows accented with black. Creams and oils were heavily scented with perfumes and applied to the body to give it shine. Among the men a brightly polished, oiled head was a thing of high fashion. Bases made of white lead and perfumed oils were sometimes applied.

The facial features of the Egyptians were quite regular, and the skin tone was light brown. The nose was often wide at the alae with large nostrils, and the bridge was long and straight. The lips were full and well formed with little or no eversion. (Make-up shades KT-2, KT-3, and KT-34 are excellent for Egyptians.)

Mesopotamians

This group comprises the ancient Chaldeans, Babylonians, Assyrians, Medians, and Persians. Natural hair was grown on both the head and the face by the men and elaborately dressed and curled (Figure 11.6). Sometimes they dyed the hair with henna to an orange-red shade. Some wigs were also worn. The women grew natural hair or wore wigs. Hair dyes were used, and they colored their lips and cheeks with rouge. They used kohl around the eyes as did the Egyptians. Both sexes drew these lines around the eyes but did not extend them as far outward as the Egyptians. The brows were also darkened with black and drawn so that they almost met over the nose. The men especially accented their brows very heavily in this fashion. (Make-up same as for Egyptians.)

Greeks

The men were generally bearded, although the soldiers of about 350 B.C. began to shave their faces clean. The hair was curled and often almost as short as the modern male cut. Men seldom used cosmetics as did the women, but various oils and pomades were used. The women dressed their long hair in elaborate fashions, using many ringlets and decorating the hair with ornaments of gold and jewels (Figure 11.7). Many Greeks had naturally blonde hair, and dyed hair was also seen. Reddish and even bluish shades of hair dusted with gold or colored powders were in fashion. Some bases were made with white lead and the cheeks and lips

FIGURE 11.6   *Mesopotamians.*

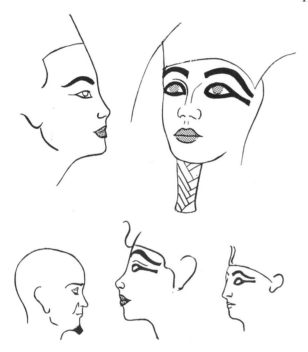

FIGURE 11.4   *Egyptians.*

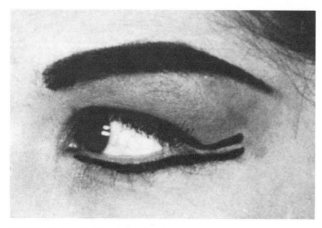

FIGURE 11.5   *Egyptian eye make-up.*

FIGURE 11.7   *Greeks.*

painted red. The skins of the ancient Greeks were about the same tone as the modern Mediterranean Europeans, being olive for the women and a darker olive or bronze tan shade for the men so that the regular charts (pages 55 and 71) can be followed for mediums for make-up.

Romans

Most Roman men had short haircuts very similar to those worn today, and their faces were cleanly shaven. Some of the older men and philosophers wore their hair and beards long and flowing. Although some men used facial cosmetics, they were mostly confined to the women (Figure 11.8). A white skin was very desirable in Roman fashion so bleaches and bases were used by the women. Cheek and lip rouge was also used as were hair dye and some wigs. Hairstyles were quite distinctive from the earlier Greek styles. For make-up, follow the charts (pages 55 and 71) and use the medium shades of bases.

Hebrews

The ancient Hebrew men were always bearded and wore their hair rather long (Figure 11.9). The women very seldom used cosmetics so should always be made less painted when they appear with Greek or Roman types on screen. Wigs were not worn and the women's hair was quite long and naturally dark. In general the charts (pages 55 and 71) can be followed for base shades (also see the section on Jewish national types in Chapter 10).

DARK AGES TO PRESENT TIME

The remarks made under this heading apply to Europeans and later, of course, European settlers and their descendants overseas.

The Dark Ages

The men of this era wore beards and moustaches and cut their hair moderately short. Women's hair was long but dressed close to the head (Figure 11.10). However, in northern Europe from the Roman conquest to the twelfth century, only the nobility wore beards and long hair. The other men cut their hair rather short and were cleanly shaven as were the Roman soldiers (Figure 11.11). From this period until the present day, there are no important make-up differences with this exception of the hairstyles on the men and women and the shape or absence of facial hair grown by the men. Follow the charts and the descriptions of the straight, age, and national types to ascertain the correct make-up shades to use.

Renaissance

The Renaissance in Europe brought about many changes in hairstyles for the women, but the men, although shapes varied, still were bearded in most countries.

FIGURE 11.8 *Romans.*

FIGURE 11.9 *Hebrews.*

Male wigs were quite in fashion, and baldness was covered in polite circles (Figure 11.12).

Seventeenth Century

Figures 11.13 through 11.15 show many examples of seventeenth century styles and faces. The Cavalier men wore they hair shoulder length and curled. Beards became smaller and the moustache took on more importance as it was curled, trimmed, dressed, and waxed or pomaded. The Puritan men cut their hair short and

FIGURE 11.10   *Byzantines*.

FIGURE 11.12   *Renaissance*.

FIGURE 11.11   TOP AND UPPER CENTER: *Saracens or Moors*. LOWER CENTER AND BOTTOM: *Medieval*.

round and were closely shaven. Wigs or perukes were also in style during this century in France and later in England after the Restoration. The women's hairstyles were again very elaborate, and false hairpieces were added to the coiffure to give it height and size. Small black gummed silk patches came into use during the last half of the century. These were cut in various designs such as stars, crescents, and other geometric shapes and applied to the cheeks or skin as facial ornamentation.

Eighteenth Century

Beards disappeared from almost all the men's faces in Europe in this century, but some moustaches remained on the military men, especially in the Germanic countries. The pigtail or queue wig and hairstyle lasted

FIGURE 11.13   *Seventeenth Century.*

FIGURE 11.14   *Seventeenth Century. Walter Slezak in a period costume with a lace front wig, moustache, and beard on lace with the edges of the beard section hand laid for a natural effect. (Courtesy of R.K.O. Pictures.)*

almost throughout the century and was especially worn by the military men. The fashion of powdering the hair and/or wig with white powder or wheat flour rose to its height during this period for all the upper classes and for soldiers of all grades. Other hair powders were pinks, blues, and grays. The lower classes often had their hair queued or left the hair full in the back without a queue.

Wigs were worn by many women as in the previous century for balls and court affairs. False hair was also used with the natural hair. Powdering the hair was also practiced. Face painting was done in France by most of the women of the court, but the English restricted most of their cosmetics to creams, lotions, and other beauty preparations (Figures 11.16 and 11.17).

Period paintings are also excellent sources for the study of characters (Figure 11.18). The British military had two styles during the 1770s and 1780s that predominated (Figure 11.19). One was the whitened hairdress shown on the left, for those who wore hats, showing the clubbed fold at the back. While the officers generally had a half-inch velvet ribbon, the other ranks had a leather cord with a centered small rosette. The right side of Figure 11.19 shows the braided and end-tied hairstyle of the flank companies (grenadiers and light infantry) for those who wore caps instead of hats.

Nineteenth Century

Short hair for men came into vogue after the French Revolution as wigs and long hair had been the mark of the aristocrat. Early in this period sideburns and moustaches began to appear, and throughout the century they continued to grow in size and shape until full beards became fashionable for men by the 1860s. At the very end of the century and into the beginning of the next, most beards were removed, leaving only the flowing moustache with the curled ends, known usually as the "handle-bar" variety. By 1910 all but the older men were clean shaven as is the general custom today. Hair powdering disappeared, as did patches, early in this century, and wigs were mostly relegated to official or legal functions in the British courts (Figures 11.20 and 11.21).

Steel cut engravings were widely employed during the nineteenth century, and they were especially good for illustrating hair and beard styles. Fine facial details were also possible with these engravings as noted in the upper illustration of Figure 11.22. Sideburns so popular in that period were popularized by General A.E. Burnside. Other hair and facial styles can be seen in these portraits of military men who typify the male styles of the day.

Twentieth Century

Until about 1910 to 1915, the hair and beard styles for men were the same as for the last part of the previous

FIGURE 11.15 *Faces of the seventeenth century (steel cut engravings).*

George Monk, Duke of Albemarle.

Henrietta Maria.

Increase Mather.

Captain John Smith.

John Winthrop the Younger.

Governor Winthrop.

Oliver Cromwell.

King Charles II.

Sir Edmond Andros.

century. Then, from the World War I period through the 1960s, beards were seldom seen, although finely trimmed moustaches were in vogue, and soon after we saw longer sideburns come in and then many beards. Young people grew long hair and beards as a symbol of protest, it seems, and this lasted until the 1980s when shorter head hair and considerably less facial hair was grown by youth. Also, some more trimmed and better kept facial hair growth was retained by a limited number of men during the 1980s. What we can expect in the future is anyone's guess!

Women's make-up and hairstyles have run the gamut of very long hair dressed up and over padding forms to very short hair to very curly styles or just long and hanging. Older women tend to retain a hairstyle for a longer period than younger or possibly more stylish women (of all ages). Make-up went from little or no make-up in the 1910s through the painted features of the flapper period of the late 1920s and early 1930s. Then the great influence of the glamour girls produced by the film industry in the 1930s and 1940s began.

FIGURE 11.16 *Eighteenth century.*

Models discovered false eyelashes and eyelining in the early 1950s and overdid it until the 1960s when a no-make-up look almost dropped all color from lip-colors and the foundation became lifeless pale for high fashion effects. Eyes were made Garbo-ish with brown and pearl, and false lashes were still *de rigueur.* During the protesting 1960s, we saw a lack of attention paid to skin care and appearance by the young, while the older began to discover the art of cosmetic surgery and skin care by machines and aestheticians.

In the 1970s the natural look became important, and by the end of the decade, red lips and more sensibly painted eyes contributed to the effect. The 1980s will certainly see its changes and nuances as well.

Certainly black-and-white film techniques influenced the American woman and, indeed, most of the world as well, and the use of highlights and shading were made popular by the famous Westmore Brothers of Hollywood. When color film and color television invaded the entertainment world, the make-up methods and shades also greatly influenced the styles of make-up for the commercial manufacturer. Visual

FIGURE 11.17 *(A) Early eighteenth century. (B) Mid-eighteenth century (steel cut engravings).*

FIGURE 11.18 *The Mozart family painted by De la Croce in 1780. Such paintings are excellent for authentic hairstyles.*

communication hit a new height with color and will continue to have its effect on fashion.

Twenty-first Century and Beyond

Although this heading sounds facetious, the make-ups for the many science fiction and fantasy films being produced today assuredly open a wide avenue of imagination for future styles for men and women in costume, hair, and make-up. Futuristic films have spawned the new and controversial special effects make-up field in addition to everything else, and examples of such are given in Chapter 15.

SPECIAL HISTORICAL CHARACTERS

Some persons of history have become identifiable by the mere mention of their names, and performers have been playing roles that recreate segments of their lives. In U.S. history, George Washington, Abraham Lin-

FIGURE 11.19 *Standard British military styles of hair dress.*

FIGURE 11.20 *Nineteenth century.*

FIGURE 11.21 *Male faces of the nineteenth century.* TOP LEFT: *President U.S. Grant (steel engraving).* TOP RIGHT: *President Abraham Lincoln (Brady photograph).* BOTTOM LEFT: *Louis Pasteur.* BOTTOM RIGHT: *General Robert E. Lee.*

coln, Generals Grant and Lee, and others bring a facial familiarity to mind that must be exactly matched to be creditable. Actors including Henry Fonda, Raymond Massey, and Gregory Peck have depicted Abraham Lincoln and are compared as standards for any other recreation of this famous face.

In the history of other nations, royalty and political and military personalities and their spouses have be-

come facially familiar to many of us, and many good make-ups have been done on performers to illustrate this. Here picture files will aid and often new research must be done. Libraries are a fertile source for books and prints, and after consulting these, a thorough study of the performer's face who will play the part should be done. Tests with instant color film are a great aid to seeing how the make-up may be progressing or for

FIGURE 11.22
American military men of the nineteenth century (steel cut engravings).

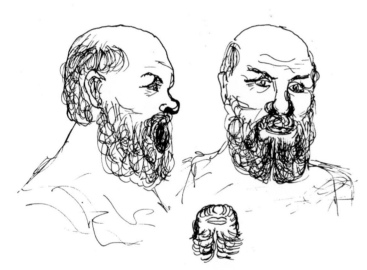

FIGURE 11.23   *Rough sketches of sculptures. A bust from Rome. At age 70 Socrates was considered to be an ugly man, with a large bulbous nose, bald head, short stature, and quite stout. Insert drawing shows beard dressing suggestions.*

the finished result. It is always good to try to pose the performer in the same manner as the portrait, photo, or such that is being used as a reference to see just how close one appears as the other.

Quick pen or pencil sketches of sculptured busts are an aid in researching a character when paintings or photos may not exist or be readily available (Figure 11.23). Many fine heads of Greek and Roman sculpture are available from museums.

All in all, this book can give you methods and materials, but one must practice the art to perfect it. No one can guide *your* hand in the execution of a make-up—only innate talent and repetition will make one make-up better each time it is done. The greatest of the make-up artists *never* stop learning, experimenting, and training themselves in their art.

CHAPTER 12

Special and Popular Characters

MYTHOLOGY, MODERN AS WELL AS ANCIENT, HAS INFUSED a myriad of popular conceptions and fantasies into human thought. Some are stylized, popular, and happy characters that have become part of our life as well as entertainment, while others have drawn upon our dark imaginations and hidden fears to instill a different element to which many are drawn out of curiosity and desire to see the unknown.

HORROR CHARACTERS

Perhaps the misty world of spirits and demons offers the greatest scope for the imagination, and make-up

FIGURE 12.1 *Lon Chaney, Sr., in* The Phantom of the Opera *in a make-up that inspired many young make-up artists of the 1940s to attempt to copy. This was a classic horror make-up done with old nose putty, facial distortion, false teeth, a bald front wig, and strong highlights and shadows. (Universal Pictures.)*

artists' conceptions have opened avenues of thought, sometimes heretofore inconceivable in frightening aspects and elements. Certainly, advancing technology and research in the field have produced more realistic and startling effects and creatures year after year, and it seems that the public not only dotes on these imaginative films but also supports the trend by contributing huge box office returns to the producers. Motion pictures have expanded the scope and depth of the horror style of entertainment far beyond what was possible on any stage in the past. Today, and in the future, we will see even more thrillingly terrifying films with the use of electronic and other spectacular tools of the engineers as well as the visual creations of the make-up artist. Special effects have entered many fields including not only make-up but also lighting, filmic superimposition, musical accompaniment, and other related crafts and arts.

Silent films introduced Lon Chaney and his many characterizations and special make-ups (Figure 12.1). Lon Chaney, Jr. told the story that his father kept his make-up/dressing room closed to everybody, and when he had to make repairs or changes, filming stopped and he went alone into his sanctum, emerging only when he was satisfied that his appearance was as he wanted it to be. Father passed no secrets down to son, and much of his art and craft died with him.

By the 1930s, studios had formed make-up departments that were mainly concerned with beautifying the performers until Jack Pierce of Universal Studios was faced with creating a Monster for the film *Frankenstein* (Figures 12.2 to 12.5). His creation commenced a trend that still leads the box office: the fantasy-horror film.

Although the early "horror" make-ups were done laboriously with old-time materials, later re-creations employed latex pieces that are far easier on the performer and can be pre-prepared for use. For example, *The Mummy* make-up had a complicated cloth wrapping on the body covered with a glue and dusted with Fuller's Earth to simulate a decaying body, while the same effect was later created with a grayed latex base paint covering a pre-made costume.

Other horror style make-ups such as *Dracula* were done with a gray-green foundation, very dark red lips and the eyes outlined to appear deep and penetrating

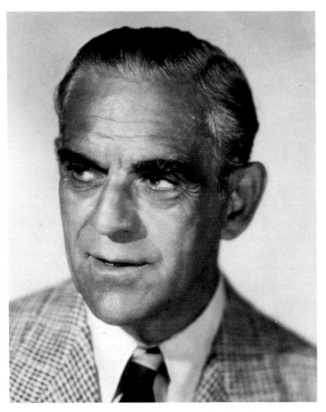

FIGURE 12.2   *Mild-mannered, gentlemanly Boris Karloff became the newest screen sensation in 1931, not as the Englishman he was but as the fearsome Monster in the Universal Studios film,* Frankenstein. *(Universal Pictures.)*

FIGURE 12.3   *Universal's make-up artist Jack Pierce created the make-up with a headpiece made of cloth, cotton batting, spirit gum, and collodion. (Universal Pictures.)*

FIGURE 12.4   *So strong was Karloff's portrayal that audiences began to call the Monster "Frankenstein" rather than the evil doctor who created the creature in the film! (Universal Pictures.)*

(Figure 12.6). *Mr. Hyde,* the alter ego of Dr. Jekyll, has gone through many phases from the John Barrymore one, to that of Frederic March, Spencer Tracy, and Louis Heyward; and to the more modern ones often seen in revived productions today (Figure 12.7).

Two of Rick Baker's horror make-ups include the decaying corpse sequence in *An American Werewolf in London,* which was done with foamed latex appliances on the actor (Jack Stave), showing the destruction on the side of the face and neck by the claws of the

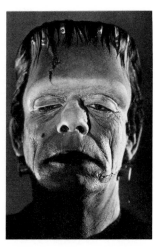

FIGURE 12.5   *The evolution of the Monster make-up from the Pierce-Karloff cotton and collodion head* (LEFT) *to the easier-to-apply and wear latex one evolved for Glenn Strange* (RIGHT) *when he took the role in the 1940s. (Universal Pictures.)*

To create this effect for the Paramount Pictures film *The Elephant Man* starring John Hurt, Christopher Tucker studied the actual photos and even a cast taken of the head and shoulders of Merrick that are the property of the London Hospital. It was found, however, that the actual distortions of the jaw and cheek lines were so radical that if the prostheses that he would make for the actor were to match the actual features, they would be impossible to apply on a normal face. Tucker therefore produced distortions so that the actor could still articulate his lines, and yet retain the patheticism that was required for the audience to have sympathy for the poor creature rather than be just horrified by the effect.

In designing the project, Tucker had to break down the make-up appliances into sections. Due to the thickness required in some facial areas where joins are normally made, he designed the make-up in two levels. That is, he had appliances that were *under* the outer ones.

On a face cast of the actor, he first sculpted his concept of the completed Elephant Man. Then, on a separate facial cast, he modelled the underlying sections, made molds, and cast these in silicone rubber.

Outer sections were molded separately, overlapping each to match the skin texture. The head was made in two sections, the face in fifteen, and the right arm was made in three—hand and wrist, elbow joint, and the upper arm. The upper torso was built in nine sections, some of them being up to four inches thick in places. Due to the exigencies of the shooting schedule, the molds were made in triplicate, and so it took long hours for eight weeks to prepare them.

Although most of the appliances were made of

werewolf that had attacked him (Figure 12.8). The slush molded latex mask made for the same film for one of the Nazi demons shows the detail of the sculpture to demonstrate the decay the mask was supposed to convey (Figure 12.9).

Although not really in the horror make-up category, distortions of the human body that are startling to the eye, such as the recreation of the terribly disfiguring disease known as neurofibromatosis that afflicted a man named Jon Merrick during the last part of the nineteenth century, can be grouped in this classification. When Merrick died at age 27, his body and face were so distorted by the ravages of the illness that he was quite unrecognizable as a human being.

FIGURE 12.6   *Other famous make-ups were those created by Jack Dawn: (A)* Dracula *for Bela Lugosi; (B)* The Mummy *and (C)* The Wolfman, *both for Lon Chaney's son (who hated to be called "junior" after his father died!), all in films for Universal Studios, which seemed to monopolize the horror film field at that time. (Universal Pictures.)*

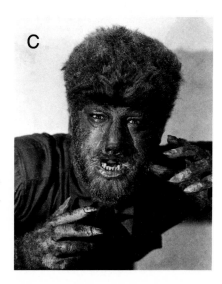

FIGURE 12.7 *Frederic March as Mr. Hyde in the Paramount Pictures production (1932) of* Dr. Jekyll and Mr. Hyde, *the most exciting version of this much filmed story. Note the painted nostrils, overpainted mouth, false teeth, and shadows and highlights make-up along with extra brows and a low-browed hairpiece. (Paramount Pictures.)*

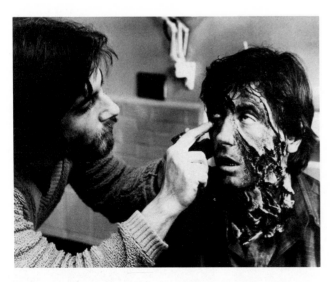

FIGURE 12.8 *Rick Baker's horror make-up for the decaying corpse sequence in* An American Werewolf in London. *(Polygram Pictures, 1980.)*

foamed latex, the two head sections were made of polyurethane foam with a latex skin into which hair was inserted, hair by hair, and, being more durable, were used a number of times (see Figure 12.10). A urethane foam made by Dow Corning was employed for a number of the sections (see Appendix G).

The make-up took some six hours to apply during the production, and due to the long make-up time as well as the long shooting day, John Hurt worked only every other day. The actor's patience as well as the make-up artist's skill resulted in a classic make-up and a remembered film. The point was made once more that regardless of how skilled a performer may be, unless the actor "looks" the part, the audience loses much of its empathy for the character being portrayed (see Figure 12.11).

SKULLS, DEVILS, AND WITCHES

Always related to horror is the fear of necromancy, demons, and devils that pervades all cultures to some degree. Bones and skulls induce the fear of death and destruction, and the image of the Devil might take

FIGURE 12.9 *A slush molded latex mask made for one of the Nazi demons in* An American Werewolf in London. *(Rick Baker photo.)*

FIGURE 12.10 *Christopher Tucker applying the prostheses on John Hurt for* The Elephant Man *(Paramount Pictures).*

many forms, but all of them are basically evil. Figure 12.12 shows a skull make-up done with black-and-white foundations in a carefully drawn replica of the human skull on the face. A black, hooded costume completes the overall effect.

Figure 12.13 shows conventional devil make-up with horns, small moustache and pointed goatlike beard, and upturned eyebrows and eyes. The hairstyle or a skull cap provides a standard effect as do the pointed

ears. Facial colors range from red to green or even a natural shade.

Witches are weird, mysterious, not-quite-human women that are portrayed in many ways in films and television (Figures 12.14 and 12.15). Shakespeare's Three Witches in *Macbeth* are classic examples of these creatures that make excellent study vehicles, and make-up artists should always attempt to make them as different as possible (see Figure III.4). Such have long hooked or flat, shapeless noses and chins; thin, wrinkled, or overly distended lips; missing or blackened teeth; straggly brows and hair; and the skin from a dull yellow or greenish tone to possibly purplish color often covered with moles, hair, or dirt. Cracked and blackened or long, curved, sharp fingernails on thin, bony, ugly hands with large knuckles are generally in order.

The beautiful variety of witches or those portraying the evil women in Dracula's castle should be cold but exotic, forbidding yet compulsively exciting, and always sensual to the point of which no one can refuse whatever they desire of one (Figure 12.16). Blood-red lips set in pale, translucent skin with green or purple eye make-up and long hair can add to their filmy costumes and gliding walk.

Mephistopheles, in Gounod's opera *Faust*, is one of the most popular conceptions of the devil character. He has been portrayed in many different ways and

FIGURE 12.11 *(a) and (b) John Hurt, before and after make-up. (Photos courtesy Christopher Tucker.)*

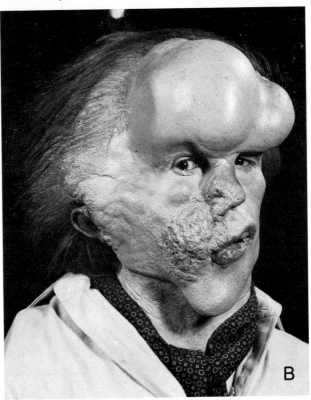

A

B

with various foundation colors—such as greenish tones, red, or even a fluorescent make-up have been used. The overall feeling of the make-up should display a diabolical intent and extreme maliciousness unless the script calls for these qualities to be hidden until the moment of exposure. Hands may have long nails in black or green, with the color being the same as the face. They might be shadowed and highlighted to appear long and delicately bony. Human nature is steeped in the lore that devils and witches are evil in intent, and the make-ups may be asked to reflect this.

ANIMAL MEN-CREATURES

Some of these creatures are created to be horrible, like the Wolfman of Jack Pierce, while others are the benign Lion created by Jack Dawn on Bert Lahr for the MGM production of *The Wizard of Oz* (Figure 12.17). One of the first major uses of foamed latex for appliances, the make-ups in this film have become works of enduring excellence. Also see page 154.

Bob O'Bradovich designed the make-up on Richard Burton, shown in Figure 12.18, for the NBC TV production of *The Tempest*. The clip-on fangs were never used because they impeded the actor's speech. See sec-

FIGURE 12.13   *Devil make-up.*

FIGURE 12.14   *Witches. Three witches with make-up by the author on three older actresses. The facial effect was enhanced by asking them to remove their dental plates. (Carl Fischer Studios/make-up by the author.)*

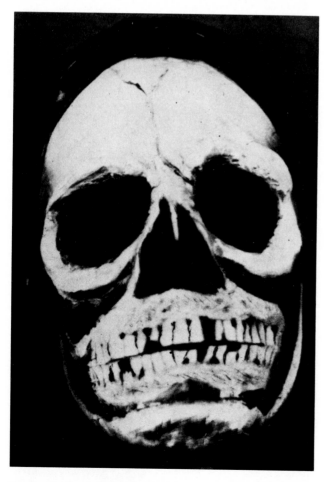

FIGURE 12.12   *A skull make-up.*

FIGURE 12.15   *Witches. The hooked, warted nose and chin, straggly hair and brows, wrinkled features and discolored skin typify the concept of the ugly witch. The eyes can be bleary and bloodshot or small and piercing. Blackened teeth also help.*

FIGURE 12.16 *Witches. Beautiful witches also have their place in fantasy films and can have the over red lips and uptilted brows of those shown, to the less stylized form of just straight make-up.*

FIGURE 12.17 *The Cowardly Lion of Bert Lahr in MGM's* Wizard of Oz. *(Courtesy MGM Pictures.)*

tion on Tooth Plastics (page 207) for tooth casts and construction.

John Chambers's make-ups in *The Planet of the Apes* made an unusual script idea into a spectacular production with his chimpanzee, gorilla, and orangutan creatures (Figure 12.19). Although the extras had slush molded heads, the principles and *speaking* apes had foam latex pieces that consisted of sections for the upper face, lower lip, chin, and so on. Many of the pieces were precolored before application, and foundation was used to blend in the edges. Much facial hair and head hair also aided in disguising the edge of the appliances. The performers' faces had their own teeth and lip areas blackened out, and many had false teeth (of the ape creatures) inserted into the appliances. This was a first in making performers into truly believable animal creatures, and what could have been a laughable B-grade picture, turned out to be a film classic solely due to the design of the appliances to allow the performers a means of making expressions that were almost human on animal features. A foamed latex from the Goodyear Rubber Company was used in making the pieces on many of the characters.

However, the man who carried the ideas of making a man to animal to unusual dimensions is Rick Baker. As a very young man, he was obsessed with the idea of producing a gorilla creature that was completely believable and ended up in the animal suit himself to make the characterization complete. He even played the most famous gorilla in screen history in the color film version of *King Kong* as well as many other commercial and feature films.

Before the project of making a full ape culture for

FIGURE 12.18 *Richard Burton as Caliban in* The Tempest.

the film *Greystoke,* for which enormous undertaking he had a complete lab-workshop-studio built in England—a project that took him two years of incessant

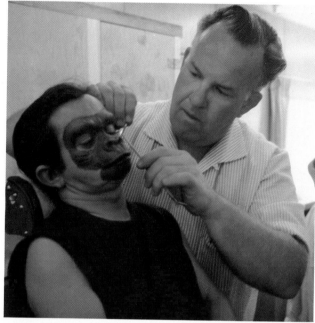

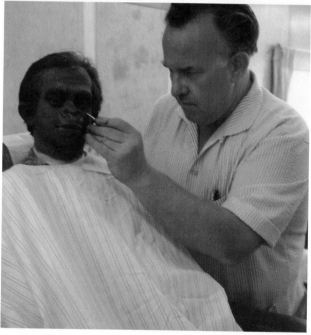

FIGURE 12.19 *John Chambers's extraordinary make-up for* The Planet of the Apes *(20th Century Fox—1967) gained him a special Academy Award for making the apes so lifelike and completely practical in use for the performers to create characters the audience could believe fully. Without this credibility, the effect could have been ludicrous. (Courtesy John Chambers.)*

work to complete, aided and encouraged by his wife Elaine (who runs all his foamed latex for him)—he had accumulated a vast knowledge of the primates and other animal creatures and won the first regular Academy Award for make-up for his spectacular *An American Werewolf in London* (see Figures 12.8 and 12.9).

One of the projects that worked out many of the methods for producing facial expressions on oversize heads and body conformations and ease in replacing fragile parts for the suits of hair was for the film *The Incredible Shrinking Woman*—again, with Rick playing the gorilla and his wife pulling the strings (Figures 12.20 and 12.21).

The operation of the facial expressions in the gorilla mask were far beyond what a performer could do with his own face pressing against any part of the mask, so a series of devices was required to activate these. Again,

it becomes evident that even though a performer actually does the role *in make-up* (even though it is a full or partial covering of the face at times), if mechanical devices improve the performance, are they not a part of the *make-up?*

In *The Incredible Shrinking Woman,* Rick played the gorilla Sidney and we see in Figure 12.18 the complexity of the costume he constructed. In the center is the knotted hair suit with its sponge rubber chest. The heads in fron of this are the close-up head, the "hoot" head, and the "roar" head—designed to form the various expressions that gorillas employ—while to the left is a slush molded head, and above it a fake fur suit for stunt men to wear. In the boxes are the dextrous mechanical extension hands (four pairs in case of damage to the hands during filming), while surrounding are various hand and foot positions to be exchanged on the suit for various attitudes of the appendages. Extra heads, hands, and feet for replacements are also seen as are various simian skulls in the upper left that were used for research to sculpture.

For *Greystoke,* Rick improved the simian suits to an amazing degree of reality (Figure 12.22). First, he drew sketches of the apes, giving each its own facial features and personality traits distinguishable from each other. Next, sculptures were done over the performers' life masks to follow the sketches. Note the similarity

FIGURE 12.20   *Complexity of costumes constructed by Rick Baker for* The Incredible Shrinking Woman. *(Courtesy Universal City Studios, 1980.)*

FIGURE 12.21   *On set with wife Elaine and special effects technician Guy Faria, Rick is in the Sidney suit with the close-up head. Note the multiple cables that produce the movements in the headpiece. As the feet were not being filmed here, Rick wore comfortable shoes! (Courtesy Universal City Studios and Rick Baker.)*

of the drawing of the character Figs with the sculpture at the lower left.

For the film, these ape characters were developed extraordinarily by Rick Baker, as can be seen in the ape Silverbeard (played by Peter Elliot) as he first ap-

FIGURE 12.22   LEFT: *Drawing of character "Figs."* RIGHT: *Clay sculptures of characters. "Figs" lower left. (Photos courtesy Warner Bros. Pictures and Rick Baker.)*

pears and then later in the film showing an aged character (Figure 12.23). This is an unusual instance where make-up has been employed to age an animal character played by an actor.

Each ape character in *Greystoke* had four heads: a close-up head that was cable operated and capable of changing expressions from, for example, eating to licking lips (however, the performers supplied the jaw movements and the eyes were covered with full scleral lenses); a "roar" head to snarl and show teeth activated by the performers' movements; a "hoot" head that made a pursed lip expression; and a stunt head of slush latex filled with polyfoam (Figure 12.24). The suits were hand knotted onto a lycra spandex base body stocking that was stretched over a muscle suite made of lycra spandex with varying densities of foam muscle formation, rib cages, and other bone structures. As well, there were various hands and feet that were interchangeable (Figure 12.25).

Once again, the art of make-up and the genius of those who can achieve the likenesses of both humans and animals becomes the entire point—of the plot, the story, the script, the action, the characters, and direction of that portion of the film. Without the make-up, it would make so much dumb show by competent performers—and no one has union-organized simians as yet to perform quite as such!

Not to be outdone in the animal-men gender, Stan Winston came up with his "Manimal" transformation for the television series of the same name (Figure 12.26). His unusual method of going beyond the simple facial cast of a performer to sculpting a full head in plastalene of the expression he wants to depict demonstrates his artistic ability and desire to make his creations as lifelike as possible. Working with a series of close-up photographs and measurements in addition to a life

FIGURE 12.23   *The aging of the ape Silverbeard, played by Peter Elliot, as he first appears in the film,* left, *and then later,* right, *showing an aged character.*

mask, he inserts teeth and eyes into the sculpture to visualize his conceptions. In the case of the "Manimal," this snarling expression is seen reflected in all the stages of the transmutation from man to black panther (Figure 12.27).

FANTASY TYPES

Fairies, elves, sprites, nymphs, and other fanciful characters can be done in an impressionistic style. Fairies can be made up straight and pretty in light shades of foundation, while elves, sprites, and the like should be done in light shades of brown or green to resemble forest colors. They can have small pointed noses and ears. Somewhat similar are leprechauns, although the foundation shade can be normal while the ears and nose can be pointed. Upturned brows and a small unkempt wig will add to the effect (Figure 12.28).

Within the fantasy types one can include such makeups as might be found in *Alice in Wonderland* (Figure 12.29), Shakespeare's *A Midsummer Night's Dream, The Wizard of Oz* (Figures 12.30 and 12.31), and similar productions on stage or for the screen (Figures 12.32, 12.33, 12.34, and 12.35). Again, foamed latex or urethane appliances will see wide use to depict the characters.

Also related to fantasy types might be included humanoid robot creations like the one Stan Winston devised and made for the film *Heartbeeps,* with Bernadette Peters and Andy Kaufman playing the two characters (Figure 12.36). To achieve the necessary translucence and artificiality and to produce a new look for the

make-up, Stan Winston employed a gelatin mixture for the appliances and intrinsically colored them with gold for Aqua, the female, and bronze for Val, the male robot. The latter had 12 separate pieces, and Aqua had 6. The coloration was delicately and carefully done using silver as highlights, so no extrinsic color or foundation was required on the appliance, only on the exposed skin areas. Matte adhesive was used to

FIGURE 12.24   *White Eyes is shown with his roar head, showing the fine detail in the facial portion of the foamed latex head, the teeth in the mask, and the scleral contact lenses on the actor's eyes.*

FIGURE 12.25 *Figs on set in the foreground with Kala, Tarzan's ape-mother, and on the left, a tiny latex rubber Tarzan. The other apes are background characters whose suits and heads are as carefully made as the principles' costumes. Figs is seen with his relaxed-mode feet and dextrous mechanical arms that extended the performer's arm proportions but still allowed him to manipulate the fingers and pick up things.*

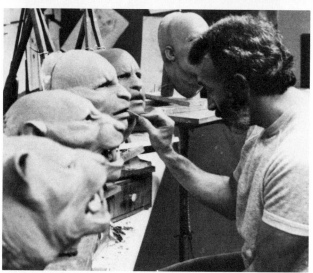

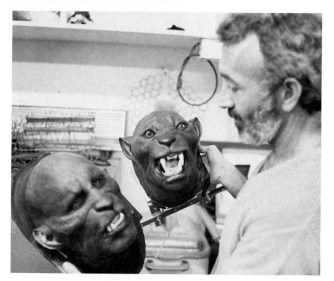

FIGURE 12.26 *Stan Winston, sculpting the Manimal progressions from man to black panther. Final sculptures are on right.*

adhere the pieces, but being the material they were, they required constant attention. The make-ups were nominated for the Academy Award as being innovative and unusual in concept.

DOLLS, TOYS, AND TIN SOLDIERS

Within the range of fantasy and children's stories are found playthings come to life (Figure 12.37). Light shades of foundation, with stylized cheekcolor circles, drawn red lips, accented eyelash make-up, and drawn brows are suitable here. Any make-up done for chil-

dren's stories should avoid any effect that might be frightening and should induce children to cuddle the toy rather than be afraid of it.

STATUARY

A marble effect can be obtained by first covering the skin with a white foundation and then carefully giving the skin reflectivity with a pearl foundation. A plastic or latex cap is used over the hair or a very stiff wig, sprayed with white and the same pearl for translucence.

Gold or bronzed statuary can be simulated with

FIGURE 12.27 *Stan Winston's* Manimal *sculpture of the commencing facial expression of Simon MacCorkindale taken from the photo below it on the left, and a stage in the transformation with make-up appliances on the right showing the progression of the idea and the expression. The on-screen transformations included the use of various bladder inflation effects and film superimposition as well as make-up changes in tight close-up from man to black panther.*

FIGURE 12.28 *Roddy MacDowall as a leprechaun in CBC-TV* General Electric Theatre. *(Make-up by the author.)*

FIGURE 12.29 *Bobby Clark as the Duchess in NBC-TV's* Alice in Wonderland *with a complete foam latex face by Dick Smith.*

FIGURE 12.30   *Ray Bolger as the Scarecrow in MGM's classic* The Wizard of Oz, *with make-up created by Jack Dawn. Although the face appears to be covered with cloth, it is actually a latex skin with a cloth impression. (Courtesy MGM Pictures.)*

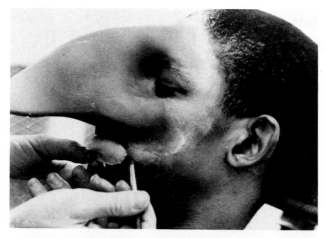

FIGURE 12.32   *Two of Stan Winston's make-ups for the film,* The Wiz, *showing the application of a stylized beak-nose prosthesis, and the completed Lion make-up, showing a different concept from that of the earlier* Wizard of Oz *film (Universal/Motown, 1978).*

FIGURE 12.31   *Jack Haley as the Tin Man in the same film. Although much of the effect is costume, the silver-painted face and spout nose was very effective along with the head and chin pieces. (Courtesy MGM Pictures.)*

gold, bronze, or copper foundations. Gold, white, and other colored hairsprays (Nestle) can also be used on the skin, or the old method is to use a mixture of gold or bronze bronzing powder, with glycerin, water, and alcohol as a vehicle. This is then applied with a large silk sponge in a thin coat on the body. Care should be taken with the latter as it can block the skin pores, so the entire body should not be covered to allow breathing space for the skin.

SURREALISTIC MAKE-UPS

The world of surrealism is wide and varied, and many different aspects and directions can be taken with make-ups to express the craft and imagination of the artist. Some employ flower petals around the eyes, others use sequins, while some design with color.

Rock singers sometimes employ such fantastic make-ups as contributory to the imagery they wish to convey. Originality in application, design, and placement of colors on the face or hair gives free reign to the imagination. However, once the concept is established, it becomes an intrinsic part of the singer or group, and to copy such exactly by another person or persons is often considered to be a breach of theatrical etiquette—just as it is for specific clown make-ups in the circus. Nevertheless, certain forms exist that seem to establish a framework or create a style indigenous to the person or group.

FIGURE 12.33 *Maurice Manson's "Santa Claus" or "Jolly Old St. Nicholas" is a plump, happy, old elf with pure white hair and whiskers in abundance. Red cheeks, a round, red nose, and laughing eyes complete the picture.*

FIGURE 12.34 *"Old Father Time" is another fanciful character that must be far more decrepit than "Santa Claus." Long, white hair, a drawn face, and a long, whispy beard typify this make-up.*

FIGURE 12.35 *Bob Schiffer of Walt Disney Studios created this pumpkin head on Jonathan Winters for a Disney Halloween Special with a slush and paint-in molded head in two sections with the inside fit made of polyurethane foam.*

FIGURE 12.36   *From the film* Heartbeeps. LEFT: *Bernadette Peters as* "Aqua." RIGHT: *Andy Kaufman as* "Val."

FIGURE 12.37   *Toy soldier.*

FIGURE 12.38   *Barton MacLane in RKO Pictures,* The Spanish Main *(1945) in a typical pirate make-up. Note the scars on the cheek and forehead, the unshaven beard effect, and the added drawing of the eyebrows along with the period wig.*

PIRATES

Most pirates should be dark, swarthy characters who seldom shave and often have scars on their faces to add to their ferocity (Figure 12.38). KM-8, KT-3, KT-4, or K-3 are foundations that may be used. Eye patches, hairy arms and bodies, and straggly, unkempt hair are typical. The pirate crew should always be interspersed with Negroid and Mongoloid types, and appropriate foundations should be used.

CLOWNS

Basically, professional clowns divide their make-ups into the *whiteface* and the *tramp*. The typical whiteface uses a clown white or Super White that is powdered with a white powder or prosthetic base powder (Figure 12.39). Red, blue, green, and black shades are used to create the designs and accentuate the features. Many clowns employ ordinary china marking pencils for the colors as these are easy to sharpen by unraveling the paper covering. Copying the face or special markings by another clown is considered to be a serious breach in circus etiquette, so unless one is copying a specific clown character, new designs are always called for in design.

Facial hair is seldom used—unless just painted on for Keystone Kops or beard effects. A skull cap can be made from a white stocking and pulled over the head. Earholes can be cut when the cap is centered. Wigs with bald fronts and red hair are often seen. Red latex noses (slush molded, usually) are special clown features and can be made in various shapes.

The tramp clown variety can be made happy or sad, depending upon the turndown or turnup of the mouth drawing. Here the unshaven beard effect is almost always used, and the face can be made up with light to medium shades of foundation to set it off. Mouths

FIGURE 12.39   ABOVE: *Tramp clowns.* BELOW: *Various whiteface clowns.*

can be white or red. The costume is always of the old clothes variety, seldom fitting anywhere and exaggerated whenever possible.

There have been many famous clowns from the Barnum and Bailey Circus; among them were Emmett Kelly and Paul Jerome as tramps and, in whiteface, Felix Adler, who always wore a rhinestone on the end of his nose (he called it an Irish Diamond, or a sham rock) and always led a baby pig on a string for his walkaround (Figure 12.40). Other famous whiteface

FIGURE 12.40   *Whiteface and tramp clowns.* LEFT: *Felix Adler in the whiteface make-up he made famous.* CENTER: *Paul Jerome as The Tramp.* RIGHT: *A classic whiteface clown, Jackie Le Claire.*

clowns include Jackie LeClaire, Al White, Ernie Burch, Al Bruce, Arnie Honkola, and many others.

Another phase of this type of make-up is practiced by mimes. However, this is mainly a white face with few other markings, and the neck is seldom painted, leaving a rather masklike look to the make-up.

OLD-TIME OR EARLY FILM TYPES

Both male and female performers in the old-time films seem to have rather pale shades of foundation (due to the old orthochromatic film) (Figure 12.41). Eyes were often heavily shaded and drawn, and the lips dark with overly accented cupid bows for the women. Brows were quite thin and drawn. The heroes were always clean shaven, and the villains were always bearded or unshaven. The latter often did not wear a foundation so that their skin did not have the smoothness of the heroes'.

FIGURE 12.41 *Heroes, heroines, villains, and character make-ups generally based on stage make-ups of the early nineteenth century.*

Conclusion to Part III:
Dedication

DEDICATION IS A DEVOTION TO AN IDEAL OR CAUSE, AND within the sphere of the make-up profession it is a necessity for one who wishes to accomplish good character make-up. The study is far more complicated and involved than that of simple straight beauty make-up and requires a mental commitment and passion to create, analyze, consider, probe, investigate, and reason out visual effects that will elicit believability and entertainment from the audience. Only dedicated people will become good character make-up artists.

Part
IV
Laboratory Techniques

While some enjoy applying make-up appliances,
others would rather make them.

ALTHOUGH MAKE-UP LABORATORY WORK AND KNOWL-
edge are not part of the requirements of various make-
up unions' entrance tests, the developments and principles form the modern extension of character make-
up. Even those make-up artists who have no particular
desire or even skill in the lab should know, as part of
their complete knowledge of the present field, the
language and techniques of this advanced branch of
the profession. We shall try to survey briefly the salient
points here for reference.

To begin with, the work area of the make-up lab
should be separate from the make-up room due to the
fact that the procedures of the lab often require the
use of materials that create considerably more debris
and mess, and at the same time the equipment and
products employed are different, requiring their own
work and storage space (Figure IV.1–IV.3). The lab
room should be well lit and ventilated (with a fan-
exhaust hood if possible) and should have some strong
work tables for both sitting and standing work levels,
a linoleum-covered floor, plenty of electrical outlets,
and substantial shelves for books, molds, casts, storage, and so forth. Great care must be taken to clean
sinks so the drains do not become clogged with plaster
or other residue. For a small lab, a safe way is to use
a siphon or spigot system on a 5-gallon water container
to a small sink whose drain is simply a rubber hose
that leads to a 5-gallon bucket. The latter can be
emptied carefully when the residues settle in the bottom, and water can be separated from the solid masses
(which can then be thrown in the trash bucket). Large

labs sometimes have special filter systems built in to
their sinks.

As for personnel, one will find that there are always
many neophytes interested almost solely in sculpting,
mold making, prosthetis making, experimenting with
new items, and doing make-up research rather than

FIGURE IV.2 *The author's developmental lab.*

the (to them!) more mundane daily straight make-up
application that takes up at least 80 percent of the
make-up artist's normal career. Almost all the new
make-up labs in California (where the bulk of the work
is done in this field) have a stream of young hopefuls
who want to work in that area of make-up, but competition—and talent—is quite fierce and extensive.
Certain specialists have evolved whose work has been
outstanding, but there is always more room for the
best. Almost every procedure in lab work is enhanced
and aided in time and effort by more than one person

FIGURE IV.1 *The author's small appliance laboratory. It illustrates the
water arrangement with a 5-gallon bottle siphoning water through glass
and plastic tubing to a small sink with a rubber hose drain to a 5-gallon
bucket. In this way, no insoluble plaster goes down the regular plumbing,
yet sufficient water is available for most molding and casting.*

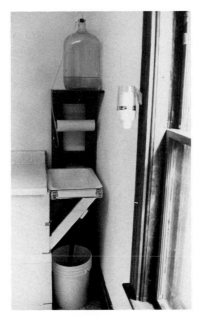

FIGURE IV.3   *Three views of Tom Burman's Studio where "monsters are made" from molds and casts by the ton.*

doing the particular job, so a crew system has been developed, and therefore, the finished product may be a blend of the work of many hands and minds.

As in all cases where enormous talent has affected the output, there are those in the production field who have no concept of the time, energy, and effort that must be put in to produce a laboratory creation. Many directors and producers have never understood or even taken the time to see the involved procedures that may be necessary to devise a prosthetic appliance, special make-up effect, or such and thus often set time frames for completion of lab work that is unrealistic. The more complicated and the more innovative make-up effects become, the more competition for exciting and new visual results will be asked for—almost as a matter of course.

However, in today's rapidly advancing technologies, it does not seem that a day goes by without some new laboratory method, material, technique, or discovery being made that improves and adds to our knowledge of make-up magic. As such, make-up artists are closely akin to magicians and illusionists in the sense that once a new trick is invented or produced, someone is bound to improve upon it.

It should be borne in mind that the products and tools mentioned herein are the author's recommended ones solely because they have been found to achieve the purpose he has desired. There are, no doubt, other methods and procedures as well as products and items that will accomplish similar techniques or parallel ones so one should always look for other ways to do the job. Artists who rest on old methods, regardless of their tried-and-true qualities, will more likely than not be left behind by those willing to research and test further to improve any make-up procedure or item. Look for the both the unusual *and* the commonplace when searching because, as with any concept or working method, both innovation and adaptation will solve most problems. As well, every make-up lab technician has run into problems, breakage, and disasters, so plan every phase carefully and follow directions closely, but don't be discouraged if something goes wrong. Redo it, and most times, whatever you do will be better.

Appendices A and B provide listings of many equivalent products in addition to addresses of suppliers and manufacturers of special items.

CHAPTER 13

Casting and Molding

MANY YEARS AGO, ALL THE WORK TO BE ACCOMPLISHED on a performer's face had to be done on the face for that performance only. Products such as spirit gum, cotton batting, adhesive tape, collodion (both flexible and nonflexible), nose putty, mortician's waxes, thread, and other materials were employed. However, the make-up artist now can premake several of a needed appliance out of rubber and synthetic latices, plastics in a great variety, gelatins, and other prosthetic materials so that a fresh appliance with a finely blendable edge is available whenever required. The clarity of the image structure of the screen and the closeness of the audience in small theaters require this excellence.

To achieve this, the first step is to make a duplicate positive casting of the performer's face or the body part required with a permanent material such as plaster or stone. From this casting, the make-up concept can be sculpted in modeling clay called *plastalene* and a negative made of the newly sculpted features. Then, by various methods, a copy of the sculpture is made, called an *appliance,* or *prosthesis,* in a skinlike, flexible material that is attached to the performer to delineate a character for a specific part. Impression and duplicating materials for molding and casting can be quite varied and different, and each is employed to accomplish a particular purpose.

IMPRESSION MATERIALS

Impression materials are used in taking casts and making molds of a *positive,* or original, item by making a *negative* into which plaster or stone can be poured to form a *duplicate positive* (Figure 13.1).

There are rigid types of impression materials such as plaster, Ultracal, and dental or tool stones, and there are flexible varieties such as the thermoplastic, reversible types of *moulages,* and the nonreversible *alginates.* Each material is rather specific in use and is best suited for certain purposes. No one material will do the work as well as a combination of methods and applications.

Rigid Impression Materials

Materials that harden and form a nonflexible cast or mold include the various plasters and stones. Although there are many varieties of these, we limit our selection

to a few for ease in operation and suitability for make-up work.

Dental Plaster

This is a quick-setting (3 to 4 minutes from liquid to hard) material for use when speed of casting and setting is important. It is the weakest of the rigid materials, so it should not be used where strength of the cast or mold is a prime requisite.

To mix, add the plaster to cold water, making small mixes because the setting time is so rapid that a set will occur before a large mixture can be spatulated properly. A thin mix will take longer to set than a thick mix but will be weaker in tensile strength. Keep adding plaster to the water until the surface of the water is covered. Then immediately mix by spatulation until a smooth creme is obtained. Generally, 10 to 15 seconds is taken for mixing with water, so this leaves about 2 minutes for application to the subject before setting time is reached.

Dental plaster can be used to take small impressions of parts such as noses, chins, eye sockets, or fingers and for repair of casting plaster molds where speed and not strength is required. It makes an excellent waste mold material because it breaks away easily. It is generally sold in 10-pound cans.

FIGURE 13.1 *Tom Burman explains the complications of a special mold to Gena Kehoe.*

#1 Casting Plaster

The U.S. Gypsum Company makes most of the casting plasters and stones, and they can be readily purchased from large building supply companies in 100-pound bags. They should be stored in a dry place in plastic covered trash cans placed on a wheeled platform for ease in handling.

Casting plaster is a white material that sets in about 20 to 25 minutes and is the standard type of plaster used in making many casts. It is not suitable, however, for pressure molds or for those that require any heat curing. It is the best for any slush or paint-in latex molds as it has good absorptive qualities. To hasten the setting time when taking casts of faces or other body parts, add a teaspoonful of sodium chloride (common table salt) to each quart of water used in the mix. In this book, we shall refer to casting plaster only as *plaster.*

Hydrocal and Ultracal

These are low expansion gypsums that are considerably harder than plaster and can be used for pressure- or heat-cured molds. Although there are varieties such as Hydrocal A-11 and B-11 (two setting grades), the Ultracal 30 is most used by the make-up lab technician because it is a superstrength product. It is recommended where extreme accuracy and additional surface hardness are required, and it is harder and stronger than Hydrocal A-11 or B-11. It also has the lowest coefficient of expansion rate of any gypsum cement available, but its very gradual set and ample plasticity make it ideal for the splash casting that this work requires. It dries with a grayish color so as to distinguish casts and molds made of this material from plaster. Ultracal 60 is similar in all respects except that it has a set of about 1 hour and is designed for very large models where additional working time and the highest possible degree of accuracy are required.

Dental or Tool Stones

This is one of the hardest of the casting materials and is a somewhat heavier material than other rigid types. It has low absorption qualities so is not suitable for slush latex molding. Its setting time is similar to that of Ultracal 30, and most of the professional lab technicians are split between these two excellent materials for any foamed latex or urethane molds.

Although dental stones come in a variety of colors, they are more expensive than pink tool stone (made by Kerr), which is a similar if not identical stone. It is best purchased, therefore, from a jewelry manufacturing supply house than a dental supply for reasons of price. Both Ultracal and stones become quite hot as they harden and so should not be used to take facial or body casts.

The mixing procedures for plaster, Ultracal, and stone are similar, and the strength, hardness, and density of the cast depends upon the amount of water used in the mixing. With all mixes, a thin mixture sets more slowly than a thick one. Small quantities can be prepared in flexible rubber or plastic bowls that dental technicians use, or large mixes can be done in polyethylene kitchen wash bowls (which are available in most department stores in kitchen wares).

Before use, always loosen up the surface of the gypsum material so it will not have any lumps. Then place the desired amount of water in the container. This can only be judged by experience, but remember that gypsum materials are cheap and it is better to have too much than too little when casting. The gypsum can then be added a little at a time with a large spoon or ladle, sifting the material in so that it does not enter the water in lumps. To make a regular mix, keep adding the gypsum material into the water until the surface of the water is completely covered with *dry* material. Then allow this mix to soak undisturbed for about 2 minutes, when the surface of the material will take on the appearance of a dried river bed. Now stir with a spoon or a nonflexible spatula until a cremey mix is obtained. It is then ready for immediate use. If one wants a thin mix to make a splash coating in the mold, put less gypsum material in the water and stir right away to remove any lumps. Dental technicians often use a very stiff mix of dental stone for making positive casts of teeth or for imbedding the casting in a tray. Such requires vigorous spatulation in the bowl to get the mix smooth and workable.

A number of methods can be used to further strengthen Ultracal or stone molds, and they vary from expert to expert. As usual, you will find that whatever works best for *you* is what you will settle upon. Basic strengthening of the mold is done by Werner Keppler and Terry Smith with wet hemp fiber applied in layers. Gus Norin always used spun glass, but this material is difficult to handle. Dick Smith uses a 5-ounce burlap in strips, which he says is easier to handle than the loose fiber and can be guided more readily into the areas that require the most strengthening. He also uses a wire cloth for strengthening the outer surface of small molds. John Chambers sometimes employs an acrylic latex cement hardener from the Wilhold Company, while Dick Smith uses Acryl 60 to reinforce the strength of Ultracal 30, employing 100 cubic centimeters of Acryl 60 with 100 cubic centimeters of water and then adding 660 grams of Ultracal 30 for a sample mix. Sources for these supplies are given in Appendix B. It should be noted that these strengtheners may be found to be an aid to the longevity of molds that require baking or for pressure molds to prevent cracking. Also, the heavier and thicker the mold is, the longer the baking time will be for foamed latex.

Flexible Impression Materials

Moulage

Years ago there was only one flexible impression product, known as *moulage* (one trade name was Negocoll). This was a reversible material that had to be heated up in a double boiler and brushed on to the part of the body to be cast while it was rather hot. It did not have much tensile strength, so a heavy coating was necessary along with a *mother mold* of plaster to hold the correct shape and act as a retaining form. It could be reused many times as long as the water content was kept to the correct level. Still, it was a tricky material to use and required much preparation and care afterward.

Alginates

Today we have a number of impression materials, called *alginates,* that are simple to prepare, are easy to apply, require no heating, and form a considerably stronger gel than moulage. Dental alginates have a normal setting time of about 3 minutes, so a technician must work quite rapidly if he or she wishes to take an impression. However, these alginates have been used with considerable success by many make-up artists.

Recently the Teledyne-Getz Company, one of the major producers of the dental alginate, has introduced a Prosthetic Grade Cream (PGC), which is a colorless, odorless, fine-grade, slow-setting alginate impression material. Normally, PGC sets in $4.5 \pm .5$ minutes, but a retarder may be added to allow as much as 16 minutes of use before setting. This PGC has a long shelf life of up to 5 years when stored at room temperature in a sealed container.

The Mid-America Dental Company also makes a prosthetic grade alginate product that has a 5-minute set time. Their directions state that warmer water than 70°F hastens the gel while colder water slows it, and up to 25 percent more water can be used for easier mixing. They also recommend to maintain the gelled alginate on the subject for about 3 minutes beyond the gel time because the material gains strength during this period. Otherwise, the directions and material are similar to PGC by Teledyne.

In use, all alginates should be shaken in the container to fluff up the powder and ensure uniformity. They come with two small measures, one for water and one for the powder. The faster setting types are affected by temperature, and warm, humid weather will hasten the setting time. Some workers use ice water to extend the setting time. PGC is rated for use between 70° to 77° F and from 40% to 60% relative humidity, so it has a wide latitude for use. Alginates of all grades are measured quite similarly, with two scoops of powder for one full vial of water—that is

17.7 grams of powder to 50 cubic centimeters of water.

To mix, measure out the desired amount of water into a mixing bowl, and add the powder. Immediately spatulate for about 45 seconds, and transfer directly to the portion being cast. Once gelation takes place, a second coat of alginate will not adhere to the first unless the area is brushed with a solution of sodium carbonate (monohydrate) in water (a teaspoon of powder in 6 ounces of hot water and dissolve). This will make the surface tacky; then additional alginate mix may be added. A retarder can be made with a solution of tri-sodium phosphate or may be obtained from Teledyne-Getz (PGC Retarder) or Mid-America.

This material will take deep undercuts and can be applied in a far thinner application than the moulage. It must, however, be supported by a mother mold for most applications, and impressions made with alginates must be used to make positive casts immediately as they commence to dry out very fast. Many people submerge the finished alginate impression in water while they are mixing the positive casting material, but if two are taking the cast, one can have the positive material immediately ready for use when the alginate impression is removed from the original (face, arm, and so on). As PGC has less than 3 percent deformation, very accurate casts can be made with it. However, with the use of any extra retarder, the physical properties of PGC are reduced in value. The retarder is not recommended for the standard dental alginates as they will become lumpy and fail to set properly.

DUPLICATING MATERIALS

Duplicating materials differ from impression materials in that the latter are generally used to take primary or original casts of the subject, while the former are used in making duplicates of the original positive by making a permanent negative mold suitable for this operation.

Formerly, heavy concentration and thickened natural latex compounds were utilized for duplication as well as a heat-melt synthetic material called Koroseal. However, we now have a series of cold molding compounds in polysulfide rubber, silicon rubbers, and polyurethane synthetic rubbers that serve as very fine duplicating materials and exhibit great flexibility as well as have a variety of hardnesses to duplicate many varied materials.

Cold Molding Compounds

Although a number of firms make cold molding compounds, the Perma-Flex Mold Company and the Smooth-On Corporation make a number of flexible compounds that are used in the make-up lab for duplicating purposes. Perma-Flex CMC Blak-Tufy, Blak-Stretchy, Gra-Tufy, and UNH are polysulfide liquid molding materials cured at room temperature by means of catalysts. They are available in three- and two-component systems, and while the two-part systems are fast and produce good duplicates, the three-part systems are indefinitely stable while the others have a shorter finished product shelf life.

Blak-Stretchy is the most elastic of the types, with a 9 to 1 elongation and a durometer hardness of between 12 and 15. It has a working time of 20 to 30 minutes. The first coats can be applied free flowing, then Cab-o-Sil M-5 can be added to the mix to thicken it so it can be spatulated on. It is a two-part system so has a cast storage life of about 4 months of shape retention. Its weight is about 12 pounds to the gallon, and like other black materials, it is reinforced with carbon black. Curing time is 20 hours at 80° F.

Blak-Tufy is a three-part system that cures in 16 hours at 80° F and has a durometer hardness of 23 to 25. In-between hardnesses can be produced by mixing Blak-Stretchy and Blak-Tufy.

Gra-Tufy is a similar product to Blak-Tufy except that it has a zinc oxide reinforcement pigment instead of the black. It is useful when snow-white plaster casts are necessary in reproduction. Elongation is similar to Blak-Tufy at 4 to 1. Its weight is 15 pounds to the gallon.

UNH is similar to Gra-Tufy except that the set time makes an extremely fast cure possible. The setting time is 5 to 10 minutes, with usable molds available in less than an hour at 80° F. This is a three-part system and has a hardness of 20 to 22 on the durometer.

Regular CMC carries no reinforcing pigments or filler and is considered to be the most dimensional of the cold molding compounds. The base A polymer is a clear viscous liquid cured by additions of B and C curatives in various proportions to get work life and setting time from 1 hour to overnight. Regular is the least viscous of all the cold molding compounds and is often used as a safe thinner for any of the other polysulfide cold molding compounds. Durometer hardness is 14 to 15. Directions for use of these polysulfide materials are available with the products from the Perma-Flex Company.

Perma-Flex makes another type of cold molding compound that deserves experimentation, the *P-60* and *P-60-S* series. These are two-component systems. They are amine polysulfides whose shelf life extends to a year or more. The item to be cast can be given a two-coat layer, then a layer of open weave fibreglass cloth and another layer of compound. This will provide a strong but lightweight mold. In use, as these are all flexible molding materials, a mother mold of plaster is recommended to hold the shape of the cold molding compound. The compound needs to be properly keyed into the plaster to set correctly for duplicating. Open weave burlap can also be used for a reinforcing material.

To duplicate a plaster positive, it must be first given a couple of thin coats of orange shellac. Then it is coated with a silicon mold release for ease in separation.

Another type of duplicating material is Perma-Flex CMC *Blu-Sil*, which is a room-temperature-set two-component system consisting of a white silicone type fluid base A, and a blue-colored modified silicone catalyst B. It is designed to be mixed at the time of use into a fluid that will set to a flexible rubbery solid about 30 on the durometer. It is often used when a duplicate is to be made of a very delicate positive. Few or no parting agents are required. However, its cost is double that of most polysulfides.

Another variety of elastomer (rubberlike substance) for making molds is Smooth-On Corporation's *PMC-724*, which is a polyurethane and polysulfide synthetic rubber compound. Part A is a brown thin liquid, and Part B is a white syrup that remains pourable for at least 20 minutes when mixed. At a mix ratio of 10 A to 100 B, it has a durometer hardness of about 40. Sonite Seal Release Wax or petroleum jelly are suitable release agents, and shellac can be used to seal porous casts before duplicating with PMC-724. A Part C is available to achieve varying degrees of durometer hardness down to 6, and a Part D can be added to produce varying viscosities in the mixed uncured rubber. Consistencies can range from a thin latex paint to a grease-like putty that can be buttered on to vertical surfaces. Up to 80 parts of Part C can be added to make the softest mold, and only 2 parts of Part D need be added to make the mix very heavy. PMC-724 remains usable for at least six months from shipment date in unopened containers stored in a cool, dry location. None of these duplicating elastomers is suitable for skin use, and none has a long shelf life so order what you need when you need it rather than stocking it in the lab. Also see page 218 for its use as a prosthetic material.

SEPARATORS

Separating mediums are also known as parting agents, mold lubes, or releasing compounds and serve to ensure that two materials, similar or dissimilar, can be easily withdrawn from each other when desired. Some separating mediums are more versatile than others, and some are for a specific purpose or material.

In addition to preventing adhesion, separators must provide upon the faces of the model or mold a continuous, non-water-soluble, and smooth film that possesses little or no frictional resistance to movement across its surface. A properly selected sealer-separator combination prevents penetration of moisture from the fluid gypsum material into the pores of the mold or hair on the body. It also allows free movement of the set gypsum as it expands. The basic requirements of a satisfactory separation medium are that it:

1. Prevents adhesion of the cast;
2. Protects and lubricates the surface of the mold;
3. Spreads easily and uniformly in a thin, continuous insoluble film;
4. Will not react destructively with the gypsum surface or with the mold surface.

One of the most versatile separators in general use is petroleum jelly. This can be used to separate any gypsum material from any other during casting procedures. It is also used to coat facial or body hair when taking a cast with either a rigid or a flexible impression material. It can be used full strength or diluted with two parts of RCMA Studio Make-up Remover for a thinner mix.

Many varieties of liquid alginate dental separators can be adapted for make-up use. These are heavy liquids that can be painted on a gypsum cast and that form a film when dry (also see pages 185–186 on transferring plastalene). These separators are excellent for use with tooth plastics to gypsum (see pages 207–209). Dental technicians also employ thin sheets of wet cellophane for separation and to obtain a glossy surface on acrylic teeth in the mold. Dental supply houses carry these items as does RCMA on their Special Materials list.

Silicons are also good release agents and PMA Silicon Mold Release or Dow Corning Compound DC-7 are excellent for coating any gypsum material to separate many plastic compounds such as RCMA Molding Material, Plastic Cap Material and others. It can be used full strength or can be diluted with methylene chloride or MEK.

It is well to keep separate brushes for use with the above two separators so that they can just be wiped clean after use rather than cleaned each time in a strong solvent.

Many sculptors use a separator made from 120 grams of stearic acid and 540 cubic centimeters of kerosene. However, to avoid the odor of the kerosene, Studio Make-up Remover can be substituted for the solvent.

The stearic acid is melted and the SMR carefully added by stirring after the melt has been removed from the heat.

Another favorite sculpture separator is one made by dissolving shavings of Ivory Soap in hot water to form a saturated solution. Both the stearic acid and the soap separators are brushed into the molds until the surface appears glazed.

Alginates and plastalene do not require separators from gypsum materials. However, if the plastalene is coated with Plastic Cap Material, some technicians use a very thin coating of the silicon type mold release over the surface of the positive so that the cap material does not stick to the negative when it is separated after casting.

Sealers

Many technicians coat their porous plaster or stone face casts with thin orange shellac (dissolved only in alcohol) to protect them from the oils in the plastalene and to make separation easier. Polyurethane lacquers are used to coat molds for use with some foam latices or for those employed for flexible urethane castings.

CAST OF THE FACE

The face is the basic body area for which most appliances are made, and the *life mask* is a primary tool for the make-up lab technician. To prepare for taking the impression in PGC, sit the subject in a comfortable chair with a headrest (Figure 13.2). Make certain that the head is not in a completely horizontal position as otherwise the chin and neckline will be distorted in the cast. Some workers prefer a 45° angle while others will cast with the head almost erect. If just a limited front section is to be cast, the hair can be covered with

FIGURE 13.2 *(A) Casting a face. The subject is prepared and the PGC is applied. (B) The PGC and the plaster mother mold are being removed. (C) An Ultracal 30 positive being cast.*

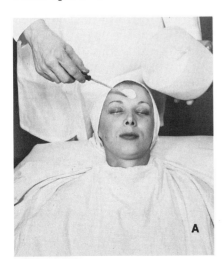 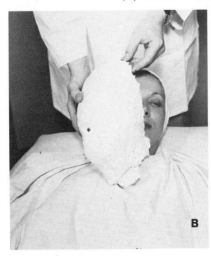 

a shower cap, but if a more extensive cast is desired, a rubber or plastic bald cap can be attached. Any hair showing—brows, lashes, moustaches, or head hair—must be coated with petroleum jelly in a liberal fashion. As well, the entire face can be given a very thin coat of petroleum jelly or mineral oil. Of course, all women's street make-up should be removed. Cover the clothing of the subject with a plastic make-up cape.

If just the upper part of the face, such as the eye and forehead area, is to be taken, the nostrils can be filled with cotton batting and covered with petroleum jelly, with the subject breathing through the mouth. However, if the entire face is to be cast, some workers use drinking straws in the nostrils while others prefer to work carefully around the nostril area. One excellent method is to cover the nostrils with PGC while the subject holds his or her breath, then with a forceful expulsion of air from the nostrils, a breathing passage will be cleared. Take care to explain this procedure carefully to the subject prior to commencing the impression so that complete confidence will be gained before the PGC is on the face.

Usually, 20 scoops of powder (177 grams) to 500 cubic centimeters of water will be sufficient for a full face. Some technicians mix the entire batch at one time, while others split it in half and start with one batch and then go to the second *before* the first batch sets. However, this means that at least two people must be doing the cast—one to apply and the other to mix. With the dental alginate (3-minute set time) this was necessary, but with the PGC, which has a longer working time, one person can do the impression.

Although some apply the PGC to the face with the hands, the material tends to set faster because the hands are warm while a brush or spatula is not. If you do use the hand method, it is well to have a basin of water close by to rinse the hands whenever necessary.

Don't throw excess alginate, plaster, or stone down the drain of a sink. It will stop up the trap and the pipes! It is best to work with a sink setup over a 5-gallon bucket. The liquid can be decanted off later and the residue on the bottom thrown into the trash bucket.

Mix the PGC as directed on page 167, and apply to the face with a 1-inch paint brush or a flexible spatula. The usual method is to start at the forehead area and work down the face as the PGC will run somewhat during application. Always do the nose and nostril area last. Also take care in the under-chin area that a coating of the PGC is brushed to adhere under the chin. Unless the ears are covered with the shower cap (or a huck towel), their cavities should be protected with cotton batting covered with petroleum jelly. To ensure that the mother mold will adhere to the PGC, strips of burlap (used for strengthening Ultracal molds—terry cloth may also be used) about 2 by 4 inches can be imbedded in the PGC surface before it

sets. These can be placed on the forehead, cheeks, mouth and chin areas.

After the PGC is set to a rubbery mass, it must be covered with a mother mold of rigid material. Some workers prefer the fast set dental plaster made in thin mixes and brushed on in successive coats until about a half-inch thickness is obtained. A layer of cheesecloth can be placed over the plaster and another thin coat brushed in for the final application. This will provide more strength to the rather fragile dental plaster. Others use plaster bandages of the type employed by medical personnel for making broken limb casts. This is obtainable from medical supply houses and, although rather expensive, does an excellent job. Follow the directions given with the plaster bandage for application. If accelerated set is desired, warm water with a bit of salt is used for the dipping of the bandage. Short, overlapping sections can cover some of the longer ones that are placed around the head until the proper thickness is obtained. Usually ¼ inch of bandage thickness is sufficient in strength for this type of mother mold. Finally, some technicians simply accelerate the set of casting plaster with the addition of a teaspoonful of salt in a quart of water, and use this for the mother mold. Again, take care around the nasal passages and allow breathing holes in the mother mold material.

· The plaster will heat up somewhat, and when it cools off, the impression is ready to remove from the face. If the strips of burlap have been embedded in the PGC, the cast will come off the face intact with the PGC stuck to the mother mold. Having the subject make a few faces will loosen the PGC from the face and facilitate removal. A mixture of dental plaster can then be used to seal up the open nasal passages on the mother mold. Be certain to coat the inside surface of this plaster with petroleum jelly so that it will not stick to the positive to be cast. One can also fill these nostril holes with plastalene if desired.

The impression is now ready for making the positive cast (Figure 13.3). This should be done immediately as the alginate will commence to shrink as it dries out. To support the cast, one can put it in a cardboard shoe box or a plastic bowl. If the PGC begins to separate from the mother mold at the edges, metal hair clips can be used to hold them together. Another method is to use a dental product made by Johnson and Johnson called *Secure Dental Seal,* which is a paper-thin, 100 percent adhesive that can be wetted and placed between the alginate and the plaster of the mother mold to form a good temporary bond that will control a thin alginate edge and maintain the shape required. It is far better than the soft dental adhesive cream material that is used by some workers for this purpose as it does not require a heavy coating that might distort the casting.

Another adhesive material can be made by dissolving a product called Vistanex in a solvent such as

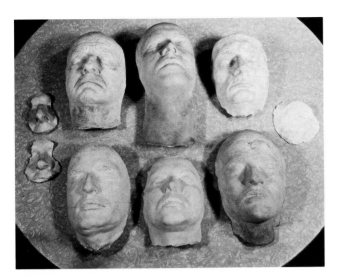

FIGURE 13.3 *Various facial casts and section casts.*

toluene. This can be painted on the outer part of the alginate and the inner surface of the mother mold. It dries rapidly and will adhere the two parts on contact when they are pressed together.

A mixture of casting plaster, Ultracal 30, or stone can then be brushed into the negative PGC mold. Blowing the thin coating around with sharp breaths will remove bubbles from the surface. More liquid gypsum can then be poured into the mold. When the mixture begins to thicken, one can spatulate the mix up the sides of the mold so that the cast will be about an inch thick all around while being hollow in the center. Unless one wants a very heavy positive, it is not necessary to fill the negative cavity completely. If more than one positive is desired, don't use the burlap strips on the PGC so that the mother mold can be removed from the PGC and then it can be carefully stripped from the positive and replaced in the mother mold. Another positive cast can then be made immediately. Otherwise, one must pry the mother mold away from the PGC (often breaking it in the process), and strip the PGC away from the positive and discard it.

The positive can then be scraped free of any small casting imperfections, the nostrils cleaned out, and the edges of the cast smoothed out with a metal plaster tool. If the cast is heavier on one side than the other, a Surform tool can be used to scrape away the excess plaster or stone. Mark the back of the positive with the name of the subject, the date, and possibly a serial number for cataloging the molds. As the positive will not have thoroughly hardened yet, a sharp pencil is best to mark in this data.

Another way to take a face cast is with casting plaster. The face is prepared the same way, and a mix of casting plaster is applied directly to the face with a 1-inch bristle brush. The same precautions should

be observed in applying the plaster to the nostril areas as were described for alginate.

This is a slower method because the casting plaster (salted to accelerate the set) takes a longer time to set, but no mother mold is required and if the subject is patient and does not move, an excellent impression can be taken. When the plaster is beginning to cool after its warming-up period is over, the impression can be removed from the face. A saturated solution of Ivory soap in water is then brushed into the impression as a separator. Use this soap solution copiously, brushing out the foam and bubbles until an even sleazy coating can be felt on the mold. Allow to dry for 15 minutes, then pour up a stone or Ultracal positive. When the stone has set and hardened, a few sharp blows with a chisel and mallet will crack the plaster negative from the harder positive. Do not attempt to separate the stone positive from the plaster negative until the stone has completely hardened; otherwise the positive may be damaged when cracking off the plaster.

Some technicians use plaster to take a cast of the entire head. Again the hair should be covered with a bald cap and the subject placed so that he or she is face down on a stack of towels or hard pillow. A piece of lead wire is stretched around the head just behind the ears to get an accurate outline of the head. This wire outline is then placed on a piece of strong cardboard and the head shape traced out. The outline is then cut out, and this form is slipped over the head to form a retaining wall for the plaster. The cardboard surface can be greased with petroleum jelly for ease in separation. The back of the head is then taken in plaster. When this impression has set, the subject is carefully turned around and up to the normal facial casting position. The cardboard retainer is then removed and keys can be cut into the plaster on each side and at the top of the impression with the end of a spatula. The impression should be at least 1 inch thick on the edge so that this can be done. The surface of the edge can then be coated with petroleum jelly and the cast of the face taken in casting plaster as previously described. Normally, the ear cavity is filled a bit more with cotton and petroleum jelly than with alginate for ease in separation when the cast is removed.

When the plaster impression of the face has set, it can be removed. The cast of the back of the head is then removed, and both halves are fitted together to make the positive. The inside of the impressions should be coated with the soap mixture and then, with the two halves held together with large elastic bands (made from auto tire tubes) or heavy cord, a positive can be cast.

To make the usual hollow cast, fill the mold about half full with a thin mix of stone or Ultracal 30. Carefully spread the material around the inside of the mold by rolling it in different directions to ensure an even coating. The excess can then be poured out and

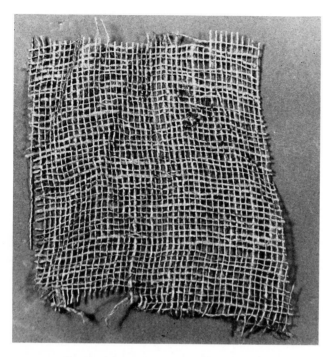

FIGURE 13.4   *The open weave burlap recommended by Dick Smith for strengthening molds and casts (see Appendix B).*

this flash coat allowed to semi-set before another thin mix is applied in the same manner. After three such coatings, the positive can be strengthened with small sections of burlap if desired, and more stone is poured into the mold and can be worked around inside with the hand to ensure an even coating (Figure 13.4).

When the positive has completely set, the plaster negative is carefully cracked away by tapping it with a small hammer to split the plaster. Some workers use

a chisel and a wooden mallet for this. This is one of the oldest methods of making a full head cast. A newer method was devised by Dick Smith and is described and illustrated in Figure 13.5 (see pages 174–177) by Tom Savini. In most cases, face casts that will be used for plastalene sculpture are coated with orange shellac or clear lacquer to seal the pores of the gypsum material and make the cast easier to clean.

SCULPTURE MATERIALS

The term *modelling clay* encompasses two types of clay-like materials. One is a water-based clay that is purchased as a powder and mixed with water to the desired consistency for use as a modelling material. Sculptures made of this material must be kept damp so they will not dry out and crack. It finds some use in the make-up lab but is not the best sculpturing material. Make-up artists find that the oil-based *plastalene* or *plasticine* serves their purposes better. It comes in many grades from a very soft #1 to a hard #4. It can be obtained in the normal dark green shade or in a white. Opinions vary considerably as to the most useful variety, but most technicians use the dark green in either the #2 or the #4 hardness. It comes in 2-pound blocks (see Appendix B).

The tools for use with plastalene modelling are usually small and wooden (generally much smaller than the clay sculpture tools fine artists use), and the small-

FIGURE 13.6   *(A) Metal tools.* LEFT TO RIGHT: *Stiff and flexible spatulas, a plaster knife, two metal plaster carving tools.* *(B) Sculpture tools.* LEFT TO RIGHT: *A small spatula; two wire tools for nostrils, lines, wrinkles; three wooden carving tools for plastalene (the smallest are the most useful); a pointed ball-head wooden tool for making pores.*

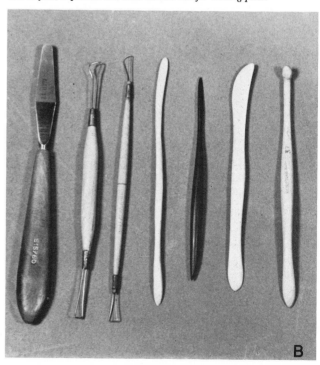

est wire tools are used as well (Figure 13.6). This is mainly because the make-up lab technician starts with a face cast in gypsum material and adds the plastalene to achieve the sculpture rather than starting with a large, roughly shaped head of plastalene and carving it to shape with tools. When adding clay to a plaster or stone positive, always apply only a bit at a time until the desired shape is built up. The wooden tools with the flat ends can do this work, while the fine wire tools can be used to diminish the sculpture a bit at a time. The surface can also be scraped with a piece of coarse burlap to remove a thin layer of clay.

Texturing the surface to make it appear to have pores can be accomplished with a small knobbed-end wooden tool, with a plastic stipple sponge, or with a *grapefruit* tool made by coating the surface of various citrus fruits with a few coats of casting latex, drying this with a hand hair dryer, and then peeling off the latex after powdering it. This will take an impression of the surface of the orange or grapefruit (hence, the name!), which has a great similarity to the human skin.

To smooth plastalene, a brush dipped in isopropyl alcohol will serve for small areas. For a larger expanse, some petrolatum jelly or mineral oil can be rubbed sparingly on the surface with the hands or a wide brush. Take care not to use too much of the latter two materials as they will soften the surface of the sculpture. Some workers will coat a finished sculpture in plastalene with three coats of RCMA Plastic Cap Material (clear type) to provide a surface that will still take some tool and texture markings to refine the sculpture.

If one uses the sponge method, they will find that the various grades of polyurethane stipple sponges work better than the red rubber ones because the latter often leave bits of rubber imbedded in the plastalene. More often than not, a combination of the grapefruit, the sponge, and the knob tool will produce the best results for the varying skin pore textures.

Another method is to make latex stamps of skin texture areas. This can be done from the life mask cast of an older person who has good forehead wrinkles, for example. Just coat the original positive with a number of coats of casting latex (depending upon the thickness of the stamp desired), dry, and then remove it. This negative impression can be then used to press into a forehead plastalene area to obtain a reasonably good sculpture of forehead wrinkles. This method cannot be used too successfully on the soft parts of the face like the eye area, but it is suitable on the bonier parts.

Due to shrinkage of latex pieces, some size allowances can be made on appliances that cover a large area. This is one reason why many technicians make multiple appliances rather than one large face piece.

As well as wooden tools for plastalene, there are

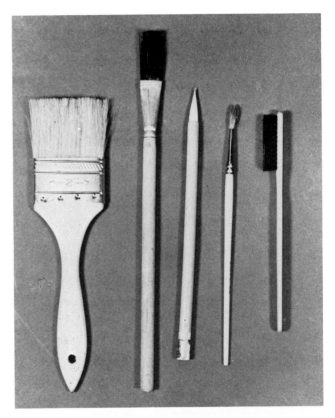

FIGURE 13.7   TOP: *Brushes.* LEFT TO RIGHT: *Soft bristle brush for cap material or other uses; long-handled bristle brush for use with plaster; Chinese bristle brush for paint-in latex work; #10R sable brush for many applications; bamboo-handled bristle brush for cleaning plastalene from castings.* BOTTOM: *Two black rubber dental bowls for mixing small batches of alginate or gypsum materials and two small polyethylene measuring cups for weighing various foam ingredients.*

various metal tools with cutting edges for trimming plaster and stone, rasps and Surform tools for removing more material, and some labs even have machine tools with coarse wheels to rapidly cut and shape gypsum casts.

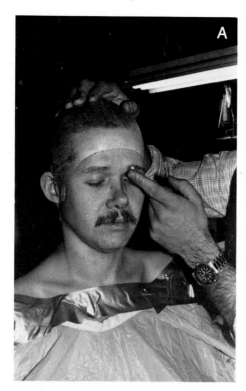

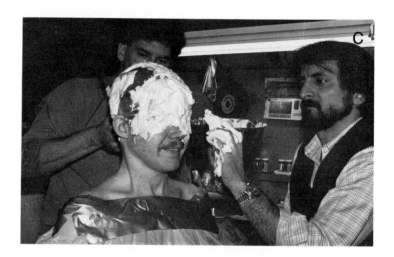

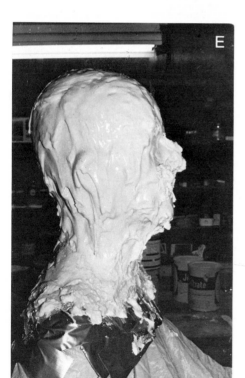

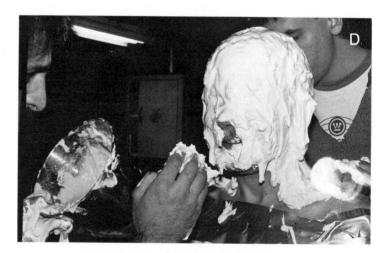

FIGURE 13.5   *Casting a full head and shoulders. (Photos and casting courtesy Tom Savini from* Grand Illusions.) *(A) The subject's head is covered with a plastic cap and adhered to cover all the hair. The brows, lashes, and facial hair are generously coated with petroleum jelly while the entire skin to be cast is also covered with a fine coat as well. (B) Tom mixes a large batch of the alginate in a kitchen mixer after weighing and measuring the proportions carefully. (C) With the aid of an assistant, he commences to apply the alginate all over the face and head, taking care around the nose section. He generally fills the ear cavity with cotton and petroleum jelly as can be seen here and, unless the ears must be part of the* cast, fills the back of them with a plastic wax material to avoid heavy undercuts. Note that the subject has been covered with a plastic sheeting, which is held to the body with gaffer tape. (D) The mouth and nose area being worked on with care. (E) The head, neck, and shoulders covered with alginate. (F) Previously cut and rolled strips of plaster bandage are soaked in water for use. (G) The first strip goes over the head (H) and down to the neck area on both sides. (I) Plaster bandages on the neck (J) and face area, again taking care not to cover the nostril vents for breathing. The face is covered to a thickness of about four layers of bandage on the front half. (K) The front section of the head, face, and*

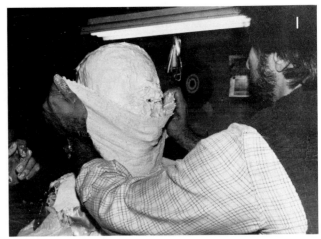

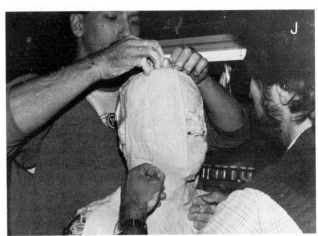

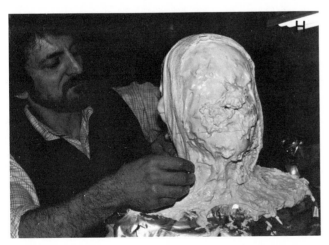

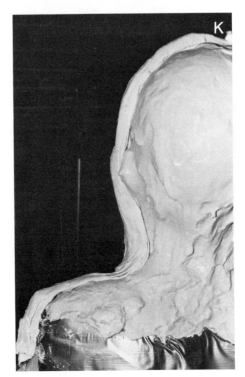

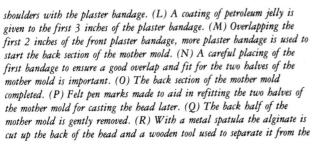

shoulders with the plaster bandage. (L) A coating of petroleum jelly is given to the first 3 inches of the plaster bandage. (M) Overlapping the first 2 inches of the front plaster bandage, more plaster bandage is used to start the back section of the mother mold. (N) A careful placing of the first bandage to ensure a good overlap and fit for the two halves of the mother mold is important. (O) The back section of the mother mold completed. (P) Felt pen marks made to aid in refitting the two halves of the mother mold for casting the head later. (Q) The back half of the mother mold is gently removed. (R) With a metal spatula the alginate is cut up the back of the head and a wooden tool used to separate it from the

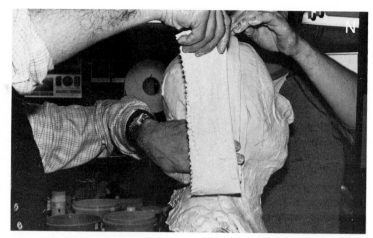

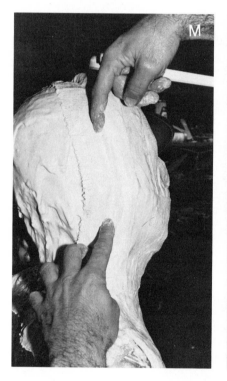

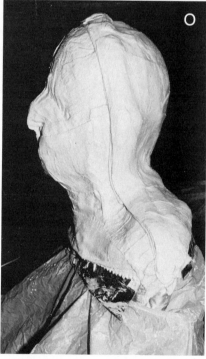

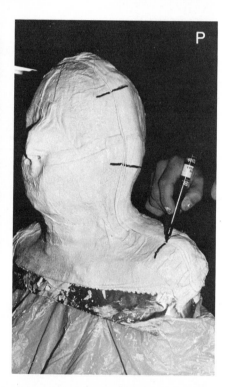

subject. (S) The head in a forward inclined position. (T) With the head forward, the alginate and front half of the mother mold are removed with care. To prevent the alginate from separating from the front half of the mother mold, he imbeds a layer of a terrycloth material into the alginate on the forehead, mouth, and chin area and down each side of the face. See page 167 for directions on adding to a set alginate surface if the alginate sets too soon before this can be done. (U) With the mold put together, the sides matching on the lines, and the alginate held to the sides of the mother mold if necessary with a false teeth gel, it can be taped, as shown, to a plastic bucket with gaffer tape and the mold held together with the same. Cord can also be used, but the tape is convenient and works well. The nostril holes can be filled with quick-set dental plaster or some plastalene forced in. A mix of Ultracal 30 is made. (V) After a splash coat is worked around the mold, the head is filled with the remaining Ultracal. Some workers prefer to make a head about one inch thick to save weight, but the usual full head is solid for strength. (W) When the Ultracal is set, the mother mold is removed and the alginate separated from the head. (X) After the usual minor clean-up around the ears and nostrils and any imperfections such as small holes or bumps are filled and removed, we have a completed perfect head cast.

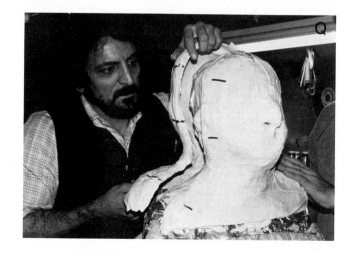

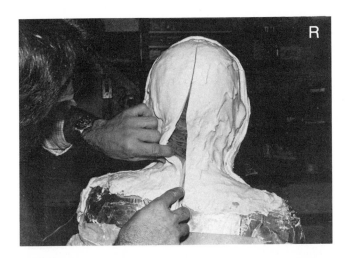

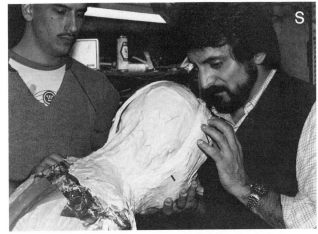

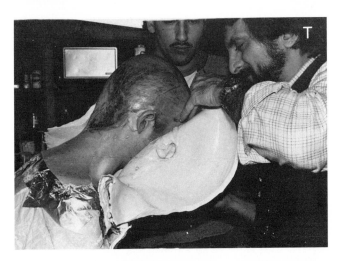

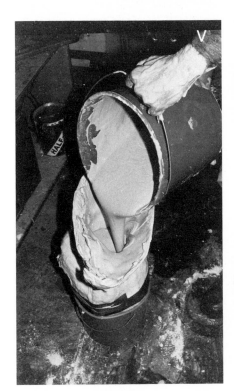

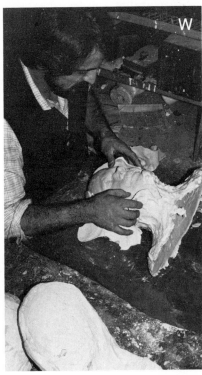

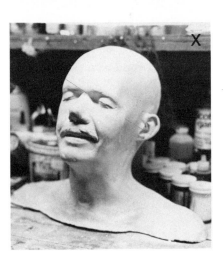

Two kinds of spatulas are used in the lab. One is the regular flexible type that is used for measuring small amounts of material for weighing, applying alginate to the face, and many other uses. The other is a stiff metal one that is obtainable from a dental supply house for mixing alginate or plaster. A strong bladed knife is needed for trimming, and a heavy screwdriver is useful for prying apart two-piece molds.

Plastic bristle 1-inch paint brushes, some with regular short handles and others with long handles (for getting into castings easier), will find much use, as will some #10 round sable hair brushes as well as some cheaper Chinese bristle brushes (for paint-in latex application) (Figure 13.7). A few wig-cleaning bristle brushes are handy for scrubbing plastalene out of plaster and stone molds as are some bristle stipple brushes of small sizes.

Sculptor's calipers for measuring head size and shape are useful for making large castings and a hand or electric drill with a ¼- and a ⅜-inch bit for drilling escape holes in positive casts for foamed latex or urethane and a ¾-inch half-round coarse reamer or rasp bit for cutting keys.

Dental sculpture tools such as dental spatulas, curved tip tools used for handling dental amalgam for filling teeth, and others will be found useful for tooth sculpture (Figures 13.8 and 13.9).

Stan Winston feels the life mask is only a guide and serves, along with many close-up photos, as a basis for sculpting in plastalene a head and face with the expression he wishes to use as a perspective for the make-up (see Figures 12.24 and 12.25 for his "Manimal" make-up). He imbeds false eyes and teeth in the sculpture to add to its lifelike qualities to aid his conceptions (Figure 13.10). He feels that many times he can capture an expression in photos that he cannot in a life mask, and the clay sculpture gives him what he seeks. P.S. One has to be a very good sculptor to do this!

FIGURE 13.8   *1/2- to 3/4-inch rotary files for use in cutting keys in gypsum materials. They work best in a good drill press or, if need be, in a hand-held electric drill.*

keys by rounding 1-inch balls of plastalene in the palms of the hands, then cutting them in half with a strand of thread. Place them about a half-inch from the corners of the enclosure. Make a thin coat of stone or Ultracal 30, and blow it in to remove the bubbles. (A sculptor's trick is to use sharp breaths of air blown in on a thin liquid plaster mix surface to remove all the small surface bubbles.) Then fill up the enclosure with the remaining mixture. A strip of burlap can be used as a strengthener if desired, but the basic thickness of the plate is normally sufficient for most uses.

When the stone or Ultracal 30 has set, remove the blocks and slide the cast off the glass plate. The plastalene in the keys can be pried out with a wooden modelling tool. The keys can then be sanded with medium then fine sandpaper so that there are no un-

MOLDS

Flat Plate Molds

The simplest of the two-piece molds is the flat plate type. With this basic casting one can make scars, cuts, bruises, moles, and other small appliances that do not require specific facial contours. When the prostheses made on the flat plate are adhered to the skin, they take on the natural curve of the area without any trouble in most cases.

To make a flat plate mold, form a rectangle whose inside measurements are about 6 by 10 inches, with four pieces of 1-by-2-inch wooden blocks (Figure 13.11). These can be held in place on a glass surface with four lumps of plastalene at the corners. The inside surface of the wood as well as the glass should be coated with petroleum jelly as a separator. Make some clay

FIGURE 13.9   *Various tools and clamps that are useful in the prosthetics lab.*

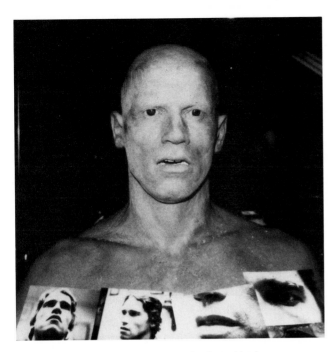

FIGURE 13.10 *Working from photographs of Arnold Schwartzenegger, Stan Winston sculpted a very accurate reproduction in plastalene as a work model for the effects he created in the film* The Terminator.

dercuts. For use, items can be modelled on the plate with plastalene, with a ¼-inch gutter left around each of the items modelled and the remainder of the plate clayed in to a depth of about ¼-inch. Petroleum jelly can then be applied to the exposed stone areas in the gutter and keys, and surround the plate with another set of greased wooden blocks (1 by 4 inches), and secure with plastalene at the corners to form a box around the plate. Stone or Ultracal 30 can then be mixed and added in the usual way to fill the cavity. Each of the plates, when completed, should be about 1¼ inches thick.

When the stone or Ultracal 30 has set, remove the retaining wall and separate the mold. The keys may then be lightly sanded after the plastalene has been thoroughly cleaned off both sides of the mold. Acetone can be used for this cleaning or TCTFE (trichlorotrifluoroethane) (Prosthetic Adhesive A Thinner is this variety of solvent). The mold is now ready for use with foamed latex or foamed urethane. The sculpted side of this type of mold can also be employed for paint-in plastic molding material as well. In the case of the latter, the smooth surface side upon which the original sculpture was done is not required as the plastic molding material uses an open-face mold (also see page 204 for these molds).

Slush and Paint-in Molds

The basic casting latex molds are those that are taken as a negative in casting plaster of a plastalene sculpture on a facial cast. Noses are probably the most commonly used prostheses in make-up work as the change of a

nasal structure can delineate a character more quickly than any other. Many make-up artists take a full cast of the face to get the correct proportions of the face when modelling a new nose, chin, cheeks, eyebags, and so on and make all their casts and molds directly from the original positive. Others take an additional cast only of the area (such as the nose) desired for use in making the necessary prosthetic mold.

This individual mold can be taken in PGC or even directly in dental plaster. The nostrils should be blocked with cotton batting and covered with petroleum jelly, leaving sufficient clearance to get a clear impression of the nostril areas. During the casting operation, the subject may breathe through the mouth as it will be free of plaster. Paint the plaster on the face with a 1-inch bristle paint brush, taking care not to get any in the nostrils or the mouth. Cover about half the eye sockets and cheeks with plaster to get sufficient edges for the final positive. Don't forget to put petroleum jelly on the eyebrows and lashes for ease in separation.

When the plaster is set, remove the cast and pour up the positive with Ultracal 30 or stone using either petroleum jelly or the soap method in a thin coat for a separator. Another good separator for plaster to stone is the stearic acid-kerosene one discussed on page 169. When the stone has set, the dental plaster negative can be easily cracked away, leaving the stone positive ready for use.

The Nose

To make a slush or brush application mold from this section mold of the nose (or using a full face cast), model the desired shape in plastalene on this positive, making certain that the blending edges are smooth so they will be imperceptible in the finished appliance. Take care that the nostrils are sufficiently filled on the positive with plastalene so they will not have any undercut that will prevent the removal of the negative cast to be taken. However, sculpt the nostrils so they will look natural and fit closely to the nose in the finished appliance (Figure 13.12).

The nose exhibits one of the more varied skin pore areas of the body. Sometimes the pores are small and almost imperceptible on someone with fine skin, and at other times the pores are large and the surface of the skin quite rough. Generally, a slight overporing is normally called for during sculpture as some of the fine details are lost in the casting of the negative on the surface of the appliance. Only by experience can the technician know what is the correct depth and extent of making pores.

For a clown nose (which is generally slush cast), no pores are called for. Thus, the surface should be made as slick as possible by dipping the fingers in water (or mineral oil) and rubbing them on the surface of the plastalene.

It should also be noted that sometimes only a nose

A

B

C

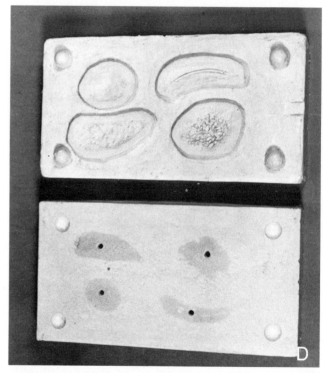

D

E

FIGURE 13.11 *Flat plate molds. (A) The sculpture of some bruises and a cut are made in plastalene. (B) After the plastalene gutters have been made, the flat plate is surrounded by wooden blocks held together with lumps of plastalene. These wooden blocks have been lacquered and coated with petroleum jelly as have the areas of the plate showing around the sculptures on the plate for ease in separation. (C) After Ultracal has been mixed and poured in, the mold set and separated, we have the two halves of the flat plate mold. (D) The plastalene is removed from the flat plate positive and 1/4-inch holes drilled into the Ultracal for vent holes. This mold was then coated with a polyurethane lacquer and employed to cast foamed polyurethane pieces. (E) The appliances (with flash removed) of foamed polyurethane.*

bridge or tip need be made so the sculpture need not be overextensive. However, the undercuts on the nostrils must be filled as well as the outer curve of the alae with plastalene to avoid undercuts.

When the modelling is completed, coat the remainder of the exposed positive with petroleum jelly as a separator. The sculpture should then be surrounded with a clay wall about ¾ inch higher than the tip of the sculptured nose and at least ¾ inch from the sculpture all around. As the casting latex builds up by absorption into the plaster negative, this amount of thickness is minimal. Larger casts should be thicker (1 to 2 inches) to allow for an absorptive surface.

Some technicians spray the surface of the sculpture with aerosol solution or Kodak Photo Flo (1 cubic centimeter in 200 cubic centimeters of water) to reduce the possibility of surface bubbles in the plaster. Paint

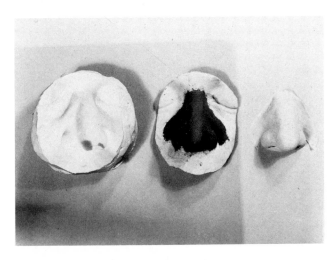

FIGURE 13.12 *Simple paint-in latex nose. A plaster negative* (LEFT) *made from a positive plastalene sculpture on a positive cast of a nose* (RIGHT). *A painted-in casting latex nose made in the negative mold.*

in, then blow in, a mix of casting plaster to cover all the sculpture and surroundings, and carefully spatulate and pour in the remainder of the mix to fill the cavity of the plastalene retaining wall. Some workers use a vibrator table to hasten the removal of all bubbles while others tap the mold on the table top to force any bubbles to the surface of the plaster.

As soon as the plaster heats up and cools, the two sides can be pried apart with the blade of a screwdriver. Any plastalene remaining in the negative should be carefully removed with a wooden modelling tool, taking care not to scratch the surface of the negative. Allow the negative cast to dry out for about an hour, then it can be cleaned of any remaining plastalene with a small bristle brush dipped in acetone or TCTFE. Rinse the negative in water before each use, and the latex appliance made in the cast will separate out easily when it is completed and dried.

If the negative is to be used for brush application, no further preparation is needed, but if a slush molded appliance is to be made, the negative requires that a plaster wall be built up in the area of the upper lip on the mold so that an even level will be achieved when the casting latex is poured into the mold.

The advantage of any slush or brush-applied latex piece is that it will fit many faces because the inner surface is not made to fit any particular face the way foamed latex appliances are. However, one cannot achieve the more skinlike quality on the outer surface of any slush or paint-in appliance that a foamed one will display (see page 190).

Whenever any additional plaster is to be added to a mold or cast, the surface should be thoroughly wetted with water before the new plaster is applied. Such additions work much better with a freshly made negative as it has a greater water content than one that has been left to dry out—even overnight.

Any repairs on the negative, like holes left by errant surface bubbles that may not have been dissipated when the cast was vibrated, can be easily made by scraping the underside of the negative with a metal tool to remove some plaster and using this to do the repair rather than making a new mix. Dental tooth filling tools are best to get into small places for these repairs.

Oriental Eyelids or the Epicanthic Eyefold

For these, a cast of the upper face from the nose to the hairline can be taken or the full face cast used (Figure 13.13). Very little poring is needed on the lids themselves. If one wishes to make a series of stock Oriental eyelids, it is well to vary the eyelid curves so they can be adjusted to fit a number of Caucasoid eyes. Again, the negative should be taken in casting plaster and made about an inch thick.

Bald Caps

Latex bald caps can be made in a negative plaster mold or painted on a positive one (Figure 13.14). To make a negative-type mold, get a small-sized balsa wood head form from a hat supply house and slice it into nine fairly equal sections by two cuts made one way, then turning the head 90° and making two cuts again. Then put the sides together and hold them in place along the neck portion of the head with heavy elastic bands. Coat the entire head with a #1 or #2 plastalene, smoothing out the surface to form a bald head sculpture. The surface can be lightly pored to give it some texture. A retaining wall of plastalene is added to the desired shape of the mold, and then it can be sprayed with the Photo Flo solution.

A good spray gadget is one made for spray painting small objects, called a PreVal Sprayer made by the Precision Valve Corporation and sold in many paint stores. It consists of a bottle for the liquid and a removable unit that has the propellant and spray tip. The latter screws onto the bottle and can be used to spray many liquids. Buy several and keep one with the Photo Flo solution ready for use.

The retaining wall should be about 1½ inches high, and the casting plaster negative taken of the plastalene positive must be that thickness all around. When the plaster is beginning to set, cover it with cheesecloth, and brush over a bit more casting plaster to strengthen the mold. When the plaster has fully set and cooled, remove the elastic bands from the balsa wood head, and carefully draw out the center section of the cut head. This can be facilitated by passing a piece of coat hanger wire through this section before putting it together to make the mold. Once this center section is out, the others will remove easily without cracking the negative. Remove the excess clay and the mold is ready for use.

The same basic method can be used for making positive stone or Ultracal heads for plastic caps. How-

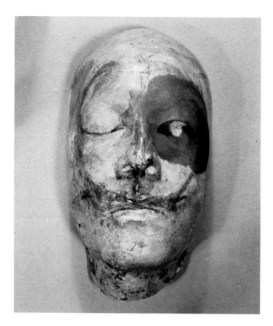
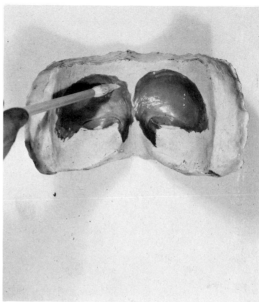

FIGURE 13.13
Epicanthic eyefold. LEFT:
*A face cast with one side
sculpted in plastalene of
an epicanthic eyefold.*
RIGHT: *The finished
negative mold made from
plaster being painted in
with casting latex to form
a large appliance that
will cover the brows as
well. A smaller piece may
also be painted-in that
will just cover the eyelid
area with the same
material. Such appliances
can also be made of
foamed urethane or latex.*

ever, do not make the casting plaster negative over ¾ inch thick, and do not use any cheesecloth reinforcement. When the plastalene has been cleaned out of the negative, coat the inside with petroleum jelly and fill the cavity with a stone or Ultracal mix. The mix should be spatulated up and around the negative so that the finished cast is about ½ to ¾ inch thick all around in a hollow shell. When the stone or Ultracal 30 has set, the plaster can be cracked away and you have a positive cast to paint or spray the cap plastic. As previously mentioned, casting latex can also be used on a positive head, but the head positive should be of plaster. Due to the separating medium (silicon grease type) used on the head molds for plastic cap material, molds so treated will not work with painted-on casting latex.

Generally, if the surface of the plastalene positive original has been pored sufficiently, the positive gypsum head will have a proper surface for painting on plastic cap material or latex as the finished cap will then be reversed for use.

To make a head mold for spraying, where the adjustment of the spray on the final coating can be made to provide a more textured coat, some technicians make the head mold as slick as possible for ease in removal, and of course, the bald cap is not reversed. John Chambers had some head molds that were chrome plated, while Bob Schiffer has made some from fibreglass, both of which are very smooth and work quite efficiently for spraying.

If one can borrow a positive head mold and wishes to make a copy, there are a number of ways that a duplicate mold can be taken. One of the basic ways is to use a flexible duplicating material such as PMC-724 or Perma Flex's Gra-Tufy CMC (see pages 167–168). Another way is to make a three-piece duplicating

mold of stone or Ultracal 30. This type of mold is the same as would be made to serve as a mother mold for the flexible impression material as well (Figure 13.15).

Basically, one would divide the head mold down the center with a plastalene wall about 1½ inches high, with the division line on side one (so that when the wall is removed, the line is still on the center line of the head). A stone or Ultracal mix is then brushed and spatulated onto side one to a thickness of about 1 inch and allowed to set and cool. (Always remember that where any gypsum product meets another in casting, a separating medium is necessary.) The clay wall is then removed and keys cut into the sides of the mold (three are sufficient), and the sides are covered with a thin coat of petroleum jelly as a separator. Side two is then taken to match the thickness of side one. When this has set and cooled, two angle keys can be cut to a depth of about ¼ to ½ inch about 1 inch apart. A cap cavity is then formed out of plastalene of the top area of the head mold and this taken in the same material (remember to use a separator). This third cast should have a flat top, so the plastalene wall should be higher on two sides, and the surface can be flattened with a board that has been lacquered and then covered with petroleum jelly. When this part has set and cooled, the three pieces can be separated and the original removed. The mold is then reassembled and is ready for use.

If a flexible material was used in making the mold, the procedures would be approximately the same for making the three-piece mother mold. Then in the disassembly, the flexible mold is removed from the original and replaced in the mother mold for preparing the negative for use. The negative is filled with stone or Ultracal, and a duplicate is made in the usual way. This type of duplicating mold will allow one to make

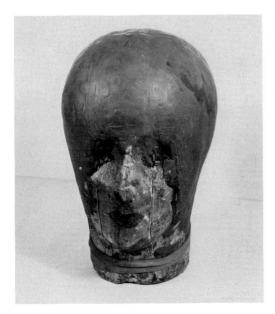

FIGURE 13.14 *Slush bald cap mold.* LEFT: *A balsa wood head (which has been previously section-cut in nine parts and held with an elastic band at the base) is covered with a coating of plastalene.* RIGHT: *A one-inch-thick plaster mold made of this form with the form removed. This mold is used for slush casting quick latex bald caps.*

a number of reproductions of the original. Check the basic shelf life of any duplicating material in the plastic varieties if you want to save the mold for future use. Of course, a stone or Ultracal 30 duplicating mold will last indefinitely with care taken in its separation from the duplicates.

Full Head Castings

Full, slip-over casting latex heads can be made with a slush cast of either one- or three-piece construction. These must be made of casting plaster and be about 2 inches thick. Such molds should be reinforced with burlap strips for strength. A mix of 20 percent Ultracal 30 with casting plaster will also add strength and not diminish the slush molding time excessively.

For use, these large molds must be securely tied together with heavy cord or rope. Take care in using clamps on plaster molds as the extra pressure exerted by the clamp may crack the mold.

Many other areas of the face and body can be cast and slush or paint-in appliances made. To get the best fit as well as the most natural appearance, however, foamed latex or urethane prostheses are far superior, especially for movement and fit.

The Two-piece Mold

The two-piece, or positive-negative, mold is required for foamed appliances or for any pressure-molded ones. In theory, the positive cast is modelled upon, with the desired shape of the appliance to be made, in plastalene. Then a negative is made of this, the clay is removed, and the mold is filled with the appliance material, placed together, cured, and separated, and the appliance is ready for use. The procedure is similar to that given for the two-part flat plate mold on page 178.

To make a mold for a foamed latex or foamed urethane small piece—say, a nose—from a full-face cast, a plastalene wall is built around the nose of the cast to a height of about ¼ to ½ inch above the tip of the nose (Figure 13.16). This cavity is then filled with PGC alginate and allowed to set. Then the clay wall is removed and the alginate separated from the face cast. This will produce a *section* negative into which is poured an Ultracal 30 or stone mix to make a positive. When set, this positive is then imbedded in another rather heavy mix of the same gypsum material. Generally, a surrounding round wall is made to hold the stone or Ultracal 30 mix out of 3-inch-wide linoleum, rubber floor mat, sheet lead, or even 6-inch plastic pipe. As the positive has still a great amount of moisture in it, it will incorporate itself into the new gypsum mix readily. Carefully place the positive in the center of the cavity, and with a dental spatula, seal the edges in so that the blend is perfect.

A good way to prepare the surface of a formica-covered table to do this procedure is to rub on a thin coat of petroleum jelly. Some workers prefer to use a glass plate or even a board with a formica top. The linoleum or other material wall can be secured with cloth adhesive tape to keep its shape.

Although with small molds there is less chance of breakage than with larger ones, many lab technicians use fibre, fibreglass cloth, or burlap in the mold. Others prefer to use a circle of wire cloth around the positive section mold, taking care to imbed this fully within the mold.

When the stone or Ultracal starts to heat up in the set, remove the retaining wall, and while the material is not yet quite hard, scoop out three or four circular keys with the round end of a spatula or knife blade. Clean up the edges and fill any defects. The desired

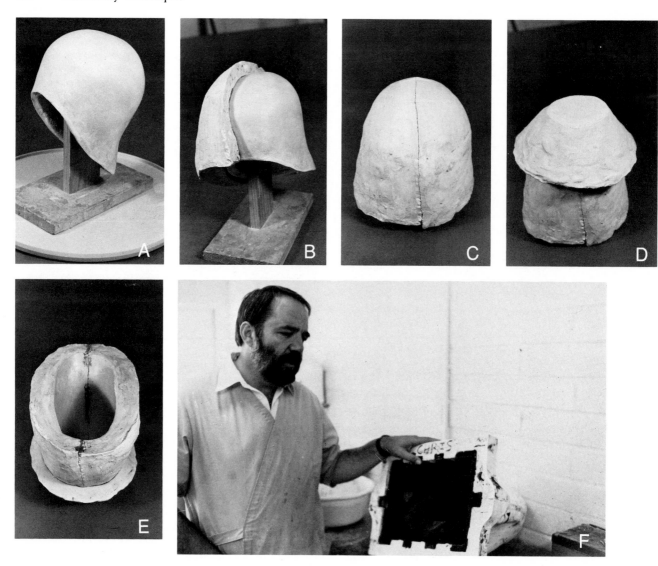

FIGURE 13.15   (A) A positive stone or Ultracal head on a stand and turntable (for ease in spraying or painting the caps). (B) Showing one side of a section cast made to duplicate this head mold. Keys are cut (C) And the other side is made (note keys cut on top). (D) Then a cap is cast to hold them together. (E) The completed three-piece mold put together. It is a good idea to coat the inside of the mold and the rim with two coats of lacquer to seal the surface. Small chipped areas can be filled with plastalene and then the mold given a coat of petroleum jelly as a separator. Ultracal 30 can then be used to make positive heads in this mold. (F) A head mold or full head mold can also be made of polysulfide duplicating material for a seamless flexible mold (see also page 167) as shown here by Tom Burman.

shape of the appliance can now be sculpted on this built-up positive form in plastalene and gutters formed for an overflow area. Clean the key areas and apply a separator on them and the overflow where the stone or Ultracal 30 can be seen. The surface can then be sprayed with the Kodak Photo Flo solution and the form enclosed in a 6-to-8-inch (depending upon the height of the sculpture) circle of linoleum or such. Heavy elastic bands or cloth adhesive tape can be used to hold the form in shape. A stone or Ultracal 30 mix is then added in the usual manner to fill the cavity, and reinforcement material is added if desired.

When the stone or Ultracal 30 is completely set and almost cool, remove the wall and separate the two sides of the mold. The plastalene is then removed and

cleaned off the mold, and the keys can be lightly sanded to remove any burrs. The mold is now ready for use.

Molds made for foamed latex prostheses should not be more than 1½ inches thick to allow good heat penetration for curing the latex. Also, if the appliance is large or deep, it is a good idea to have a vent hole for overflow as well. On large noses this hole can be drilled through the positive on the nose tip with a ¼-to-⁵⁄₁₆-inch drill (Figures 13.17 and 13.18).

Two-piece molds for foamed polyurethanes are quite similar to those for foamed latex except that the urethane molds can be heavier as no baking is required.

Large molds that might be for extensive forehead pieces or for full heads can entail a positive core head and a two-piece negative. Not only are these molds made to have vent holes, but also they may be employed as filling holes for a gun filled with foamed latex that is forced into the mold. Large molds must often be clamped together as the mold is being filled, and flanges must be made to give a purchase point for these clamps. Of course, the larger the mold, the more strengthening it must have, both with a product such as Acryl 60 cement hardener and burlap or fibre added to the molds. It is also a good idea to mark one side of the molds with a heavy marking pen or to scratch a line down the sides of both of the molds to see easily how to put the molds together during use.

Some lab technicians coat the sculpted plastalene with RCMA Plastic Cap Material when making two-piece molds so that when separated after making the negative, no plastalene will stick to it, making cleaning of the negative easier. This coating should be removed from the plastalene with acetone for reuse of the clay material. There are many methods employed for making the positive-negative types of molds, and books relative to these methods may be found in the Bibliography.

Incidentally, all molds used for foamed procedures must be thoroughly dried out before use. One good way is to put them in an oven at low heat (100°–150°F) and leave them overnight.

Plastalene Transfers

Dick Smith often sculpts a facial transformation on the life mask, then removes sections of plastalene intact and transfers them to an individual section positive for making the two-piece molds. Using a clean, fresh dental stone life mask, he brushes on two coats of alginate dental separator and dries the surface with a hair dryer. He then sculpts his appliances with #2 plastalene as usual on the life mask, finishing the detail and the edges. He then coats the plastalene with a layer of clear plastic cap material.

The entire life mask and sculpture is then submerged in cold water for about an hour, at which time some of the plastalene sections may loosen from the life mask while others can be easily pried away. Large areas like the forehead may have to be cut in half to remove easily. These sections of plastalene are then carefully transferred to section positives. He recommends coating the section positive with petroleum jelly and then pressing on the sculpture and smoothing

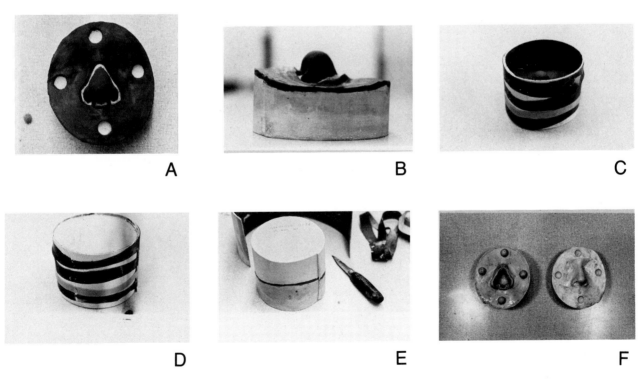

A   B   C

D   E   F

FIGURE 13.16 *Two-piece mold. (A) and (B) Two angles of an imbedded nose that has been sculpted in plastalene and the gutters made. Note that the nostrils have been employed as a run-off for the excess material when the nose is being made rather than the usual hole drilled through the positive on the tip of the nose. (C) The positive has been circled with a piece of inlaid linoleum and held together with rubber bands cut from an old auto inner tube. The inside of the linoleum has been greased with petroleum jelly. Sheet lead or rubber can also be used for this.*

(D) The negative portion poured with Ultracal 30. (E) The rubber bands and linoleum removed. A plaster knife is used to clean up any flash and to round the edges of the mold. The markings are made in pencil just before the Ultracal has set, giving the type of appliance, date, and if desired, a number for cataloging. (F) The finished two-part mold with the plastalene removed and the surface given two coats of clear lacquer to seal it. This mold will be used for casting foamed urethane noses.

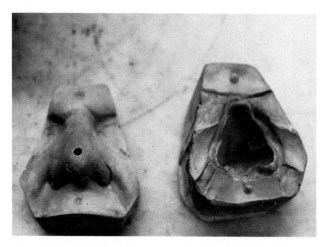

FIGURE 13.17   *A smaller foamed latex type nose mold for just a nose tip, showing the conventional escape hole bored into the tip of the positive. (Molds by Werner Keppler.)*

FIGURE 13.18   *Another Keppler mold for foamed latex for a nose tip but with channels cut for the excess rather than the hole as in the previous figure. See Chapter 14 for other two-piece molds.*

down the edges. The texture, edges, and minor repairs needed after the transfer can then be done as can the plastalene flashing for casting and the negatives poured up. In this manner one can get a better perspective of how the sculptured pieces will appear on the full face rather than just estimated on a section positive. RCMA makes a special heavy grade of alginate separator that is excellent for this use as well as for making acrylic teeth.

Molds for Teeth

Plastic teeth and caps that are pressure or heat cured are made in special metal flasks. However, for most make-up purposes, two-piece dental stone molds will suffice.

The first step in casting the mouth is to cast the upper or lower teeth (or both) to have a basic cast to work on. A dental supply house (see Appendix B) can furnish all the necessary materials and tools for this step as well as the materials to make the teeth. A set of impression plates, for both the upper and lower teeth, can be obtained in a number of sizes that will fit inside the mouth. They are available in plastic or metal. Also, it is a good idea to buy a set of rubber

dental base molds to make neat finished castings to work with.

For a cast of the upper teeth, a mix of quick-set dental alginate can be made and placed in the tray. This is inserted in the mouth and pressed up to cast the upper teeth. This type of alginate sets quite rapidly (2 to 3 minutes) so the cast can be removed as soon as the alginate is firm. A mix of dental stone should be then brushed into the negative alginate mold (while still in the tray) and then additional stone added to fill the cast. Many dental technicians employ a vibrator to ensure that the mold is well filled and packed with the stone and to dissipate any bubbles that might be formed. The rest of the stone mix should be poured into a rubber dental base mold. When the stone is of a plastic consistency, the tray is inverted onto the stone in the base mold and the join smoothed out.

When the stone has completely set, the tray with the alginate can be removed, with the alginate attached and the casting removed from the base mold. The edges can be trimmed with metal tools to form a neat casting. The lower teeth can be cast in the same manner using the lower teeth trays and the same rubber forms. As there are a number of methods for making acrylic teeth, the special molds for these are discussed in the section on tooth plastics in Chapter 14.

Latex and Plastic Appliances

AN INCREASING NUMBER OF COMPOUNDS AND MATERIALS are being used by make-up artists to produce appliances. The criterion is always that the prosthesis looks and behaves as natural as possible for the use intended and designed. Today we have casting latex to make slush or paint-in appliances, foamed latex products, foamed urethanes, gelatins, solid molded plastics, and waxes for constructions. This chapter covers each of these materials and their basic uses.

NATURAL RUBBER MATERIALS

Of all the compounds utilized for appliances, natural rubber is by far the most popular and useful. Commencing in the late 1930s make-up artists began experimenting with various forms of these products, starting with the brush-applied method into a plaster mold and later the use of the two-piece foamed latex materials.

Beyond the making of the molds and casts, the make-up artist now had to learn a bit of chemistry to make the mixes necessary for use and to understand the variety of solvents, polymers, colorants, and so forth that were part of the vocabulary of the make-up laboratory technician. Telephone calls to manufacturers, visits to the prosthetic clinics of the Veterans Administration, discussions with chemists, and finally, comparing notes with other make-up artists interested in these new and exciting directions were all part of the education of those who wanted to learn and master these challenging techniques. Times may change but avidity for learning should never diminish.

Latex

Latex is a milky white fluid that is produced by the cells of various seed plants (such as milkweed, spurge, and poppy families) and is the source of rubber, gutta-percha, chicle, and balata. Most *natural rubber* today comes from coagulating this juice and is a polymer of isoprene.

Synthetic rubberlike substances can be obtained by polymerization of some plastic materials and may also be called a *synthetic rubber,* or *synthetic latex.* The term *elastomer* means rubberlike, and there are many categories to which this term can be applied.

Natural rubber latex as employed by the make-up

professional usually refers to a pure gum latex that has no fillers and is of varying density. Some are thinner for use as balloon-type rubber (for making bladders and so forth) while others are thicker (like those employed for some foam latex formulas). When rubber latex has fillers, like zinc oxide, it can be used for *casting* as it will build in thickness when left in a slush mold or can be added on with painting-in methods. Some natural rubber latices are prevulcanized—that is, pretreated so they do not require a heat cure—while others require such heat treatment to cure them.

Synthetic rubbers and latices can be polysulfides, polyurethanes, acrylics, and so on. One thus should take care in defining a material as just *liquid latex* without explaining which type or grade it might be.

The hardness of the finished product can be determined with a Shore durometer. While a reading of 40 refers to approximately the hardness of automobile tire rubber, a reading of 10 is a much softer finished product—more what a make-up artist might wish to employ as a soft duplicating material.

For make-up purposes, three types of latex molding methods are employed: the slush molding method, the paint-in (or on) manner, and the foamed latex procedures cover the general use.

Grades of Latices

Many grades of latex compounds are available from a multitude of sources and can be employed for slush, paint-in, duplicating, or foamed latex.

Natural rubber latices that have no fillers dry with a yellowish translucence and an elastic resiliency such as might be found in elastic bands or balloons. The General Latex Corporation has 1-V-10 that can be used for dipping or making thin pieces. RCMA also supplies a pure gum latex that is excellent for bladder effects. These natural rubber latices are prevulcanized and have good stability, fast drying rates, good water resistance, and excellent aging properties and flexibility.

Slush molding latices have higher solids (due to the fillers) and higher viscosities. They can be easily colored with dyes or colloidal colors and are prevulcanized. RCMA supplies a casting latex for slush molding and brush coating. RCMA also supplies a casting filler that can be stirred into the casting latex to hasten the build-up in slush casting and to provide more viscosity for brush application (see Appendix B for other suppliers). For foamed latices and their components, see the section, "Foamed Latex," later in this chapter.

Coloration

Coloration of appliances made with latices can be done *intrinsically*—that is, with color added to the initial formulation before curing—or *extrinsically*—with color added over the finished appliance. Intrinsic coloration is best done with universal colors like those supplied by RCMA or paint stores, and a drop or so of burnt sienna color is usually sufficient to color a pint of latex as a light flesh tone. Other colors such as reds, ochres, and blues can be added for coloration, but it should be kept in mind that the color of the finished material is usually quite darker than what the liquid appears before curing. Testing on a plaster plate will show the end color before adding too much.

Dyes may also be used for coloration of water-based latices, and a 10 percent solution is usually sufficient to add drop by drop to achieve the best color. Unfilled latices will deepen in coloration considerably more than the filled types of slush or paint-in latices. Some workers use 30 percent dye solutions for intrinsically coloring a foamed latex so that the higher concentration of dye in water will not affect the water balance of the mix. Colloidal colors such as universal colors, however, are considered to be more manageable and more versatile than dyes for coloring most latex products.

Extrinsic coloration of appliances requires special vehicles for the color as some affect natural rubber or synthetic rubber appliances. Cake make-up foundations, creme-stick foundations, and a number of the creme-cake foundations in general use for make-up foundations on the skin, do *not* work as a *prosthetic base,* just as the old greasepaints did not either. The main reason is that they contain mineral oil, which has a tendency either to attack the rubber or to whiten out after application. (Note: All RCMA Color Process Foundations can be used with latex appliances.) Max Factor formerly made a product called Rubber Mask Greasepaint that was essentially a castor oil vehicle base that did not attack the rubber. However, other theatrical make-up companies have copied this foundation so it is available. Also, RCMA has devised two new varieties of foundations that are called Prosthetic Base (PB series) and Appliance Foundations that are superior to the old rubber mask greasepaint types. They also make a PB Powder that has more coverage than the regular RCMA No-Color Powder for use with these foundations (also see page 272). Powdering PB or AF Foundations with PB Powder will produce a more matte surface on an appliance than a No-Color Powder, so sometimes a light stipple of glycerin will restore a better surface halation to an appliance.

The RCMA AF or Appliance Foundation series have replaced the RCMA PB (Prosthetic Base) materials as the newer AF series are a thicker form of foundation from which the oils do not separate as is prevalent with most of the "rubber mask greasepaint" types, nor are they sticky and hard to blend as are some of the firmer ones.

The AF series is obtainable in the basic Color Wheel primary colors of Red, Yellow and Blue, that can be combined to form any of the other in-between or secondary colors. As well, it comes in White, Black, Brown (KN-5 shade) and the earth colors of Ochre-1624, Warm Ochre-3279, Burnt Sienna-2817, and Red Oxide-6205 which are some of the main basic color ingredients that are components of the majority of skin color foundation shades. In addition, RCMA makes a number of matched shades to their regular Color Process Foundations such as KW-2 and KM-2. Other useful shades are constantly being added to the line due to the wide acceptance of the AF type of make-up for a high-coverage use.

The AF series come in various sizes and also in kit form of six colors per tray which is great convenience for carrying in the makeup kit.

The AF series can be applied with brushes or sponges and provides a super-coverage of both skin and appliances made of foam latex or plastic as well as for slush or paint-in molded rubber.

The mixing of earth colors with the bright colors and the dilution with white will produce almost any shade of foundation. As most make-up artists who employ appliances in their work prefer to mix their own shades for the particular intended use, this provides an excellent basic method. The colors are easily spatulated together on a glass or plastic plate to the required shade. It is a good idea to mix more than is required so that one does not run out of a special color during the job at hand. The extra can be stored in a container for future use. It is also well to keep track of the mix by marking down the amounts of the basic materials used, such as so many spatula tips or spoonsfull of one shade to another in case additional material must be mixed.

The AF Kits are numbered as #1, #2, etc. #1 has 1624 (Ochre), 2817 (Burnt Sienna), 3279 (Warm Ochre), 6205 (Red Oxide), KW-2 and KM-2. Kit #2 has Red, Yellow, Blue, Black, White and Brown-KN-5. Other kits with regular matched Color Process shades are also being made. Special kits are available to order for specific use.

Although the viscosity of the AF series is high, its blendability and coverage is superior for any type of appliance.

A new concept of extrinsic coloration is being done with acrylic emulsion colors. Dick Smith devised a mixture of RCMA Prosthetic Adhesive B and acrylic paints that coats a foamed latex appliance with an excellent surface material. His basic formula is an equal mix of PA-B and Liquidtex tube colors. This he stipples on the appliance with the urethane stipple sponges after giving the appliance a stipple coat of PA-B and powdering. To lessen the halation, he might stipple on a coat of Liquidtex Matte Medium.

For temporary coloration (as it has a tendency to crack off with age), Craig Reardon painted the ET heads with universal colors mixed with regular rubber cement (Figure 14.1). Also, some firms make a paint for latex castings with xylol as the solvent. The latter seems to incorporate the color quite well into the latex but does not give as fleshlike a look as the Dick Smith mixture (he calls that PAX).

RCMA also makes a special line of flexible acrylic foamed latex or foamed urethane paint colors for make-up use, which are similar to the PAX material which can be painted or stippled on to the appliances. There are matting agents that can be added to produce a less shiny coating as well. A new series with FDA certified colors has also been developed by RCMA, called Appliance Paint (AP) series, and is available in colors matching the Appliance Foundation (AF) series.

pH Value

Natural rubber latices are suspensions in ammoniacal water, and the correct pH value for most systems is between 10 to 11. pH test papers can be obtained from Tri-ess Sciences Co. (see Appendix B). The excess ammonia loss of natural rubber latices can be corrected with a 2 percent solution of ammonium hydroxide in water, testing as one adds the solution to obtain the proper pH.

Fillers

Latices used for slush or paint-in casting can be adjusted for various degrees of stiffness or hardness of finished product by the addition of a filler. This filler can be prepared by adding 25 grams of zinc oxide to 100 cubic centimeters of distilled water, or it can be purchased from RCMA as Casting Filler. Depending upon the degree of opacity and stiffness required, anywhere from 1 to 10 percent of filler can be added by stirring into any prevulcanized latex for slush use.

Softeners

To make softer latex pieces, any latex compound can have a plasticizer added to it which consists of 400 grams of stearic acid, 500 grams of distilled water, 100 grams of oleic acid, 12 grams of potassium hydroxide, and 12 grams of ammonium hydroxide (28 percent). This mixture can be added 5 cubic centimeters (to 100 cubic centimeters of latex compound) at a time, testing a finished piece in between, to achieve the degree of softness required. Such products also affect the drying time of the latex as well as the durability of it.

Thickeners

Most natural and synthetic latices can be thickened to a soft, buttery consistency in a few minutes by stirring in Acrysol GS. This product is the sodium salt of an acrylic polymer and is supplied at 12 to 13 percent solids in water solution. For most applications, the use of 0.10 to 2 percent Acrysol GS (solids on latex solids) is adequate. Natural rubber has approximately 60 percent solids, and thus, 5 to 10 cubic centimeters of Acrysol GS added to 100 cubic centimeters of latex will thicken it considerably. This mix can then be spatulated into a mold to add bulk to a particular section for paint-in or slush molding. Note that thickeners retard the drying time of any latex compound.

As the Acrysol GS is a very heavy, viscous liquid, a diluted solution in water will work better for additions to latices. Therefore, if the Acrysol GS is diluted 50-50 in water, it will have half the strength of the stock solution and should be used accordingly. Tincture of green soap can also be employed as a latex compound thickener for some applications.

Latex Appliances

Slush or Slip Casting

Casting latex is poured into a slush-type mold and allowed to set for a period of 10 to 30 minutes, de-

FIGURE 14.1   *Craig Reardon with one of the ET heads that he painted.*

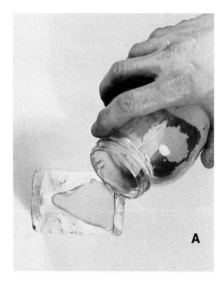
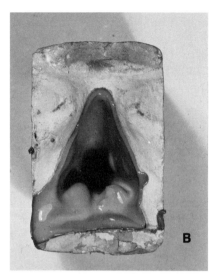

FIGURE 14.2   *Slush casting. (A) A nose mold is filled with casting latex and allowed to set until the buildup is about 1/16 to 1/8 inch. (B) The latex is poured out of the mold and the surface of the latex dried with a hand-held hair dryer. Unless oven cured in a low heat for about 1 hour, the mold should be left to air cure for about 8 hours after the latex has been thoroughly dried. (C) The nose is then removed and trimmed for use. Clown noses are usually made by this method as their edges do not* need to be thin for blending into the skin. As with all slush- or paint-in-type molds, to reuse them after they are dry (especially after oven curing), they should be rinsed in clean water before use. This restores some of the moisture in the mold and prevents the latex from adhering strongly to the plaster when the appliance is removed. Otherwise, no separating medium is required for latex-to-gypsum castings.

pending upon the thickness of the desired appliance (Figure 14.2). The excess latex is then poured back in the container and the mold turned upside down to drain. It can be force dried with a hand-held hair dryer, set under a large hair dryer, or placed in a low-heat oven. Drying time varies with the thickness of the prosthesis, but separation from the mold is relatively simple due to shrinkage of the latex. It should be noted that the appliance will have rather heavy edges and is unsuitable for blending into the face during application.

A slush cast nose is generally made for clowns as they will last longer than the delicate paint-in or foamed pieces. Full heads often found in joke and costume shops are made in this manner as well as full or partial appliances made for extras in a production where special make-ups are required but do not have to have any facial movement.

Paint-in or Brush Application Molding

This method consists of painting casting latex into the mold in successive overlapping coats with the thinnest being closest to the edge of the prosthesis. While a slush molded piece has a heavy ungraduated edge, the brush application method will allow the blending edge to be carefully controlled. This is quite necessary because a casting latex edge cannot be dissolved into the skin area as can many of the plastic types. Another advantage of a painted-in piece is it can be built up more heavily where it is necessary. For example, the alae and the bridge of the nose will require extra coats to produce an appliance that will hold the proper shape.

Sometimes workers add small pieces of paper towelling in to strengthen a piece. The towelling should always be torn into shape rather than cleanly cut to produce more graduated edges. It is then placed where desired with a brush and additional casting latex coated over the paper to make it a part of the prosthesis.

Medium-sized Chinese bristle brushes are used for this work by many technicians due to their size, configuration, and cost. Always work up a good lather of soapsuds on a cake of Ivory Soap with the brush and wipe it lightly before putting it into the casting latex. This procedure prevents the latex from solidifying or building up on the bristles of the brush and facilitates its cleaning in cold running water after use. It is a good practice to pour some casting latex in a 16-ounce wide mouth jar for painting use. The brush should always be left *in* the latex when not in use between coats so that it will not dry out. Leaving the latex-covered brush on a counter top for just a few minutes can ruin it for further use. Sometimes a coated brush can be salvaged by soaking it in RCMA Studio Brush Cleaner overnight.

Some appliances require special attention and painting—for example, Oriental eyelids where the lid area must be painted heavily enough to hold the proper shape while the upper portion that is attached to the skin must blend off in a thin coat. It is a good idea to have a series of eyelid molds (see Figure 13.13) and to paint up a sample of each for try-ons when a number of them are required. Then the eyelids can be individually fitted and the area to be painted in casting latex noted (as eyes tend to be different, some lids may be attached higher or lower on the frontal bone for the best effect).

Another case is the making of latex bald caps for extras or large casts. This method consists of pouring some casting latex into the mold cavity and turning the mold back and forth, with the latex being carried higher and higher each time and forming an edge. This procedure should be carried out carefully so that the leading edge around the forehead line of the hair receives only one or two such coatings of casting latex. Don't let the latex sit in the mold without this turning agitation because it will build in rings that might be apparent when the piece is dried out for use. Experience will show how long this procedure must be kept up to obtain a bald-effect cap that is both fine at the edges and heavy enough in the crown to hold its shape.

Pour off the excess out of the back portion of the mold, and drain fully before drying the surface with a hand-held hair dryer. Half an hour of drying is required generally, and then the mold can be set aside for another hour or so to cure fully. The cap can then be peeled off the mold, and to insure that it will retain its shape, it should be placed over a head form for about another hour so the surface can fully dry out. It is then ready for use (see mold shape in Figure 13.14).

Inflatable Bladder Effects

Many special effects transformations or illusions employ the use of some form of inflatable bladder whose effect is to ripple the surface of the skin to indicate violent changes taking place systemically (Figure 14.3). In essence, these bladders are inflatable plastic or latex balloons that can be controlled in size and flexibility by the introduction of air into them through fine plastic tubing. Such bladders may be concealed under surface appliances made of foamed latex, urethane, or plastic molding material so when they are inflated and/ or deflated, it appears that the surface of the skin is expanding as air is introduced into the bladder or contracting as the air is let out. As such, a rippling effect can be created [like that on William Hurt's arm in Dick Smith's make-up in the film *Altered States* (Warners, 1980) and Rick Baker's *Werewolf* (see Chapter 15)].

To make a simple bladder, pure gum latex can be used. Make a plaster flat plate (a good size would be 6 by 12 inches and about an inch thick for a permanent stock plate) with a smooth surface. Sketch an outline with a #2 lead pencil of the two sides of the bladder, and paint on three even coats of RCMA Pure Gum Latex, right to the edges of the outline, drying each coat thoroughly between with a hand-held hair dryer. This will normally give a sufficient thickness for the walls of the bladder, but larger-sized bladders can be made with additional coats for more strength.

Cut out a piece of heavy waxed paper, allowing about ½-inch clearance to the edge of the outline. On larger bladders, allow at least 3/4-inch (for an adhering edge). This waxed paper will delineate the inside dimension of the bladder, which of course can be made in many shapes.

Dust this waxed paper cutout with RCMA No-Color Powder on both sides, and lay it down on the latex-painted shape on one side. Take care not to powder the surface of the latex. Carefully peel up the other side of the bladder and fold it over to fit exactly the outline of the other side. Press the edges together firmly so that the latex will adhere to itself to form the two sides of the bladder. Strip off the other side of the bladder from the plaster, and trim the latex nozzle end to within about 1/8 inch from the waxed paper end.

To remove the waxed paper, push in a small rounded wooden modeling tool to force an entrance and then remove it. Insert in its place a drinking straw whose end has been dipped into No-Color Powder. Blow in the powder into the bladder cavity on each side. Then the waxed paper insert can be teased out with a pair of dental college pliers or tweezers. The bladder with a nozzle end is now complete. Clear plastic tubing can be obtained from a medical or chemical supply house and inserted into the nozzle of the bladder. This can be sealed in with Johnson & Johnson 1/2-inch Dermicel Clear Tape and then coated over with pure gum latex. The bladder is then ready to be attached and used.

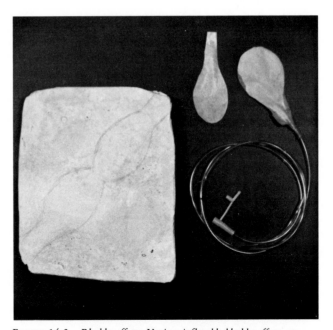

FIGURE 14.3 *Bladder effects. Various inflatable bladder effects are often seen in transformations. These can be made of PMC-724 (see page 225) or in pure gum latex. This is a plaster flat plate with an outline inscribed in pencil. A finished bladder is shown after having been attached to the tubing with Dermicel tape and a coat of pure gum latex as a sealer over it. In the center is the cutout wax paper used to keep the sides of the bladder from each other when it is folded over.*

Although the illustrated bladder has no excess edge for attachment to the skin, the 1/2- to 3/4-inch excess previously described during manufacture can be used for the adhesion area. RCMA Prosthetic Adhesive A is best for these bladders as it has excellent retention (see page 225 for other bladder uses).

Foamed Latex

The present-day ultimate in prosthetic appliance use for make-up purposes is the employment of foamed latex. It is unsurpassed for the best possible effects as well as adhesion to the skin. Since MGM's classic, *The Wizard of Oz* (1939), when Jack Dawn, then head of the make-up department, employed foamed appliances for the unforgettable characters of the Cowardly Lion, the Tin Man, and the Scarecrow as well as many others in the film, foamed latex has been the mainstay of tridimensional character work in motion pictures. In the early days of television in New York in the 1940s and 1950s, studios relied mainly on painted-in casting latex appliances for their low cost and rapidity of making them that was needed due to the lack of preparation time allowed to make-up artists for television shows. Today, foamed latex has become the standard for most appliances for professional make-up work.

Like many laboratory procedures that have been developed strictly in the make-up field, the parameters of the chemistry and mixing of foamed latex have been found to extend beyond the seeming restrictions of the expensive scientific or special equipment to ordinary, easily obtained ones (Figure 14.4). At first, special hand-made beaters, bowls, and laboratory ovens were considered to be required, but many labs today use kitchen-type Sunbeam Mixmasters with regular beaters and bowls, along with either standard electric kitchen ovens or the newer air convection table-top type. Only when full heads and bodies must be made are larger ovens required and restaurant-type Hobart mixers employed.

However, still some conditions do not vary much, and general room temperature and humidity do affect foaming procedures. The optimum is 68 to 72° F and a midrange of humidity. It is not difficult to maintain such with air conditioning and heating today. Higher temperatures will cause faster gelling and setting, while lower temperatures will extend the time. Successful foaming operations are impractical below 60° F.

Small quantities of ingredients can be weighed in plastic cups and the standard bowls of the Mixmaster used. Some artists mark the foaming volume on these bowls (five volumes is approximately correct for most appliances). All measurements given are for wet unfoamed ingredients, and a basic mix can often do quite a few small pieces. As it is rather impractical to mix very small amounts, most workers fill the molds that they are presently working on and, with the remaining material, fill some stock molds for extra pieces. Otherwise, much foamed product is wasted.

Wholesalers of foamed latex ingredients normally sell 5 gallons as a basic lot, but 1-quart sizes of both the three-part latex and the four-part Burman formula are available through RCMA. Some other suppliers will furnish 1-gallon lots. It should be noted that a lot consists of the measured amount of the basic latex material plus all the necessary chemicals for the foaming operation, although they are also available separately from most sources. European users will find a very soft and excellent foam available through Christopher Tucker in England (see Appendices B and G for other suppliers).

The Three-part Foamed Latex

1. Mix Parts A, B, and C thoroughly to be sure there is no material separated and caked on the bottom of the containers.
2. Combine 170 grams (6 ounces, or 3/4 cup) of Part A and 22 grams (1 tablespoon) of Part B in a mixing bowl.

FIGURE 14.4   *Mixing foamed latex. (Lab work by Werner Keppler.) (A) Paper cups used to measure ingredients by weight. (B) Beating foam in a kitchen mixer with regular beaters. (C) The latex mix beat to correct volume and refined.*

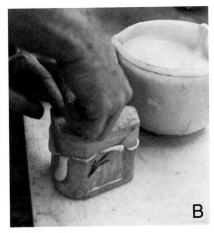
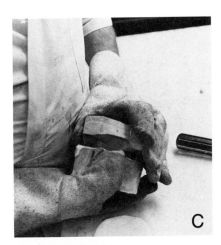
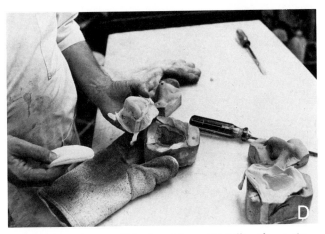

FIGURE 14.5   *A nose mold for foamed latex. (A) Filling the negative with a spoonful of the foam. (B) The positive set on and hand pressure applied. (C) After baking, the mold is separated. Gloves are used to hold the still hot mold. (D) The appliance removed. Flash or run-off can be seen. (E) Comparison with a file photo upon which the nose was based for duplication.*

3. Mix with electric mixer to three to six times the original volume, depending on the firmness desired.
4. Add 6 grams (1/2 teaspoon) of Part C, and mix until uniformly mixed (about 1 minute).
5. Pour or inject into your mold and allow to stand undisturbed until gelled. Place in oven and cure (bake) at 200° F for 4 to 5 hours. If the material gels (sets up) too fast, reduce the amount of Part C used gradually until you get the amount of working time you need. Normally it is not desirable to exceed 10 minutes after the addition of Part C.

However, one of the most knowledgeable lab technicians in California, Werner Keppler, who formerly headed the Universal Studio laboratory and is now at TBS in Burbank, has modified and simplified the foaming procedures, making small to large appliances with a minimum of difficulty. He prefers to use the three-part material and makes minor adjustments for each run if need be (Figures 14.5 and 14.6).

Using the regular Mixmaster bowls, Keppler's overall mixing time is 10 minutes, and his measurements are made in plastic cups. One cup of Part A (the latex)

in the small bowl is about the minimum, and this will do half a dozen noses, while four cups in the large bowl will suffice for a full head.

His molds are prepared by coating a fully dried out, cool mold with two coats of thinned-out clear lacquer. After this has dried, he coats them with a thin brushing of castor oil and then uses Mold Release prior to use. He also prefers the Kerr Toolstone (about 2 inches thick) with fibre (hemp casting fibre) reinforcement for making his molds.

To prepare for a run, Keppler lays out his molds side by side coated with Mold Release. The foamed ingredients are measured on a scale and poured into the mixing bowl along with a few drops of burnt sienna universal color to make slightly flesh-tinted appliances. If he decides that he wants softer than usual foam, he adds two drops of glycerin per cup of Part A. He then hand mixes the ingredients a bit and places them under the mixer, starting at a slow speed and gradually turning the speed wheel up to the desired one. Timing begins when mixing by machine starts.

With the small bowl, he mixes at a speed of 7 the measured ingredients Part A (latex) and Part B (curing agents) and with the large bowl, the speed is 12, for

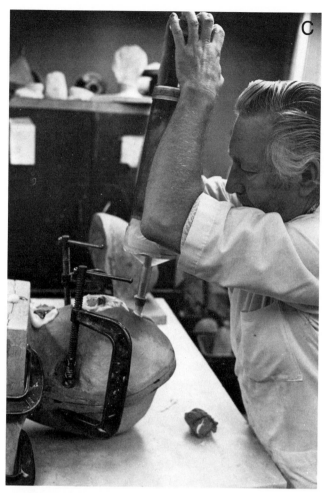

FIGURE 14.6   A full head mold for foamed latex. (A) Coating the negative with separator. (B) For a large mold with many surfaces, a special gun is filled with the foamed latex mix and injected into a hole drilled into the mold surface for this purpose. This particular mold is a three-piece affair with a front, back, and core all held securely together for casting with large screw clamps. (C) As the mold fills, two exit holes for the excess material are seen. As the foam bubbles through the one closest to the injection hole, it is stopped up with plastalene. The second hole has just been plugged with the clay material before removing the injection gun. (D) The injected molds are placed in the large oven for baking at the proper temperature and timing. (E) The cured appliance is removed from the oven and the front half of the mold opened and powdered.

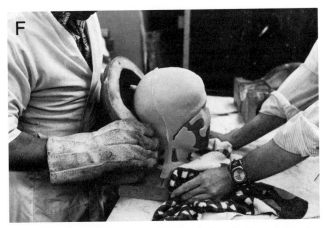

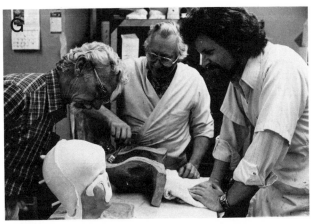

(F) The back half of the mold is removed as the latex head remains on the core section of the mold. Note that as the piece is a head and forehead appliance only, the core face portion has been cut away to remove any undercuts. Note also the excess foamed latex that has flowed into the prepared channels away from the front blending edge of the appliance.
(G) Leo Lotito, make-up department head of TBS Studios (LEFT), views the completed appliance with Terry Smith (RIGHT), the make-up designer for the production (The Last Star Fighter, Lorimar Productions, 1984). Note also the two approximately half-inch projections that are the injection

point and one of the excess flow holes of the mold. These will be removed, and the hair portion of the appliance will cover them. For a large piece like this, Werner often paints the inside of the mold with a coating of the latex used for the foaming procedure to give a skin to the surface of the appliance. Werner (CENTER) is seen recoating the molds with castor oil prior to allowing them to cool off slowly before another use. Large molds are covered with towels or blankets to slow cooling process. These molds were made of Kerr Toolstone.

4 1/2 minutes. He then tests the mix by dipping a kitchen knife into the mass. If the mix does not fall off the blade readily, he estimates that it is ready. He then turns the mixer to speed 1 to refine the foam for 3 1/2 to 4 minutes. He then adds the Part C (gelling agent) and beats for the remaining time (about 1 minute), stirring down to the bottom of the bowl with his knife blade to insure that the gelling agent is thoroughly mixed in. He also hand reverses the motion of the spinning bowl to aid in removing all the excess or large bubbles that might have formed. If the piece is a large one, he often adds the same amount of water to the gelling agent to slow down the gell time.

He then pours into his molds immediately and holds them shut until the excess material gells to the touch. Large molds are filled with an injection gun and are clamped shut during this procedure as they are filled.

Keppler bakes the molds in a 200°F oven for 2 1/2 hours for a small piece and full heads or large molds for 5 1/2 to 6 hours. On completion of bake time, the molds are removed from the oven with welders' gloves and taken apart. Small molds are opened and the piece is left out to air dry, while larger appliances are left on the positive side of the mold after being taken apart and replaced in the oven for about 1/2 hour so that their more extensive surfaces can fully dry out.

He then coats the molds, while hot, with some castor oil with a paint brush and sets them aside to cool off. Large molds are covered with towels to slowly dissipate the heat. The molds must cool off completely before another use.

He fills small holes in a finished foam appliance with a mixture of pure gum latex into which Cabosil

M-5 has been stirred to thicken it. Sometimes, when the head is a large one, he will coat the negative mold surface with a brush coat of Part A before pouring in the foam mixture, especially if the modelling is intricate and detailed on the finished piece. This "skins" the mold and becomes a part of the appliance during the baking.

Some additional notes on the three-part foamed latex are from Dick Smith, who uses it only occasionally since he seems to prefer the four-part latex mix. He does not recommend the use of the castor oil separator but does lacquer-coat his molds and uses a stearic-acid-type mold release. When beating the foam he sets his mixer on the highest speed for about 2 minutes so that the volume of the mix will be increased to about four volumes. Three volumes will make a firm foam, four volumes a soft foam, and five volumes a very soft piece. He then turns the machine down to a speed of about 3 and refines the foam for 3 minutes. This way the volume remains about the same, but the foam cell structure becomes finer.

He measures out the gelling agent with a spoon that has been waxed (the gelling agent won't stick to a waxed surface) or a small receptacle such as a cut-off paper cup that has been coated with a paste wax. A level teaspoon of gelling agent is about 8 grams, which he normally uses. If the foam mixture doesn't gel within 15 minutes after the mix is stopped, it will start to break down and collapse the foam. A good average time is between 5 and 10 minutes for gellation, when the room temperature is in the mid-seventies.

He also mixes his color in the gelling agent, using casein artist tube colors slightly thinned with water.

A light flesh tone is provided with a teaspoon of color with 1/4 teaspoon of water for a standard mix (using burnt sienna as the colorant).

Like Keppler, Smith uses a spoon to scoop out the foam from the mixing bowl and pours it carefully into the mold, taking care not to entrap any bubbles. If the negative has deep texture or pits, the foam mix should be poked into these with a pointed wooden modelling tool. Special care should be taken with nose tips.

Smith cautions that too much foam should not be placed in a small mold as it will make it too difficult to press completely closed. He recommends a preheated oven of 210° F. With a mold about 4 inches high, 3 hours' baking time should be sufficient. When the foam is believed to be cured, the mold should be removed from the oven and gently pried open. The foam should be poked in an inconspicuous spot (such as an overflow area). If it stays indented, it is not baked enough. Leave the piece on the positive half of the mold, and return it to the oven for another half-hour. Test it again before removing completely. He also recommends that all molds be wrapped or covered in old bath towels to prevent rapid cooling or cracking.

Foam can be made softer to the touch by adding 3 grams of Nopco 1444-B from Diamond Shamrock Chemical Company to the 22 grams of Part B and then adding it to Part A. If you want to beat to volume 5, also add 10 grams of water to the ingredients before beating.

As one can readily see, foamed latex mixtures can be varied by different technicians to achieve what they consider to be the most efficient or simplest method of use. Certainly there is little contradiction in the methods, and they show the versatility of the material as well as slightly different approaches to achieving the same end result.

Most lab technicians that employ dental or tool stone molds will coat them when hot with castor oil, but this method should not be employed for Ultracal molds as it often leaves the edges of the appliances gummy. Use only the Mold Separator with the latter after the mold has cooled off and before the next use.

The Four-part Foamed Latex
Some basic differences in the three-part and the four-part foamed latex procedures are noted here.

Sponge rubber will cure properly only in a *thoroughly dry* mold. When one mixes gypsum products a large amount of water is used. Some of this water goes into the molecular change when the gypsum sets up, but the rest has to be removed from the mold before it is suitable for molding sponge rubber. Separate the two halves of the mold, and bake at 200° F for 4 to 6 hours if a small mold and from 10 to 12 hours if larger.

Once the mold is dry, it should not be heated above 200° F as the water of crystallization will leave and the mold material will break down. Mold separator should be applied in a *thin* coat to both halves of the dry mold only where the modelling occurred. Do not allow this material to get into the keys. When the mold is thoroughly cold and the separator has dried, the sponge can be poured in.

Room temperature and humidity have a great bearing on the operations and results. With an optimum room temperature of 68 to 72° F, it may be necessary to vary the amount of gelling agent from between 9 and 16 grams to have the foam jell in about 10 minutes.

Using the deluxe model of the Mixmaster with the small bowl, the beater speeds and times should be carefully noted for each batch. (Note that some other kitchen mixer models usually beat too fast at their slowest speeds and so are unsuitable.) As the latex is stabilized with ammonia, the recommended speeds will assure that sufficient remains in the mix so that it will gell with 14 grams of gelling agent. If it is beaten faster or to too high a volume or the temperature of the latex or air is too warm, the mix will gell too fast—even during the mixing process. Conversely, if the latex or air is too cold, 15 minutes of beating time will not remove enough ammonia and the foam will not gell in 30 or more minutes. If, during the final refining process of slow beating, the speed of the beaters is too fast, a nice fine foam cannot result and the mix will set up too fast. Also, with some other beaters that do not fit the contours of the bowl correctly, poor mixing of ingredients will result.

To insure a very slow low speed, one can connect a light dimmer fixture (or studio light dimmer) in series with the Mixmaster. The controls of the Mixmaster are then set at 12 (the highest speed) and all the speeds set up with the dimmer. One must carefully calibrate the dimmer switch with markings to indicate the speeds of 1, 5, and 6 for correct operations. This way, the Mixmaster runs much cooler and quieter, and infinite speed control is possible.

Although the latex is simply stirred carefully to remove any lumps, the three other components—curing agent, foaming agent, and gelling agent—should be thoroughly shaken up just before weighing them; otherwise the heavier ingredients will settle to the bottom while the lighter ones will be poured off first. Thus, the first batch mixed and the last will have the same mix if they are properly shaken. Otherwise the formula will vary considerably—and so will the foaming operation. The four-part latex is considerably thicker than the three-part one, so turning the containers every 2 weeks on the shelf is quite important to avoid settling and extend shelf life.

To prepare the foam mix, pour 12 grams of curing agent into the mixing bowl and add 30 grams of foaming agent and 150 grams of the latex, and mix thoroughly with a rubber spatula. Adjust the beaters so they fit to the sides of the bowl and then:

Begin the mixing and foaming cycle.
1 minute at speed #2 (mixing)
7 minutes at speed #8 (whipping)
4 minutes at speed #4 (refining)
2 minutes at speed #2 (ultra-refining)

Then add 14 grams of gelling agent (shake bottle first!) taking 30 seconds to do so and staying at speed #2. Finally, continue mixing at speed #2 for 1 more minute, turning the bowl back against the beating action to assure a complete mix. This gives a total beating time of 15½ minutes. (Timings are for 70°F.)

The mix can then be poured into the prepared molds. Roll the molds around to cover the necessary surfaces, then slowly set on the other half of the mold and add weights to close it tightly. The remaining foam can be poured on a glass or formica sheet where it should set to a solid mass in 10 to 30 minutes. When you can press it down with a finger and a permanent indentation is formed, you may put the molds in the preheated oven and *bake 5 1/2 hours at 200° F.* When the time has elapsed, the now vulcanized piece can be removed from the molds and the molds put back into the oven to cool slowly. Don't forget to turn the oven off! To prolong the shelf life of the latex, it should be kept as close to 70° F as possible.

Those who have used this formula and material for some years note that a new mold should be first prepared for use by spreading castor oil on both sides to clean it and to prevent the latex chemicals from attacking the mold. A new mold should be given a good soaking coat of castor oil, and after each use, while the mold is still warm after being removed from the oven, it should again be coated with oil and then wiped with a clean towel. These latter directions preclude leaving the molds in the still-hot, turned-off oven to cool. Instead, after coating the hot molds with the oil, set them aside wrapped in towels to cool. After the mold has thoroughly cooled, and before the next use, both sides of the mold should be covered with the recommended mold separator for foamed latex.

Another suggestion is to add the color directly to the latex, tinting a gallon at a time, and then storing the latex in 1/2-gallon brown glass wide-mouth bottles until ready for use.

Most workers agree that overcuring in an oven of no more than 200° F will neither ruin the appliance nor the mold, but careful watch should still be paid to *all* timings for foamed products. Large-sized batches might require double the beating time—for example, batches that cannot be handled in the large Mixmaster bowl and that require the large restaurant kitchen-type Hobart mixers.

Regarding the injection by gun of large molds, breathing holes or exit holes placed along the line of the injection flow are necessary, and as the foam begins to escape from each successively, they can be plugged with a blob of plastalene until the mold is filled and the material has gelled. These blobs of clay can be removed before the mold is placed in the hot oven.

Carl Fullerton, who often works with Dick Smith, considers the four-part foam to be superior and experiments constantly with it to get better results. In a move to save the molds from calcining due to excessive heat, as well as the shock of going to and from a hot oven, he does not preheat his oven, but turns it on just before putting his filled molds into the oven. In this way, the cold mold goes into a relatively cold oven. He turns the oven up to 200° F, or just under sometimes, and bakes for only 1 to 1 1/2 hours. He then turns off the oven and leaves the mold to cool for about 3 hours. At that time both the oven and the mold should be cooled off. He then takes the mold out and separates it. His theory is that the shorter baking time not only saves the mold but also makes a stronger appliance. Again, personal experimentation will allow the lab technician to devise individual methods and techniques based on the experiences of others as well as researching new ideas.

Some of the main differences between the two systems, the three- and four-part formulas, is the lacquer sealing of the molds for the three-part foam that is not recommended for the four-part method; the amount of beating time—for example, only 10 minutes for the three-part formula and 15½ minutes for the four-part one; the speed recommendations for each; the baking timings; and, of course, the basic chemistry of each formulation. As well, the three-part materials carry a somewhat greater latitude for storage as well as use temperature than the other. Nevertheless, in all cases, the final product is the ultimate criterion, and whichever one the make-up artist decides upon to use, he or she might also be smart to test the other system to have a means of comparison on a particular project. Also, although simplification of any manufacturing process leads to faster and less complicated procedures, the final outcome should not suffer unless the compromise will accomplish the same result.

In this vein, the newer flexible polyurethane foams might be considered. Certainly, with more experimentation and possibly even more improved products, these materials will gain more favor and use by make-up artist technicians as their two-part formula does not require any beating with a machine or baking in an oven as well as the piece being completed in less than an hour from the start of the entire operation.

Repairing and Cleaning Foamed Latex Appliances
When an appliance first comes out of a mold, it may have small defects (especially in a large section or a full headpiece). These can be repaired by mixing some pure gum latex with Cabosil M-5 (see page 210), or

the inside can sometimes be strengthened by brush coating in a thin piece of netting or nylon stocking.

Rips or tears made when the mask or large piece is removed from the performer can sometimes be repaired by using Prosthetic Adhesive A on each side of the tear, drying a few minutes, and then contacting the sides for adhesion to take place. Then both surfaces, inner and outer, can be stippled lightly with some pure gum latex to disguise the line.

Normally, any foamed latex appliance can be cleaned of most adhesives and make-up coloration by immersing it in acetone and squeezing the solvent through the foam. Two baths will normally take out all the color and adhesion, but the piece will shrink. Shrinkage can be somewhat controlled by a method recommended by Dick Smith in which he adds 1 part of light mineral oil to 6 parts of Ivory Liquid or Joy detergent in a bowl large enough to immerse the piece. The liquid should be squeezed through the foam until it is saturated—about 2 minutes. Then rinse out the detergent from the appliance with about 10 to 20 rinses of cold water. The foam should have expanded about 10 percent. If the detergent mix is decreased to about 4 parts with one of mineral oil, this results in about a 20 percent enlargement. The appliance can then be placed in a warm oven to dry. He also states that this procedure is not exact and to make tests before trying it on a valuable appliance. Generally, such methods may be employed when a fresh prosthesis is not available but are not normal practice.

Basic Procedures for Handling Latex Appliances

As both natural and synthetic latices will stick to each other and the edges may curl and be difficult to straighten out, it is standard procedure to powder any appliance before removing it from any mold. This also can be said for plastic caps, plastic molding materials, and the like. RCMA No-Color Powder or unperfumed talc can be used because they will not impart any color to the appliance. Also, when the prosthesis is removed from the mold, it is a good practice to powder the outside or face of it as well as the back.

PLASTIC MATERIALS

Plastics are widely used by make-up lab technicians for a variety of uses and appliances. For such they can be divided into foamed urethanes, solid urethanes, cap materials, molding materials, tooth plastics, special construction plastics, and gelatin materials.

Foamed Urethanes

A number of these products see increasing employment in the lab. The semiflexible types can be used for filling or supporting certain molds or constructions, while the flexible types can replace some foamed latex applications.

FIGURE 14.7 *Foamed urethane heads. (A) A commercial plastisol full head form* (LEFT) *from a beauty supply firm (RMCA can furnish these) that is suitable for some applications. It fits on a standard adjustable wig block stand that can be clamped to a table top.* (RIGHT:) *The latex filled-with-foam heads made by Werner Keppler for beard laying (see Chapter 15). (B) A self-skinning type of urethane foam head made by Gary Boham with a plastic PVC pipe insert that will also fit on the wig block stand.*

Semiflexible Types

What industrial chemists think is a *flexible* type of urethane may not be quite flexible enough for the make-up artist. *Rigid* urethanes, in the strict sense, are those with little or no flexible qualities, but the semiflexible types do have *flex* to them but might not be as resilient to the touch as what a make-up artist might class the foamed or sponge rubber appliances. There are, however, some new skinlike urethanes with which we will deal later.

One product we might put in the semiflexible class is made by the Hastings Plastics Company and is excellent as a filler product when a slush molded item must be made that will hold its shape. Werner Keppler has made some heads using this technique that are suitable for laying on premade beards (Figure 14.7). Using a face cast in casting plaster, he cut out all the undercuts around the nose and removed part of the forehead area, only keeping intact the area of the face where the beard normally grows. The ears were also

FIGURE 14.8 (opposite page) *Making a foamed urethane face block. (Photos and lab work courtesy Gary Boham.) (A)* (TOP) *The pipe sections are shown.* (MIDDLE) *The elbow covered with fiber glass material.* (BOTTOM) *The finished support with the extended elbow finished to fit a wig block stand. (B) A plywood board with the gasket channels cut and the holes drilled for the support rod and for pouring. (C) The plaster mold sealed to the board with plastalene. (D) RTV 700 flexible mold is made. The keys shown at the sides were made of cut foamed urethane and sealed into the flexible mold with the material. (E) The base of the plaster mother mold must be flat so that it will set straight upon inversion to make the heads. Rigidity is maintained with the particle board box. (F) Placement of the support rod. (G) IASCO two-part self-skinning flexible urethane foam and the flesh colorant material. (H) The mixed foam material is poured into the closed mold. Note the positioning of the clamped pipe support and the stage weights. (I) The foam having set, the mold is opened. (J) The flexible duplicating mold is stripped from the foam head still attached to the board. (K) The foam head pulled away from the board. The finished head can be seen in Figure 14.7(B).*

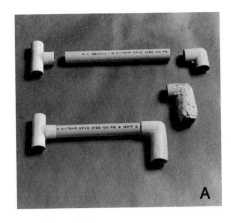

A

B

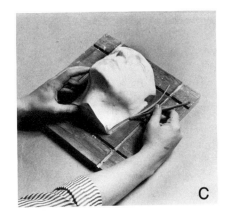

C

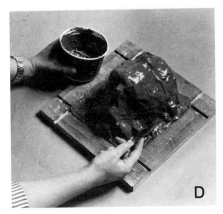

D

E

F

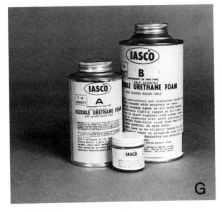

G

H

I

J

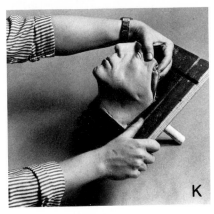

K

cut down, leaving only an indication so that sideburns can be laid in the correct area. He then made a negative mold of this face in casting plaster about 2 inches thick. Separating the two, he had a negative of the face. Into this he poured a lightly filled casting latex and let it build up to about 1/8 inch in thickness. He then poured out the latex and *almost* dried the slush piece (so that it would not shrink too much). At this point, he made a mix of the Hathane polyurethane 1640C-54 and filled the cavity. As it foamed up, he laid a silicon-greased board and weight over the top of the mold so that the foam would not spread but be confined. It is necessary to determine exactly how much foam mixture to add as it does expand. Also, the mold must be so made that it has a flat top so that the board can be put on top of it. When the foam has set, the head can be removed and a hole drilled up through the bottom and fitted with a plastic tube so that the form can be set on a head form support (see head form in use in Figure 17.2). This particular polyurethane comes in a range of densities from 4 pounds per cubic foot to 16 pounds per cubic foot. Many companies make numerous varieties of this type of foaming product (see Appendix B).

Gary Boham makes his beard block as well as full heads for ventilation hair with a flexible self-skinning urethane foam that has a 5-pound density and sets rapidly (Figure 14.8). The support rod is made of 3/4-inch PVC tubing. A T-connector is cemented to one end of the pipe and imbedded in the foam during manufacture of the head. A special elbow joint is made by using a regular PVC joint covered with fibreglass resin. The tapered portion of a wig clamp is covered with masking tape to serve as a release agent and then inserted into the open end of the 90° 3/4-inch elbow joint. A thin layer of fiberglass resin is applied and then wrapped with fiberglass cloth or tape. The continuous-strand type gives slightly more strength than the cloth. After curing, the wig stand can be removed by twisting and pulling and the completed fiberglass connector sanded as necessary.

The head form can be any ear-to-ear cast of a male head of average size that is either cut to shape and flattened on the top and bottom portions to form a block of the desired size. Although the one illustrated in Figure 14.7B does not include the forehead, it is a good idea to do so if the block is to be used to make any eyebrows. Another way is to make a section mold of a face cast using alginate for duplication (if one does not wish to ruin a good original face casting).

A wooden base is made as illustrated in Figure 14.8B with plywood or 3/4-inch particle board that is at least 3 inches longer and wider than the plaster mold. Center the mold on the board and trace its outline. Using a router or table saw, cut gasket channels all around the four sides of the mold, 1 inch away from the widest portions. These channels should be about 1/4 inch wide

and 1/4 inch deep. One inch below the center of where the mold will fit on the board, drill a hole about 1/4 inch larger than the outside dimension of the plastic pipe to serve as a support rod for the head form. Then 1 inch up from this hole, drill a 1 1/2-inch hole for pouring in the foam and to allow an exit hole for the excess during the foaming procedure. Seal both sides of the board with two coats of a polyurethane lacquer. Then center the head on the board, and seal down with plastalene.

To make a flexible duplicating mold, a number of products can be used, but the one recommended here is GE's RTV 700 with Beta 2 curing agent (see Appendix B for suppliers). The surface of the original plaster mold should be sealed with lacquer before use in most cases. This is a high elongation and tear strength material with a variety of curing agents available for room temperature use. No heating is required. An amount of 10 parts of base to 1 part of curing agent is recommended, and mixing is done by hand using a spatula or paint stirrer. Avoid rapid stirring so as not to induce bubbles into the mass, and mix for about a minute; then apply to the mold with a spatula. Any bubbles should be broken by passing the spatula over the surface. The pour time is about 1/2 hour and the work time about an hour more. Allow to stand at room temperature until the mass has completely set, but with this curing agent, a minimum of 3 hours is required. Take care that the molding material extends just beyond the gasket channel. A minimum thickness of about 1/4 inch is desirable so add additional layers as required. The edges may be reinforced with gauze to prevent any tearing when removed from the mold. Make certain to shake the curing agent well before mixing it into the base. The finished mold has a Shore A Durometer reading of 30 and an elongation of about 400 percent. Shrinkage for this use is minimal. A three-piece mother mold of plaster is recommended, and it can be confined in a box made of particle board as shown in Figure 14.8E. The flexible mold is stripped from the plaster head, the mother mold is inverted, and the flexible mold is set in. Keys can be made to ensure proper and correct fitting. The wooden portions of the mother mold, the base board, and all tools can be coated with paste wax for a separating medium, as well as the surface of the flexible duplicating mold. Trewax is a good product for this.

The support rod is placed inside the negative flexible mold and passed through the board, which is carefully fitted into the gasket portion of the mold. The correct height of the plastic pipe can be maintained with a spring clamp as shown (Figure 14.8H), lifting it up about an inch from the surface of the flexible mold to be embedded there. Stage weights (wrapped in plastic wrap as a separator) can be employed to hold down the spring clamp and board during the foaming procedure or a clamp arrangement can be devised.

The foam material used was IASCO's two part self-skinning flexible urethane foam using 240 grams of Part B to 80 grams of Part A. This can be tinted with fleshtone plastisol pigment if desired, mixing it into the Part B before adding the Part A. Mixing time is about 20 seconds before the foam will start to rise when the material should be poured into the large hole of the cast. Expansion and exotherm action start immediately, so steady the support pipe and apply some body weight to the mold to aid in weighting it.

The foam will set in about 1 minute, and the mold can be opened in 5 minutes. The flexible mold can be removed and the head stripped from the base board. The excess can be trimmed with a sharp knife or shears. The block can be cleaned with acetone to remove any excess wax, and the special elbow joint can be glued with PVC cement to the end of the pipe support. The length of this pipe support can be varied to suit, but 6 to 8 inches is normally long enough.

With experimentation, full heads, prosthetic arms or legs, and other items can be made with this same self-skinning foam. Imbedding wire forms in it is also possible to govern the position of appendages. A burnt sienna shade of colorant will be found to be useful for most Caucasoid appliances, while Negroid or Oriental skins can be simulated by the addition of a burnt umber or ochre shades, respectively. This foam has excellent solvent resistance for normal usage and is rather easy to use overall.

Flexible Polyurethane Foam

A unique material is made by BJB Enterprises called TC-274 A/B, which is a two-component flexible foam system specifically developed for low density molding, and like most of these polyurethane foams, it is just a matter of mixing A with B, noting how much time is allowed to mix, then how long it takes to foam, and finally, how long the cure time is to complete the project. This particular material has a density range about 3.5 to 4 pounds per cubic foot, but it is very soft. It has a cream time of 90 seconds at 75° F and a cure time of 15 to 20 minutes at room temperature, depending upon part size and cross section. It offers low oral, skin, and eye toxicity; low vapor pressure; and good storage stability. Shelf life is six months at room temperature, but may last considerably longer.

For molding, a regular two-piece mold similar to that employed for sponge rubber is used, although one can make it a trifle heavier to last longer (Figure 14.9). The mold should be prepared by coating it with a polyurethane lacquer to seal the gypsum material. It should then be warmed slightly and coated with BJB Mold Release #86 or RCMA MR-8 (which is a wax in solvent) two or three times, drying in between coats. A single nose will take about 8 grams of Part B and 2 grams of Part A mixed and then poured into the negative. The positive is then pressed on and the mold

clamped for about 20 minutes. Test the overflow foam with the finger, and keep clamped until the foam is not tacky any longer. The molds can then be separated and the piece removed for immediate use.

Considerable research has been done on this product

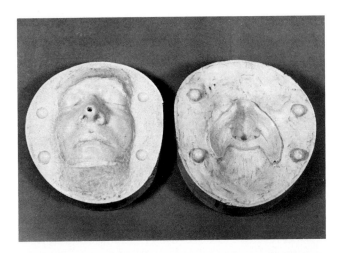

FIGURE 14.9 *Flexible urethane appliance.* TOP: *A two-part mold for making flexible urethane appliances. Note that it is constructed in a heavier form than the normal foamed latex molds for a similar-sized appliance.* MIDDLE: *The male portion of the mold.* BOTTOM: *A finished piece showing the fine edges possible with this versatile material.*

by Werner Keppler and David Quashnick of California, and some very fine foamed work has been done with it. The Part B can be intrinsically colored with universal colors or RCMA Color Tint Light (1 drop per 10-gram mixture) before mixing. It will be noticed that the manufacturer recommends a 4-to-1 (B to A) mix, while experimentation may prove that a 5-to-1 mix will produce a softer product more suitable for make-up use.

Werner Keppler makes gang nose molds for the TC-274 mix with five different noses (Figure 14.10). It takes a 30-gram B and 6-gram A mixture just to fill the cavities and give sufficient overflow. These appliances can be attached and made up like any other. The TC-274 A/B polyurethane foam material is available from RCMA as is the MR-86 and MR-8 Mold Release.

Solid Polyurethane Elastomers

It is sometimes convenient to make a core mold or an item in a more solid compound that has some elastomeric qualities. BJB makes a multipurpose elastomer numbered TC-430 A/B that is suitable for a variety of applications (available from RCMA in small quantities). This material has a high tear strength, ability to accept coloration, and being odorless, a high acceptance in the medical prosthetic industry. It can be used to make both positive casts and negative molds, and most thermoset liquid plastics and gypsums can be readily cast in molds of TC-430 A/B.

TC-430 A/B is available in two versions: TC-430 A/B for casting and TC-430 A/B-10 as a brushable grade. The standard versions of these are white, but they also come in clear (which yellows with aging).

Equal weights of the two parts are mixed together and have a work life of 30 minutes at 75° F in a 100-gram mass. Cure time is 24 hours at room temperature and 48 hours for full properties. Heat cure of 200° F only accelerates demold time, which is 4 hours at room temperature and 2 hours at 200° F. Again, good ventilation is recommended during its use and avoid skin contact with the liquid materials.

To increase the hardness, use 60 percent Part A and 40 percent Part B. To make the end product softer, use 40 percent Part A and 60 percent Part B. This will make a difference of ± 10 durometer hardness.

When TC-430 A/B is used to reproduce molds from gypsum or other porous materials, a silicon release #1711 is recommended. Shrinkage is very low and the material has good tensile strength and elongation, plus with a 50/50 ratio mix, the durometer A hardness is 50 to 55. A note about this durometer A reading: This is a means of comparing the relative hardness of a mass with a scale of 0 to 100. Soft, very flexible materials will be 6 to 10, while a stiffer product (such as auto tire) would be 40 to 50 (also see page 187 for further information).

Some general basic notes on urethanes of all varieties will help those who handle these materials. Storage in

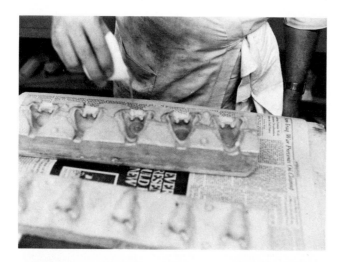

FIGURE 14.10   *A gang mold for noses.* TOP LEFT: *Werner Keppler pours flexible urethane into a gang mold.* BOTTOM LEFT: *The mold is clamped shut. Note the excess coming out of the overflow holes, which were made in the nostril area of the noses.* ABOVE: *The separated mold, showing the noses in a row still in the negative portion of the mold.*

an area between 70° and 90° F is best. During colder weather the resin should be inspected to assure that there is no crystallization. If the resin appears cloudy or the hardener becomes gummy, the component should be heated to 120° to 160° F and stirred until the material returns to its proper smooth liquid consistency.

Use only metal or plastic mixing containers and spatulas as paper cups and wood stir sticks have been known to contaminate the ingredients as they are porous and can absorb moisture in storage.

Weigh all parts accurately and always mix thoroughly. Polyethylene mixing containers are good because the cured urethane will not adhere to them.

Molds should always be quite dry and made nonporous. A slightly warmed mold is preferred. The sealing process must be done with a material that will withstand the exotherm generated when the urethane cures and must not melt at the peak temperature.

There will be some shrinkage as in all rubber or urethane materials, but this depends greatly upon the thickness (cross section) and configuration of the casting.

Always use any urethane with adequate ventilation, and avoid skin contact with the uncured ingredients. Protective barrier creams are recommended for the hands, and brushes and equipment may be cleaned with MEK if the solid urethanes are being mixed and have not hardened. General cleaning can be done with soap and water or a 1-to-1 mixture of toluene and isopropyl alcohol.

RCMA can furnish 8-ounce quantities of both foamed and solid urethane materials for laboratory work.

Cap Material

Bald front wigs and fully bald head appliances have been made with a number of materials and in a variety of ways, but the only really good bald cap material is made from a vinyl resin of the Union Carbide Corporation called VYNS. There are a number of formulas in use employing MEK or acetone (or a mixture of the two) as a solvent plus a plasticizer to control the degree of desired softness. A basic formulation is 25 grams of VYNS to 75 grams of MEK and adding 10 to 25 grams of plasticizer.

There are many plasticizers on the market, among them, dibutylphthalate (DBP) and dioctylphthalate (DOP). The Monsanto Company also makes a series of plasticizers called *Santicizers,* and their #160 is used where DBP is called for and #711 for DOP. The DOP will retain the soft quality of a cap material and be less affected by ultraviolet than DBP.

After dissolving all the solid in the liquids, cap material can be tinted by using some of the RCMA PB Foundation shaken in. To make a bald cap one requires a positive mold made from plaster, stone, metal, or even fiberglass. One of the production methods to make thin caps is to spray the cap material with a Pasche L head airbrush with a medium (#3) nozzle. John Chambers sprays about five heads at a time, one after the other, with about six coats of spray. In this way, the first head sprayed will have just about set when the second coating is applied after spraying head number 5 in the sequence. Careful guiding of the spray, plus a rather rough final coating spray is best as the caps, when removed from the molds, will not be reversed. As such, head forms for spraying are made quite smooth for ease in removal of the plastic cap from the form.

A more tedious method, but one that allows full control over the thinness of the blending edge wherever one wants it and a heavier weight of material on the hairline or the pate, is done by painting the plastic cap material on a head form that has a slight pore effect on the surface. When completed, this cap will be reversed to show this pored surface.

The best way to make caps is to have a ventilated spray hood for either the spraying or the painting due to the fumes of the solvents. Otherwise, this operation should be carried out in the open air on a calm day or with a fan blowing the fumes out a window. The cap plastic can be brushed on with a 2 1/2-inch oxhair or sable hair brush in a flowing manner rather than brushed back and forth the way one would paint. The brush should always be well wet with material as it will evaporate very rapidly. Most metal molds do not require any separating medium, but others should be lightly rubbed with a silicon mold release and wiped almost dry for ease in separation. Head molds can be lightly marked in pencil in the general shape of the hairline so that a reference point is available when graduating the coats to make a fine blending edge on the cap.

For a full, regular thickness cap, brush on 5 coats over the entire head form. Then brush on 25 coats just up to the hairline marking, graduating the coats to form a thin edge. Finish off with a few more coats over the entire head. Coats should take about 10 minutes to dry between each one, so again, doing a number of heads one after the other will save overall time.

Always remove any stray hairs immediately if they fall out of the brush while painting the cap on the mold. A small sable brush is best for this operation. The same precautions about dust and dirt in the air should be observed about painting caps as with any fine varnish or lacquer. Never sweep the room or in other ways disturb any settled dust while caps are being sprayed or painted as dirt specks will ruin the caps.

Allow the cap to remain on the mold overnight to air cure. Caps can also be cured and much shrinkage prevented by placing the finished cap on the mold in an oven at 150° F for 5 minutes.

To remove the cap from the mold, cut a line with a sharp knife or razor blade around the base of the head mold through the plastic material, and carefully strip the cap from the mold. To prevent edge turnover or the cap sticking to itself, a good powdering before removing the cap from the mold and on the other side after removal is best (No-Color powder is good for this). Some very shiny (chrome-plated metal) molds need only the edge to be started and then the cap can be easily rolled off.

Another way of making a partial bald cap is to use a try-on cap on the performer and carefully mark out the hairline that must be covered and where the hairpiece is to be added on top of the bald cap area (see Figures 9.11 and 14.20 for the use of this type of cap). As the cap will be made quite heavy, it is often not necessary to make a blending edge in the back of the neck area as it will not be adhered there. The top or pate area can be made quite heavy with 45 to 60 coats, depending upon the desired weight. Also, when the cap is made for one particular performer, the edge can be painted with a finely graduated painting that becomes heavier and thicker more quickly than on a stock cap.

Cap material can also be used to make thin sections that will be used to cover eyebrows. Generally these can be made by painting cap material on a flat plate and building edges and sufficient weight for covering the hairs. Some brows will require heavier buildups than others to cover the coarse hairs (also see page 256).

Clear cap material is also employed by some sculptors to cover their plastalene modelling. Additional sculpture can still be done over the coating, or fine lines can be refined (see page 173). RCMA supplies two grades of cap material: one tinted for cap use and the other clear for sculpture. It is fast drying and can be thinned with acetone if it thickens during use. For a slower set, thin with MEK.

Molding Plastics

Molding plastics can be divided into two categories: those that are simply painted into a negative mold in layers to form a piece and those that are pressure molded in a two-piece mold.

Paint-in Plastic Molding Material

RCMA furnishes a Molding Material in Light (KW shade), Deep (KM shade), and Dark (KN shade) that can be painted into a negative mold in successive coats to build an appliance. This Molding Material also is available in a thicker, quicker drying form in a tube under the name of Scar or Blister Making Material by RCMA. Essentially, they are methacrylates in a solvent with plasticizer and are excellent for temporary appliances. Due to the manner in which they are made and the basic qualities of the material, appliances should be produced only a day or so before use. Otherwise there is a tendency (of all such molded plastics) to lose

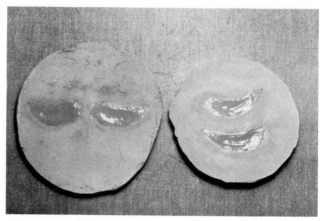

FIGURE 14.11 *Plastic molding material.* TOP: (LEFT) *A set of eyebags sculpted in plastalene on a plaster flat plate.* (MIDDLE AND RIGHT) *Finished plastic molding material appliances.* BOTTOM: *The appliances painted into mold before powdering and removal.*

some of the sculptured surface during long storage periods. However, appliances made with this material are soft and flexible and are easily adhered to the skin with Prosthetic Adhesive A. (Formulas for similar materials can be found on page 205).

Negative molds should be coated with RCMA Silicon Mold Release or silicon grease before use, taking care not to brush too much into the fine sculpturing. This type of material is especially good to make eyebags when time and budget do not allow foamed ones to be made. The liquid Molding Material should be put in a small wide mouth jar with an easily cleaned stopper or cap and thinned out a bit with acetone. Make certain that you stir the stock bottle to disperse all the coloration that might have settled to the bottom during storage.

Eyebags can be sculpted on a flat plate and a negative mold of stone or Ultracal 30 made (Figure 14.11). After the silicon separator has been applied, give the bags (don't overpaint for a blending edge) 10 coats of the thinned-out Molding Material, allowing them to air dry (don't use the hair dryer as it will bubble the material) between coats. Slightly heavier coats can then be applied as the cavity of the sculpted eyebags begins to fill up. Dry between coats. Finally, lay in a heavy coat of material so that the level of the eyebag is built up higher than the mold surface, and set aside to dry overnight. As the solvent evaporates completely, the level of the built-up portion will shrink down to slightly below the surface level of the mold so the appliance can be attached flat to the skin. To remove, powder the surface and tease an edge up with a sharp tool and then peel out the appliance. It is then ready for immediate use.

This material can also be used to make interesting forehead pieces for gunshot wounds, veins that pulsate, and such. For example, for a gunshot wound, take a negative cast of the forehead area in salt-accelerated casting plaster (Figure 14.12). The surface can be sealed with the RCMA Silicon Mold Release agent and the coatings built up by brushing the Molding Material into the form to cover the entire forehead area, making a good blending edge just above the eyebrows and at the point where the nose meets the forehead, as well as all around the hairline edge. The thickness of the appliance can be increased near the center of the hairline where either the performer's own hair or a small hairpiece will cover this edge.

After painting on about 15 coats of thinned Molding Material, a small-diameter polyethylene plastic tube can be imbedded with its end about in the middle of the forehead. This tubing should be about 4 feet long so that the other end will be off camera with a blood syringe attached. Hold the tube in position and carefully paint in a few coats of material to imbed it into the surface of the appliance. Adhesive tape can be used to position and hold the tubing where you want it to stay. Then add more coatings of the Molding Material

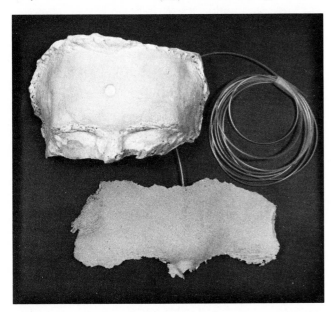

FIGURE 14.12 TOP: *A plaster cast of a forehead with a bullet hole cut into the center to show where the tubing must end.* BOTTOM: *After painting into a negative mold, an appliance with a tube imbedded into the material. The hole is cut with a sharpened metal tube and the plug saved for later insertion.*

over that to fully seal and cover the tubing, graduating the coatings away from the tubing so that it won't be seen as a bulge when the appliance is attached to the face. Thicker coats can be applied until the desired amount is reached.

Let the appliance dry on the mold overnight, and strip it off, taking care that the tube remains firmly imbedded in the appliance. Always select a color of Molding Material that will be closest to the skin coloration of the performer so that little foundation will be required for coverage. Next, with a sharp, round punch, about the size of the bullet hole, cut out a circle *just* below the end of the imbedded tube. The hole in the tube can then be cleared with a sharp instrument. A needle about the tube size can be imbedded in a wooden dowel point first. The protruding head portion makes a good tool for this. Save the cut-out portion of the plastic as this will be used as a plug. Attached to a length of clear filament, this plug can be pulled out as part of the bullet-hit sequence before the blood flows through the tubing.

Dick Smith has imbedded removable strips cut in the shape of veins in this material and, in use, filled the spaces with a fluctuating air stream to simulate the pulsation of forehead veins. Small bruises, cuts, scars, moles, burn tissue, and many other marks can be made similarly with this type of Plastic Molding Material. One of the main advantages of the Molding Material is that its blending edges are soluble in acetone. Thus, it can be washed in to be imperceptible.

Press Molding Material

This is a polyvinyl butyral plastic that is employed to make appliances with a method devised by Gustaf Norin a number of years ago. Although it is employed relatively seldom, the procedure is an interesting one that produces an appliance whose edges are soluble in isopropyl alcohol (Figure 14.13).

The original formulation consists of 4 1/2 *fluid* ounces (by measure) of Vinylite XYSG (a white powder, polyvinyl butyral) dissolved in 32 fluid ounces of isopropyl alcohol. Then add a plasticizer mixture of 25 cubic centimeters (cc) castor oil, 25 cc DBP, and 50 cc isopropyl alcohol. If a harder appliance is desired, use 10 cc castor oil, 10 cc DBP, and 25 cc alcohol. Mix well and keep tightly capped until use.

To prepare the plastic for molding, pour some of the compound into clean, cold water, and with clean, well-scrubbed hands knead the then coagulated mass until it stiffens into a ball (Figure 14.14). This squeezing will press out all the solvent alcohol into the water and should take about 3 minutes (for a piece large enough to form a nose). Remove the plastic from the water and it will now be a semitranslucent white mass. Stretch a clean, lintless dish towel, often called a *huck towel*, over the knee, and carefully roll the plastic back and forth for about 15 minutes to dry out all the water from it. It is important that, when ready to color and place in the mold, no water content remains in the plastic as this will cause shrinkage.

When the mass is dry to the touch, dip the roll of material into some dry calcium carbonate to give it some opacity, and knead this well into the material. A slight coloration can be imparted by rubbing the material lightly on a cake of a pale shade of dry cheek-color and kneading the color into the material. To hasten the drying out period, a stream of warm air from a hand-held hair dryer may be used, taking care that the material does not stick to the fingers due to the heat. It is well to keep dipping the fingers into No-Color Powder to prevent sticking during this part of the operation. Do any coloration gradually, a bit at a time, and do not color deeply as the plastic has a propensity to deepen in shade as it cures and, again, a pale shade is easier to cover with make-up than a dark one.

The material is now ready to place in the mold, which is a pressure type made of dental stone (see page 201 for pressure-type molds). Grease the female side with petroleum jelly, and *lightly* wipe the male side with the same separating medium. Place the material on the negative female side, spreading it out into the general shape of the desired prosthesis, and put the two sides of the mold carefully together. Place about 200 pounds of pressure on the mold for about 15 minutes. Large sections may require up to 400 pounds of pressure to get a fine blending edge and correct construction of the material. Twenty-five-pound iron stage weights are very good for applying this pressure. Build them up by crossing two over two at right angles

FIGURE 14.14 *Press molding.* TOP LEFT: *Plastic poured into water and coagulated.* TOP RIGHT: *Opacity and color kneaded into plastic.* BOTTOM LEFT: *Stage weights used for pressure on mold.* BOTTOM RIGHT: *Finished appliance on the male side of the mold after pressing and trimming edges.*

until the correct weight is achieved. Sitting on the built-up weight platform will increase the pressure if necessary.

After the required pressing time, remove the weights and open the mold carefully. The negative side, which was greased, should release quite easily and leave the completed prosthesis on the positive male side. If the appliance is well made with no bubbles or defects, let it stand overnight on the open mold. However, if the appliance has defects such as thick edges or lack of completeness, add a bit more molding material where necessary and repress the material for about 10 minutes. Separate and check once more. If perfect, leave the mold open overnight, but if still defective, repair in the same manner. Sometimes rearranging or replacing the material is necessary to get a perfect pressing.

The prosthesis can be removed from the mold the next day by lightly brushing under the edge of the appliance with a small flat brush covered in No-Color Powder or very fine talc. Then the overflow excess plastic material that forms in the gutters during the pressing can be cut away from the appliance, leaving just enough of the thin blending edge to adhere the appliance to the skin. This blending edge can be thinned out to a very fine edge against the positive mold with a flat sable brush that has been dipped in isopropyl alcohol. It is best to keep the appliance on the male mold or another positive cast of the same until ready

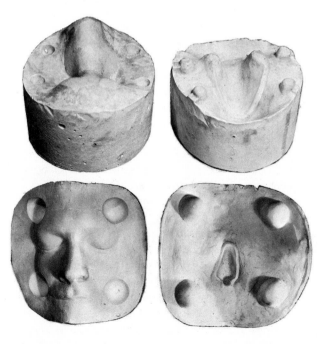

FIGURE 14.13 *Comparative casts.* TOP: *A foamed latex nose mold.* BOTTOM: *A pressure casting or foamed urethane sized mold.*

for use. It is also a good idea to make up plaster duplicates of the male side of the mold for this purpose so as to leave the actual mold free to do another pressing if necessary.

To adhere this type of appliance to the skin, cover the area where the prosthesis is to be placed with a coating of the liquid plastic molding compound and dry thoroughly. Powder this area with No-Color Powder, and center the appliance in place. With a small brush carefully touch under the appliance with isopropyl alcohol. This will immediately dissolve the appliance into the adhesive coat of molding compound, attaching the prosthesis firmly to the skin. Carefully wash the edges of the appliance into the skin with the alcohol, and powder the appliance and the edges with No-Color Powder or very fine talc. Do not use the so-called translucent or neutral powders for this type of appliance as these powders do have color and fillers that will affect the coloration of the piece. Extrinsic coloration can be done with a high-pigment creme-type foundation like RCMA Color Process Foundations or similar types.

Tooth Plastics

Dental acrylic resins of methyl or ethyl methacrylates are supplied as a *monomer,* a colorless liquid, and a *polymer,* a fine powder that comes in a number of shades. For additional coloration, there are liquid stains that can be painted over any finished denture material.

When the monomer and the polymer are mixed together, the polymer is partially dissolved so that the mixture becomes a doughlike plastic mass. The monomer in the mixture is then polymerized to form the solid acrylic resin. Two types of acrylic resins are used by dental technicians and dentists. In one, the polymerization is activated by heating the mixture, and in the other, polymerization is activated chemically. The latter type is self-curing and is generally the variety employed by make-up lab technicians for teeth. One of the best textbooks on oral anatomy and techniques, titled *Dental Technician, Prosthetic* (1965), is published for training prosthetic dental technicians in the U.S. Navy and is available through the U.S. Government Printing Office (NAVPERS 10685-C).

The Lang Company makes a number of materials for dental use that can be adapted by make-up lab technicians:

Jet Tooth Shade Acrylic A very fast-setting, self-curing powder and liquid combination available in shades 59, 60, 61, 62, 65, 66, 67, 68, 69, 77, 81, 87, and Light and Dark Incisal.

Jet Denture Repair Acrylic Strong and fast setting, this self-curing acrylic is colorfast in Pink, Fibred Pink, Translux, and Clear and is used for simulating gum tissue.

Colored Jet Acrylic Self-curing resins that set hard in less than 10 minutes. Available in colored powders with clear liquid monomer or with clear powder and colored liquids. Tans, greens, reds, blues, yellows, violet, and black are available in sets.

Flexacryl-Soft Rebase Acrylic An ethyl methacrylate formulation that is plasticized to remain resilient in a cushionlike consistency for a number of months. This material is nonirritating and dimensionally stable.

Warm weather and humidity will affect these self-curing polymers during mixing. Working time can be extended with chilled mixing containers for large batches.

In most cases, the polymer powder is wetted with the monomer to make the casting. Many workers coat the negative cast with a brushing of monomer and then add a bit of powder to start, then drop powder and liquid a bit at a time to form the mass. As long as it appears glossy, it can be worked and brushed into a form.

To make teeth, one can employ a number of techniques. Each one is a bit different and more designed for make-up use than regular dental application. The simplest way to make flat-form teeth for imbedding in gum material is to fill preformed plastic crown forms with tooth shade acrylic and allow them to set. Odd-shaped or special teeth can be made by spatulating additional material on these formed teeth. One can also obtain various preformed teeth from supply houses that serve dentists to identify tooth shades and shapes. Sometimes dentists have a set of discontinued artificial teeth that they will give you if you ask for them.

One method is described by Lee Baygan in his book on prostheses for make-up, *The Techniques of Three-Dimensional Make-up* (Watson Guptill, 1982), where he sculpts the desired teeth in plastalene on the positive dental stone cast. Using petroleum jelly as a separator on the gypsum portion of the cast, he takes a negative of the sculpted teeth in alginate in the dental tray. When the alginate is set, he removes it and, since the next step may take some time, immerses the alginate negative in water to prevent any shrinkage due to air drying. The plastalene is removed from the positive (or negative if any becomes imbedded), and the positive is given a coating of petroleum jelly. The negative is then removed from the water and the surface quickly dried (don't use heat). Self-curing tooth shade acrylic polymer powder is added to the tooth cavities. The liquid monomer is added with an eyedropper, and the mass is mixed with a small narrow tool. Adding a bit of each at a time will fill the teeth and allow some extra for flash over the gums (which he has allowed for in the plastalene modeling). The positive is then placed carefully over the negative and the two pressed together and held for a minute or two until set. Put aside and let the acrylic fully harden for about 30 minutes.

The two sides can then be separated and the teeth pried from the positive cast. The casting of the teeth is then trimmed with a Mototool and polished so that no rough edges are felt. The teeth may be colored (or discolored) with acrylics and then given a coat of 5-minute epoxy. This can be yellowed for some effects. The epoxy will gloss the teeth to a natural shine. Baygan then recommends that a bit of flesh pink be added to an epoxy mix and painted on the gum section of the acrylic to simulate the gums.

Tom Savini's *Grande Illusions* (1983) on special make-up effects shows a method of making a set of vampire teeth by building up a wax covering over the teeth in the stone cast and then imbedding some artificial teeth into this wax in front of the teeth on the cast. He also makes some long incisors by mixing the polymer-monomer into a heavy paste and, wearing surgical gloves, rolls the material to get the shape of the desired tooth. Filing and grinding produce the finished teeth. Using a medicine dropper, monomer is used to wet the area above the teeth, and some pink gum shade polymer powder is sprinkled on. The process is repeated until the proper amount is built up to surround the teeth in front. The wax is removed after the acrylic is set (about 10 minutes), and the back of the teeth are treated in the same manner to form around them. Small dental tools can be used to sculpt around the gums and teeth.

When the material is set, it can be removed from the positive form, and you will have a dental plate that fits directly over the subject's own teeth. They can be finished with acrylic lacquer if need be to restore the shine. Dental powder or cream can be used to hold the teeth in for use in the mouth.

If someone has teeth with a space between, a set of snap-on tooth caps can be made by making a cast of the front teeth and then imbedding them in a heavy stone-reinforced mold or a dentist's flask (Figures 14.15 and 14.16). Make a mix of the proper tooth shade, and after coating the positive side with an alginate release agent, put on a layer of the mix and place a sheet of cellophane separator on top of the plastic. Close the mold and put it in a dental press for a minute or two. Remove the mold from the press, separate the two halves, lift the cellophane separator sheet, and using light pressure, with a sharp scalpel trim off the excess flash material from the appliance, then repress the appliance in the mold press for a minute or so. Repeat this until no excess is extruded from the prosthesis. Replace the cellophane sheet with a new one each time a repress is made. When the appliance is complete and perfect, leave the mold halves separated until the material sets. Then ease the appliance off the positive cast and clean and polish it with burrs, abrasive paper discs, rubber cups, and felt wheels (with

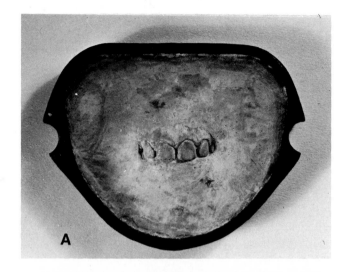

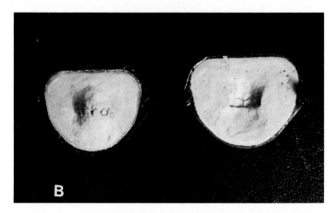

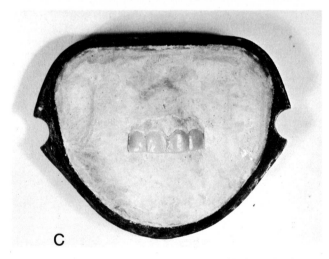

FIGURE 14.15   *Tooth caps. (A) Flasked male mold of actual teeth. (B) Male and female mold before pressing. (C) The finished caps on the male mold.*

a paste of chalk or tin oxide and water) in that order. These very thin caps should then clip on to the real teeth and look perfectly natural. By the way, they are not made for eating and are rather fragile, so keep them in a small plastic box when not in use.

Variations of these methods include sculpting the teeth in an inlay wax material rather than in plastalene or using a Flexacryl-Soft Rebase Acrylic for making the pink gum material for teeth, which may be easier on the wearer. Colored acrylics can be used to make teeth of any shade for certain effects, while the Flexacryl-Soft material can be used alone to make an appliance that would cover all the teeth for a toothless gum effect.

It is not a good idea to apply any of the polymer-monomer mixture directly into the mouth to form teeth as it may firmly attach itself to the subject's teeth. Work only from a tooth cast, and always trim the finished plate so that the performer's speech will not be impaired—unless that is the desired effect. If too much acrylic is formed on the inside portion of a tooth prosthesis, it may produce lispy speech patterns.

Special Construction Plastics

Certain special effects require construction plastics such as epoxies, fiberglass, polyester resins, and other such industrial materials often employed by auto repair shops (see pages 220 and 233). Most of these are rigid materials although some technicians are also using the flexible urethanes and polysulfide products as well. None of these is for application to the skin, and they represent mostly substances that are being adapted to make the dummies, figures, puppets, and simulations of humans or animals that are usually mechanically operated that serve to replace a living thing to achieve situations that might either be physically impossible or that might prove to be dangerous to a living person or animal.

Sometimes these materials are used as part of an overall make-up effect and the finished piece attached to the body. See Tom Burman's special make-up on page 216. In general, the special construction plastics and other materials serve as a stimulation to new make-up effects, and in the future, more industrial products designed for other uses will find their way into the laboratories of the creative lab technician/make-up artist.

Again, great care should be taken with researching not only the finished qualities and restrictions that might affect the human skin on all these products but also whether or not any dangers exist in the use of the raw materials to laboratory personnel. None of these materials is classified for cosmetic use so great care should be taken whenever their use is contemplated. Also see the section "Duplicating Materials" earlier in this chapter.

One use of two-part epoxy is for glossing over latex or plastics to make false eyes. One method is to paint ping-pong balls with artist's enamels in the shape and color of eyes and then giving them a coating of the epoxy. This material comes in a handy two-barreled ejector type of dispenser that can be purchased in hardware stores. This fast-drying epoxy can also be used to put a gloss on teeth in latex masks.

Polyester resin used by auto repair shops with fibreglass cloth comes with a can of resin and a tube of curing agent plus some open fibre or mesh cloth to soak into the compounded resin. Also available from the same source are the fillered polyester resins that are used as auto body fillers. Both types set hard and can be worked with files and sandpaper to make various constructions.

Gelatin Materials

In addition to synthetic processed materials, one can also include animal gelatin for the manufacture of special constructions (see page 156). A basic formula recommended by Tom Burman is as follows:

20 grams of gelatin powder,
12 cc of distilled water,
100 cc of glycerin.

Set this mixture aside to saturate for 45 minutes. First, heat to 140° F in a water bath to melt the mass, then pour into a plastic bag (Ziplok type) and flatten out to cool. For use, this gelatin compound can be made opaque with talc and titanium dioxide, and water-soluble vegetable colors can be used for various shades. Dry colors could also be ground into the dry mixture of talc and titanium dioxide, while kaolin can be used as a filler if required. Tom Burman uses this formula

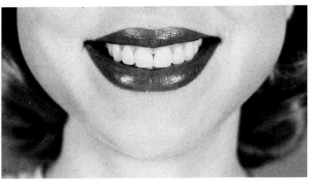

FIGURE 14.16 *Tooth caps.* TOP: *The natural teeth without caps.* BOTTOM: *Appearance after adding temporary caps. Bonding can produce the same effect on a more permanent basis (see page 87).*

gelatin with a red coloration for blood effects for certain projects. Nonmetallic powders were incorporated into the gelatin mix by Stan Winston for his robot make-ups in Figure 12.34.

Softer constructions can be made with 15 grams of gelatin and 25 grams for harder ones in the given formulation. Another formula calls for 50 cc of Sorbitol and 50 cc of glycerin with 1 gram of salicylic acid as a preservative or a few drops of oil of wintergreen.

The molds used are similar to any type of pressure mold that is well greased with petroleum jelly as a separating medium. In use, the gelatin material is melted, fillered, and colored and then poured into the negative, or female, mold. The positive is then placed on and weights or clamps applied to hold the molds together. When the material is cold, the molds are separated and the appliance powdered and removed.

ADHERING APPLIANCES

Books on make-up seldom stress the importance of selecting the proper adhesive to keep the appliances firmly attached to the skin or one material to another so that the overall effect is never destroyed by an appliance, hairpiece, or other special material coming loose. Oftentimes, re-attachment takes more time and clean up than the time that should have been taken to attach the item properly in the first place. No matter how well the molds were made and the appliances formed, if they are not properly adhered, the entire effect is lost.

There are many varied adhesives used for make-up work, some very versatile and others very specific in use. The directions for use must be accurately followed. Certainly new adhesives will be found in the future, and make-up personnel should research and experiment with any new ideas they come across. However, take care with any adhesive that it can be *removed* from the skin before using it for any make-up purposes or application of hair goods. Some of the fast-drying superglues used for pottery, metal, and other materials (such as Cyanoacrylate types) should *never* be used on the skin because they are very difficult to remove. Therefore, a good adhesive should be easy to apply, set rapidly, be dilutable for use, and be removable with a solvent that is not harmful to the skin. In addition, the adhesive should be a proper one for sticking one surface to another without adversely affecting either one. Finally, never use any adhesive on the skin without testing it first—preferably on yourself—and if any adhesive causes any skin irritation on a performer, discontinue its use and seek another for the job. For a review of various adhesives, see pages 37 to 38.

Matting and Thixotropic Agents

Many materials, among them adhesives and sealers, can be made to set with less shine or be made thicker for certain applications by the addition of special clays or earths or with microsilicas that are made in many grades and particle sizes. The Cabot Corporation makes a number of grades of Cab-o-sil, which is a fumed silicon dioxide used to increase solution viscosity. They also will reduce sheen but do not produce a truly flat or matte effect as will some of the larger-particle silicon dioxide types. Cab-o-sil, like all the others of this type, is a very fine white powder that is quite light in weight. M-5 is the lightest weight grade (2.3 pounds per cubic foot) that the make-up technician might employ, and it can be dispersed in a solution under low rates of shear. The MS-7 (4.5 pounds per cubic foot) is almost twice as heavy and is best dispersed with high shear equipment. Both will thicken many liquids (such as latices or resin gum solutions) by forming a network of particles. A nonionic surfactant such as Triton X-100 (Rohm and Haas Co.) can be added (0.5 percent) to further thicken some solutions. The nominal par-

FIGURE 14.17 *Application of small appliances.* TOP LEFT: *Kenneth Smith before make-up for the NBC production of* War and Peace *in which he played General Kutuzov.* TOP RIGHT: *A foamed latex nose is attached with Matte Plasticized Adhesive as is a slush molded blind eye appliance.* BOTTOM LEFT: *A scar is added on the cheek with scar material along with a wig and sideburns, and the eyebrows are grayed with some hair added as well.* BOTTOM RIGHT: *The completed make-up as designed by Dick Smith and applied for the production by the author.*

ticle size of Cab-o-sil is 0.014 microns, and it can be used in combination with larger-particle silicon dioxides to promote suspension and reduce hard settling. A variety of this material is the processed silica like Tullanox, which makes the product superhydrophobic.

The larger-particle Syloids (Grace Chemical) or Degussa's TS-100 have more matting effect. Syloid 244 (4 pounds per cubic foot) has an average particle size of 3.3 microns diameter but can settle out of solution—even after dispersal by high shear—unless suspenders are added to the mix. TS-100 varies in particle size from 2 to 10 microns with an average value of 4 microns in diameter. Its particular quality is that it can be added to a solution under low shear and will increase in thickening effect at about the same rate that the flattening resultant increases. For example, 1 ounce of a basic resin adhesive can be matted with 1 gram of TS-100 to produce an adhesive that can be finger tacked to remove the shine, while 2 grams of TS-100 will produce a gell that dries matte without tacking.

Clays such as kaolin and the Attapulgus varieties of AT-40 or Pharmosorb produce a flattening if added to resin adhesives, but although they do settle out readily, they are easily shaken back into solution. However, an excess of the latter has a tendency to whiten or gray upon full drying after application. It takes almost 3 teaspoonsful of Pharmasorb in 1 ounce of resin adhesive to give an adequate matting effect but, unlike some of the silicas, does little to add to the adhesive quality of a resin gum solution. Also see pages 37 and 38 for further explanations of the use of matting materials in RCMA adhesives and sealers.

General Applications

Most appliances, slush or paint-in latex, foamed latex or foamed urethanes can be attached with Matte Plasticized Adhesive. This type of adhesive is a Matte Adhesive with special plasticizers added to provide an adherent that sets with more tack and does not crystallize like ordinary spirit gums.

With small prostheses, apply a coat of the adhesive to the edge of the appliance, and allow some of the solvent to evaporate (Figure 14.17). Then carefully place the appliance on the desired area and press into place. Take care that adhesive is painted on the blending edge but not over it. If some of the adhesive is left on the surrounding skin, carefully remove it with a Q-Tip lightly dipped in alcohol. Never paint over the edge of an appliance with adhesive after it has been adhered to seal the edge as this excess adhesive will darken with the application of any foundation. Instead, to seal the edge, use a material that will seal but not discolor. Some make-up artists use RCMA Matte Plastic Sealer, while others prefer a latex-type eyelash adhesive. Either one of these products can be lightly

stippled on with a small section of stipple sponge held in a pair of curved dental college pliers. Before using an edge sealer, check that the appliance is fully attached right to the edge. If there are any loose spots, lift the appliance with the college pliers, carefully brush a touch of adhesive under the edge, and press down the appliance to the skin with a damp huck towel.

Some make-up artists prefer to attach the pieces with eyelash adhesive and to stipple over the entire surface of foam appliances with the same adhesive to seal the surface. Others use Prosthetic Adhesive A or B. These are contact-type adhesives and so both surfaces, the skin, and the appliance must be coated. As neither one of these adhesives fully dries to the touch, it should be lightly powdered with No-Color Powder on the surface of the adhesive coatings before centering the piece on the skin, then pressed in to complete the adhesion.

On a large piece, many artists adhere only the center portion of the appliance by coating just the skin area, leaving about half an inch of blending edge adhesive free. A brush with a small amount of adhesive is then passed under the edges, and the piece is pressed in to the face or body part with a damp towel. Take care that the blending edges of the appliance do not roll over to form a visible demarcation to the surrounding skin area. Such folds can be eased out with a brush dipped in alcohol and re-adhered if Matte Plasticized Adhesive is used.

Coloration may be done with any RCMA Color Process Foundation that has been slightly thinned with Foundation Thinner, or any type of PB or AF Foundation may be used. Some of the latter are heavier and stickier than others, but all are designed to give a heavy cover to both the skin and the appliances. Both the polyurethane sponge and brushes can be used for application and then the surface powdered with either PB Powder for a matte effect or No-Color Powder for one with more halation. The latter does not absorb the oils in the PB Foundations as well as the PB Powder, however. If the surface appears too dry, appliances can be overstippled with a bit of glycerin or RCMA Tears and Perspiration to restore some surface shine without affecting the foundation. Some make-up artists make a mixture of glycerin and a bit of isopropyl alcohol for a thinner mix for the stippling.

Foamed urethane appliances can be attached in the same manner, but if the outer skin of the foam is porous, it can be coated with RCMA Matte Plastic Sealer before applying the foundation. Either Color Process Foundation, AF, or PB Foundations will work over this type of appliance; liquid or water-based cake foundation does not. RCMA also makes a very flexible acrylic-base foundation called Acrylid, specially compounded for use with foamed latex or flexible urethane foam appliances in a number of basic colors (AP series).

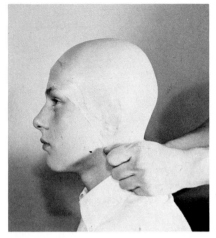 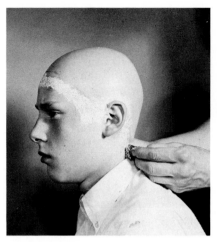 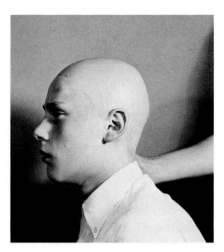

FIGURE 14.18   *Applying a bald cap.* LEFT: *A plastic bald cap is marked at the hairline with a brown pencil and is stretched to estimate the fit on the head.* MIDDLE: *After cutting and adhering, the edge is being stippled with eyelash adhesive or edge stipple to further disguise the blend and seal the edge of the cap.* RIGHT: *The completed attachment of the cap.*

Latex and Plastic Bald Caps

Latex bald caps are employed solely for mob scenes or extra work but are still effective except for close shots. Films on American Indians in the eighteenth and early nineteenth centuries when the scalp lock was worn can use latex caps, and the front blend edge is often disguised by a streak of warpaint. These caps can be generally applied with RCMA Matte Plasticized Adhesive or Prosthetic Adhesive A. Plastic bald caps are much thinner and more fragile than the latex variety but can be applied to have an imperceptible edge. The same adhesives as recommended for latex caps can be used.

Applying a Latex Cap

Slip the cap over the performer's head, and mark line with a brown make-up pencil about a half-inch from the edge of the front of the hairline, close around the ears, and allow about 1 inch below the hairline in the back. Remove the cap and cut along the line, then replace it on the head.

The front edge can be adhered by passing a #10 round sable brush dipped in Matte Plasticized Adhesive under the blending edge and pressing out the excess with a damp huck towel. The cap can be adhered around the ears and back in the same way, then the edges stippled with eyelash adhesive and dried with a hand-held hair dryer.

The cap can be made up with AF or PB Foundations or with RCMA Color Process Foundation on a sponge dampened with Foundation Thinner. If the head is to appear with a scalp lock, the rest of the hairline should be stippled with a dark gray beard stipple to simulate stubble growth. For some bald effects where the cap does not fit well in the back of the head, costume design can sometimes cover this defect.

Applying a Plastic Bald Cap

As plastic bald caps are generally thinner and more translucent than the latex variety, it is easier to trace the hairline for cutting the excess cap material (Figure 14.18). When the cap has been refitted to the head, start attaching the forehead area first by passing the adhesive brush under the edge with some Matte Plasticized Adhesive. Avoid using too much adhesive so

FIGURE 14.19   *A partial bald cap used to cover the hairline of Sarah Churchill as young Queen Elizabeth I for a Hallmark Theatre production. This cap was not attached at the back and a wig was used for the hairstyle that covered the back of the neck as did the costume (hairstyling by Jack (M) LeGoms).*

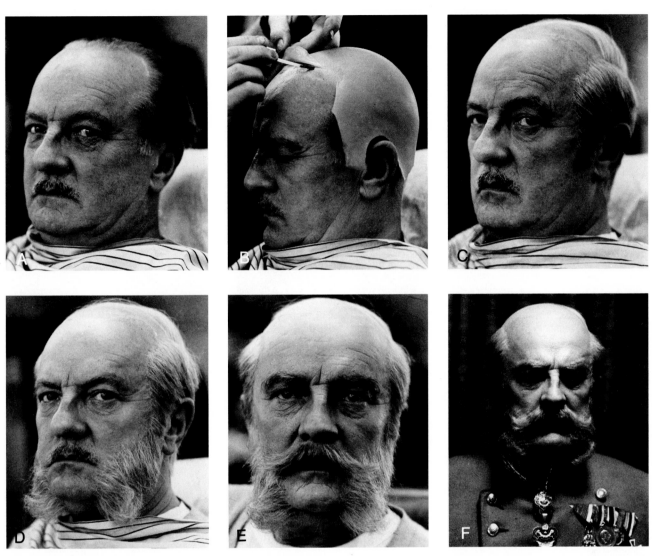

FIGURE 14.20 *Partial bald cap. (A) Basil Sydney before make-up. (B) A heavy (50 coats painted) bald cap has been made to fit the actor's hairline exactly and is being attached with Matte Plasticized Adhesive. (C) The wig and sideburns are added with RCMA Special Adhesive #1. (D) Added mutton chops lengthened sideburns. (E) After the addition of the moustache, the face was covered with a light coating of foundation and* *then blended carefully. As the Special Adhesive #1 was used on the lace and attached to the bald cap, the foundation was blended right over it. Shadows, highlights, and age lines were then added. (F) The completed make-up for Emperor Franz Josef of Austria. (Make-up by the author for an NBC production with make-up design by Dick Smith.)*

as not to dissolve the edge of the cap. Press the front down with a slightly dampened huck towel to make the edge smooth. Then stretch the sides downward by the sideburn tab, smoothing out the front and side so that no wrinkles appear, and pass the adhesive-dipped brush under the edges and press them in place with the towel. It is easier to attach a cap with two make-up artists working together since then one can stretch one side while the other side is being stretched and attached so that both sides will have approximately the same tension. However, one person can still do it. The cap area at the back of the ears can be adhered to the head and along the hair line in the back.

The back of the head is the most difficult to attach if the hair is not cut close. If the costume will cover the back of the head, don't adhere it so the wearing

of the cap will be more comfortable for the performer. If the rear of the cap is to be attached, then the back of the neck where the adhesion must take place should be cleanly shaven; otherwise the hair will prevent positive adherence. If the cap must be applied to a woman or man with long hair, carefully pin the hair as flat to the head as possible, taking care that no bobby pins point outward to puncture the cap. It is also a good idea to place the crown of a previously used cap over the top of the head prior to application of the outer cap to protect the cap further from bobby pins and to smooth out the rough surface created by the pinning down of the hair. This extra inner cap need not be adhered as the pressure of the outer cap will hold it in place.

Sometimes when just the front of the head need be

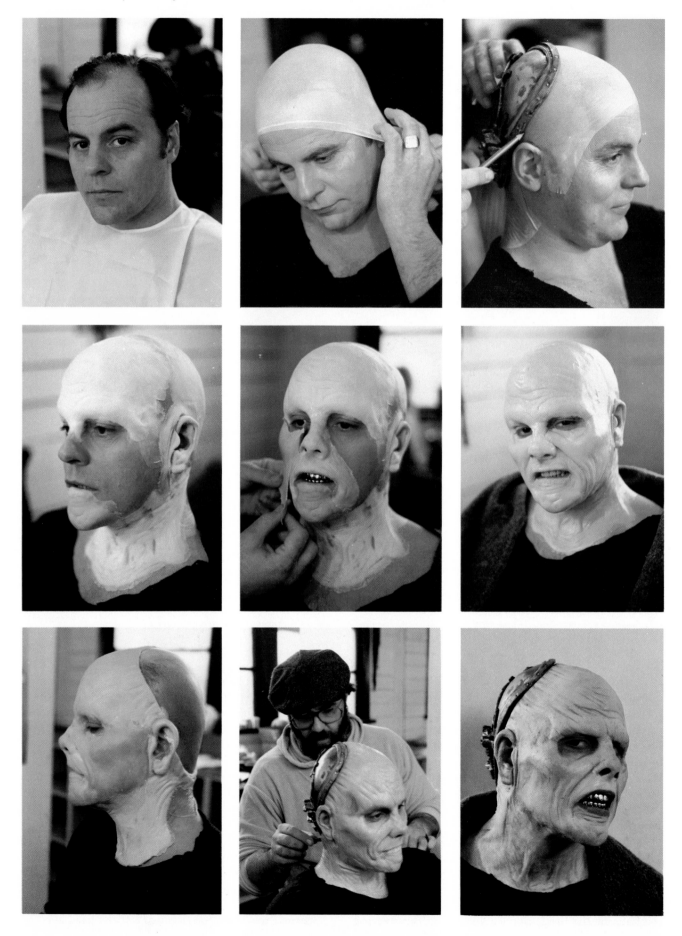

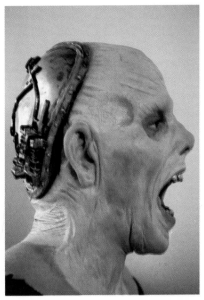
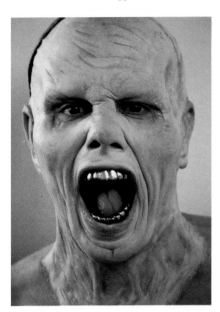

FIGURE 14.21   *Character creation for Michael Ironside in*
Spacehunter. *(Make-up and photos by Tom Burman.)*

bald, there is a Velcro type of hair bandage that can be wrapped around the forehead area and then slid up to flatten out and push back the hair (Figure 14.19). Prosthetic Adhesive A can also be used to attach a bald cap by cutting the cap completely to shape and size and then coating the inner surface of the blending-attaching inside edge with a half-inch strip of Prosthetic Adhesive A. This will dry quickly and it should then be powdered with No-Color Powder. Then place and center the cap on the head, and roll back the leading edge on the forehead line and carefully coat the skin with Prosthetic Adhesive A, dry it, and lightly powder it. This adhesive does not really dry to the touch as it is a contact type, so the light powdering simply takes out some of the tack. Then carefully roll the edge back on to the forehead, and when perfectly centered, roll the sides back one at a time and repeat the process. When this is done start pressing the blending edge of the cap firmly into the skin with a dry towel or a polyurethane sponge. This will complete the contact, and the cap will be securely attached. Unlike the Matte Plasticized Adhesive, once the bond is made, the cap cannot be repositioned by lifting an edge and regluing.

Do around the ears and back of head in the same manner. The blending edges can then be washed into the skin with a brush lightly dipped in acetone. Take care in this operation that you do not get much acetone on the cap as it will eat right through the thin edges and ruin the blend. The edge can then be additionally disguised by stippling the blend area with RCMA Matte Plastic Sealer, or eyelash adhesive may be used. The stipple should be thoroughly dried with a hand-held hair dryer before applying any foundation. Color Process Foundations, or AF, or PB Foundations may

be used, or the cap may be stippled with the Prosthetic Adhesive B and tube paint mixture described on page 188 or RCMAAP series.

Heavy Partial Bald Caps

The heavier, painted-on bald cap described on page 213 can be attached with Matte Plasticized Adhesive by placing the cap on the head in the required position (Figure 14.20). The edge is then turned up, first on the forehead, and a coat of adhesive is applied to the skin, allowed to set for a few minutes, and then pressed down on the skin. The same process is repeated on the sides. The cap can then be blended into the skin with acetone and pressed with a small-bladed dental spatula to smooth the edge. Stipple on Matte Plastic Sealer or eyelash adhesive as previously described. The Prosthetic Adhesive A method can also be used.

Lace hair goods can be attached to the cap with RCMA Special Adhesive #1. This has a heavy matting material in a solution similar to cap material and must be well shaken before use. The hairpieces are held in the proper position, and the adhesive is applied *over* the lace. This adhesive dries very quickly and totally matte, incorporating the lace into the cap so that it is imperceptible. Other adhesives may dissolve the cap material or produce an excessive shine in the lace area. This Special Adhesive #1 is only for use on the cap, so Matte Adhesive can be employed where the lace meets the skin.

For coloration, some make-up artists stipple the entire cap area with Matte Plastic Sealer that has been colored with foundation to the desired shade. Any RCMA Color Process Foundation can be used for this and the color worked into the sealer with a dental spatula on a glass plate.

The incorporation of a number of appliance elements is used in the make-up shown in Figure 14.21 by Tom Burman on Michael Ironside for the film, *Spacehunter,* where he plays a character being kept alive by a mechanical support system controlled by his brain—hence, the special steel cranial plate with its attached circuitry. First, the performer's head was covered with a plastic bald cap with an extra long neck length attached with RCMA Prosthetic Adhesive A. Next, foam latex appliances on the forehead, chin, nose, nasolabial folds, ears, neck, and cheeks were adhered with RCMA Prosthetic Adhesive B. The edges of the latex appliances were blended together with a mixture of eyelash-adhesive-type latex and the PA-B. The foundation and other coloration are prosthetic bases of Tom's manufacture, with shadows and highlights to complete the effect. The cranial plate was made of vacuum-formed sheet butyrate and painted with a silver lacquer. The rivets and other circuitry were made of clear dental acrylic cast in alginate and tinted with a gold-colored lacquer. The whole plate was then assembled and adhered to the head with Prosthetic Adhesive B (so as not to dissolve the cap material base).

The special teeth were made by taking an impression of the actor's teeth, sculpting the desired look in dental wax, and taking a cast with a polysulfide mold. The bases of the teeth were made in Flexacryl-Soft acrylic, and a veneer was made of wax. This was then turned over to dental lab technician, Armando Ramos, who cast the teeth in chrome and polished them. Due to their close fit, the teeth were simply held in place by suction. The final photos show the extreme versatility allowed by the soft foam appliances as well as the actor's feeling for the role.

CHAPTER **15**

Special Effects

SINCE THE 1950s WHEN SOME NEW AND DIFFERENT TEXtures and effects were introduced in *The Technique of Film and Television Make-up* (Focal Press, London, 1958), the advances and research into the complicated field of make-up special effects has expanded in many and various directions and is only bound by human imagination. Entire make-up laboratories, far advanced from those previously envisioned, have become the spawning places of more blood and guts, monsters, and extraterrestrial creatures through the talents of a small band of make-up artists that has gone far beyond the usual. To anyone interested in this field, the workshops of Dick Smith, Stan Winston, Tom Burman, Rick Baker, Rob Botin, and others are wondrous places indeed where the illusions of these artists come to life for the screen.

Although the Special Effects Departments and personnel of the old Hollywood studios performed marvels of the time, an increasing trend toward the transition from the human face and form to what might be termed *puppets,* whose actions are controlled by wires or cables, both manual and motorized, bladders of air, and other means of manipulation, has bridged the visual imagery in seemingly magic ways to lead the audience down the path of deception and false impression when combined by clever editing of film or tape. The demands of directors have spurred the almost overreaching thoughts of these special effects make-up artists into creating a new form of entertainment that combines the talents of the make-up artist with that of magician or creators of visual illusions. This trend has brought forth a number of magazines (with very high-quality color pages) such as *Cinefantastique, Fangoria,* and *Cinemagic,* which are distributed worldwide to the enthusiasts of this genus and are contributed to and enriched by the work of these artists in careful coverage of their films.

The combination of make-up skills with those of sculpture, casting, latex and plastics, painting, and making mechanicals produce the learning level (and all in this area will agree that they are constantly researching and learning) that pervades in their results. All the good artists are ready to share their knowledge with others in the field as one person or one laboratory or workshop cannot experiment with or discover everything that might be the best for the job. Gone are the

secret rooms that Lon Chaney or Jack Pierce locked themselves into to create their make-ups, and an opening of sincere talent and sharing makes the effects grow. Granted, like all artists who are immersed in a project, make-up artists seek closed facilities to focus their concentration, but with their peers and workers, such is respected and understood because they wish the same for themselves. Here are a series of special effects make-ups that go from the human body or face to exact image counterparts (when the human form could not stand the distortion or materials required for the effect), showing the details and directions that served in each case.

BIDDING ON A PRODUCTION

Films that entail mostly straight make-ups have their expenses in make-up covered under personnel and production costs for supplies. However, when the entire premise, scope, and effectiveness of the film depends strongly on its make-up effects, such as in the *Planet of the Apes* series by John Chambers, Rick Baker's *American Werewolf in London,* Dick Smith's *Exorcist* and *Altered States,* Tom Burman's *Invasion of the Body Snatchers,* Tom Savini's *Creepshow,* and many others, then the budgetary factors related to make-up begin to soar.

With the many private laboratories for creating these effects concentrated mostly in the Los Angeles area, it makes it easy for producers and directors of this type of film to begin to shop around for someone to do the very special make-up that is often required. They seek bids on the projects, and like everything else done on that basis, the lowest bid may not be the truest. More and more, this is *not* due to competition between the labs but because the production people start changing their minds as to the effects as each bid is received along with a composite breakdown of the project as explained and discussed with one lab.

As the concepts change, and with the variance of effects directions vacillating, the unions start to place obstacles of craft crossovers, the writers begin to alter their ideas of story direction, the directors get brainstorms injected by the input of fresh elements, and what was minor becomes major. Whereas a bid of $150,000 for the make-up might be the lowest as compared to another of $200,000 or more, the over-budget ideas can climb to $400,000 just as easily unless a strong hand controls the artistic flights of fancy often indulged in by production people sometimes unfamiliar with the creative minds of the make-up artist when they are let loose to produce an effect. Terry Smith once said to a producer who was beginning to allow his mind to wander beyond the agreed-upon concept of an effect allowed for within the budget, "Sure, I can change it. You pay for it and I'll make anything you want. It only costs money. But don't expect it to be contained within the present budget, because it's not there!"

There are other factors that also occur after the filming is in progress. First, the production can have personality problems and the director is taken off the picture or quits. The new one hired has an entirely different idea of what direction the make-up should take and changes everything. The budget goes out the window in this act. Make-up labs have a mound of canceled molds, casts, effects, and so on and make many trips to the dump to clear their storage space for the next foolish change.

Next (and this is often the worst) is the performer who decides to be uncooperative and does not want to wear the make-up in which he or she was hired to perform. When a make-up artist has a professional performer to work with who is cooperative, he or she is more than pleased. Tom Burman said of Michael Ironside (see Figure 14.21), "Michael was absolutely the perfect actor to support the make-up. It made me feel good to be instrumental in making this part possible and in making it come alive."

Finally, there are cinematographers who will not change their camera angles or lighting to aid the aspect and effectiveness of the make-up. This adds to the difficulties of producing the best appearance for the screen and sometimes seriously detracts from the illusion. As one can readily see, only with sincere coordinative efforts can any special make-up effect be at its optimum all around.

In this regard and because of past disappointments, Stan Winston now often insists upon directing and editing any so-called special make-up effects, illusions, or *transformations* (as he calls them) that he feels require his guiding hand to achieve the result that he has envisioned—and has been hired to deliver! In this way, the proof of the effectiveness of the make-up or illusion is in the audience reaction, and its reflection is the satisfaction of the creator.

Due to the horizontal work structure of some unions in opposition to the creative singularity of some lab technician/make-up artists, the majority of make-up labs is off-premises to the film studios where union jurisdiction and control are contractual. Most of those who have these outside make-up labs are free to hire and train their own specialists at this time, but only the future holds the answers to the controversy of whether or not the special effects that are being designed and executed in the make-up labs come under the union job descriptions and contracts of the myriad of directions that are being taken to produce these illusions and transformations today. It would have been sad to have unions disputing whether Leonardo da Vinci was an art director, scene painter, plasterer, construction grip, or such while completing his work in the Sistine Chapel. Present-day make-up art must remain unfettered by organizational politics and restrictions and our Leonardos allowed full rein in their quest for viable and exciting ideas and their fruition. Openmindedness in art is mandatory for the human spirit.

In addition to the special character make-up by Tom Burman shown in the last chapter, the following are just a few examples of the varied use of special materials by a few of the make-up artists engaged in this very creative and challenging work.

RICK BAKER

Make-up artists who love to do character work are always fascinated with on-screen, progressive, facial and body changes because they offer a different type of challenge and require a variety of approaches to achieve the effects. In recent years, the most outstanding transformation on-screen was the one devised and executed by Rick Baker and his company of talented technicians for the Polygram Pictures production, *An American Werewolf in London,* by the use of special devices that went beyond the physiological functions of the human body to make the change from man to wolf-creature. He called his body parts "Change-Os" to define them from static reality. The first spectacular effect was two stages of a Change-O-Hand made from Smooth-On PMC-724 with hair imbedded in the arms (Figures 15.1 and 15.2).

The Change-O-Heads used for the facial progressions were also made of PMC-724 (about 1/4 inch thick) over a mechanically controlled skull section made of fibreglass and dental acrylics that was operated by technicians to achieve the facial extensions so graphically portrayed in the film (Figure 15.3). Rick used a series of appliances on David Naughton to effect the commencement of the transformation and brought the make-up up to the stage of the first Change-O-Head (Figures 15.4 to 15.6).

The Change-O-Back effect again demonstrates the number of technicians required for this change (Figure 15.7). A number of bladders and syringes that acted as pneumatic rams for the effect were employed to ripple the spinal cord.

There are a number of ways that camera work aids effects:

1. A special camera can be employed that runs at high

FIGURE 15.1   *Rick holds Change-O-Hand #1 that shows the metal rod extension that controlled the wrist movement. In his other hand is the device that controlled the individual finger movements, and the tubing connected to the pneumatic rams controlled with air pressure to extend the hand in the palm area. (Photos courtesy Rick Baker for Polygram Pictures.)*

FIGURE 15.2   TOP: *In Change-O-Hand #2, the palm extension was continued further with the action of the pneumatics as the palm section was fluctuated with bladders controlled with plastic tubing and a syringe. All the hair was added as the filming progressed and was laid by hand on the arm and the chest (as the actor, David Naughton, had little chest hair). A small appliance nose commences the facial transformation that was the next progression in the make-up.* BOTTOM: *Due to camera framing, as the actor looked at the hand (seemingly his own), the changes occurred and he reacted to them. (Photos courtesy Rick Baker for Polygram Pictures.)*

speed to give a slow motion effect when played back at normal projection speed.

2. This process can be reversed and the camera run more slowly, which speeds up the screen action.

3. The action is performed in reverse for a normal camera run and a print made that again reverses *that* action. For example, a section of skin (made from PMC-724 or a similar product or even a foamed one) can have long hairs implanted. As the camera is running, these hairs, attached to a material under the false skin, are withdrawn until the skin is smooth. The reverse printing makes the hair appear to grow from the skin in the proper pattern.

4. A diffusion disc or filter can sometimes be employed to soften certain effects, or sometimes colored filters can add to an effect. Proper lighting to *see* the effect is essential!

A thorough discussion of all special effects in a film by department heads will aid the final concept as some-

times a minor contribution to an idea will make it work better. However, a make-up artist who does these special make-up transformations or effects should have a thorough knowledge of what the film camera can do as well as be familiar with all the new advances in electronic techniques or special effects. Of the latter, make-up artists should be aware that some directors get carried away with the sight of electronic changes and employ them to superimpose over make-ups. This may add to the overall visual effect or, in the case of some overuse of such, destroy the make-up concept. Films like *Altered States* displayed an excess of electronics or light-created effects *over* a splendid concept designed by Dick Smith for the transformations. The creative illusions of the make-up artist can be *aided* at times with such visuals but should not overpower them with their inherent artificiality.

Editing also plays an enormous role in creating visual effects. For example, if one videotapes a film and then plays it back frame after frame, the methodology

FIGURE 15.3   *The skulls, shown here in (top) frontal and (bottom) profile views, show the jaw and teeth portions that could be extended out, along with an acrylic nose section. Space was left in the forehead area for another acrylic section that could extend the movement. Ear movement was controlled by the tubing devices seen on the sides of the skull, and the cheekbone extensions were being constructed (see frontal view). (Photos courtesy Rick Baker for Polygram Pictures.)*

FIGURE 15.4   *Sequence photos of Change-O-Head #1, showing facial distortion activated by the devices of the skull that stretched the outer plastic skin in the mouth, nose, cheekbone, and forehead areas. Facial hair was individually embedded in the face mask and a wig added to complete the effect. The body hair also increased during the sequence. (Photo courtesy Rick Baker for Polygram Pictures.)*

FIGURE 15.5   *During the filming on set, note the various cables, tubes, connectors, boards, and other devices operated by the technicians activating the Change-O-Head devices. (Photo courtesy Rick Baker for Polygram Pictures.)*

FIGURE 15.6   *With Change-O-Head #2, Rick made the sculpture asymmetrical, with the mask's left side more human and in pain while the right side is more wolflike. (Photos courtesy Rick Baker for Polygram Pictures.) (A) A test with the head, note area where teeth will be set. (B–D) The extensions continue and the face becomes more and more animalistic during the change. Note that the eyes remained closed during the Change-O-Head sequences so no false eyes were required.*

FIGURE 15.7   *The Change-O-Back photo shows the number of technicians required for the change to activate a number of bladders to ripple the spinal cord by means of various syringes that acted as pneumatic rams for the effect. The hands, made of slush molded latex and mounted on long sticks, were moved on each side by technicians. With the camera shooting downward, the hands were in their proper place in the frame. (Photos courtesy Rick Baker for Polygram Pictures.)*

the various Change-O effects. In these, first the actor's face is made up with various appliances (during cutaways), then the first Change-O-Head is used to its full extension, another cutaway to another part of the body changing, and then back to the face that is now Change-O-Head #2 that is starting its extension at the point that Change-O-Head #1 left off.

Body positioning to show just the head and neck, with the rest of the body a simulation, is sometimes used, as with Tom Savini's arrow-through-the-throat effect. Here only the head of the actor is seen through a pillow, but the neck and shoulders are an appliance. In *Werewolf,* Rick Baker used this same type of effect when the wolf-body transformation is taking place, by having the actor's upper torso showing through a hole in a false floor, and the wolf body attached (which, due to its conformation, could not possibly be done on a human body) is a dummy simulation.

Within the framework of these cooperative illusionary visuals the audience is fooled into believing what they *think* they see on the screen. The mind connects these illusions into a pattern of thought that shows the effect but is deceived by the camera work and the film editing during the sequence of the transformation, change, or final result in addition to whatever bladders or mechanical devices are utilized in the make-up to produce the changes.

CARL FULLERTON

Although in the production of Paramount's *Friday the 13th: Part II* much of the gore and effects that Carl Fullerton made for the film were cut in the final editing due to a clamp-down on explicit grisly make-up effects, there remained some good foam latex appliances. The character of Jason from the original film *Friday the 13th* has grown up into a deformed, retarded youth so Carl constructed a foamed latex head, along with false eyes, upper and lower dentures, and laid-on brows and beardline to go on the actor's face (Figure 15.8).

The decapitated head of another character from the earlier film is also shown, and this is a prop made with an epoxy substrate and covered with a skin of a slush molded latex (Figure 15.9). Epoxy eyes along with a set of acrylic teeth, a wig, and some dried blood effect all add to the realistic figure. The coloration was done with acrylic paints.

In *The Hunger,* which Carl worked on with his mentor, Dick Smith, he made a number of mummified figures. One of them, seen in Figure 15.10 with Carl, was a very tall and very thin man for which a full body suit (that zipped up the back) was made. Various latex products were employed to construct the suit.

Those mummies that had to crumble into bones and dust were largely made of a fragile wax compound with a very friable polyurethane foam as an interior

of the sequence is revealed. Take for example Tom Savini's throat-cutting sequence in *Friday the 13th.* One sees a frightened girl up against a tree and the back of a figure approaching with a knife. We see the knife gleam on the left side of the frame and sweep toward the girl. As it does, the figure effectively blocks out the scene for a few frames. Here the cutting of the film takes place. Then we see the girl lift up her head, revealing a cleverly concealed throat-cut appliance, the blood pumps out as the knife gleams on the right side of the screen, appearing to be a continuation of the slash. For safety, even the knife blade was a rubber copy made by Savini to look realistic. In this momentary film cut, the audience is fooled by a visual illusion of a continued action.

This same type of illusion of continued action is also accomplished by cutaways. In the case of the *Werewolf* transformation sequence, reaction shots of the actor's face are intercut with changes of appliances for

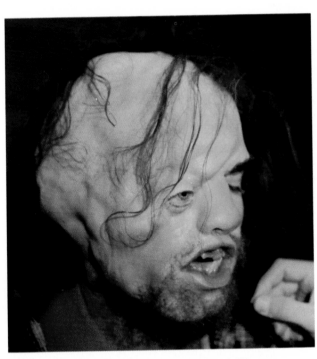

FIGURE 15.10   *Carl with one of his mummified figures.*

FIGURE 15.8   *Jason from* Friday the 13th: Part 2 *(Georgetown Productions, 1981).*

FIGURE 15.11   *Two mummies that were cast from human figures and constructed in the wax material that crumbled with very little agitation.*

support medium. Figures 15.10 and 15.11 show two examples of figures, one an appliance bodysuit make-up on a performer, and the other, puppets or props of similar characters, and both were designed and executed by the make-up department for the film to give continuity to the overall concept.

FIGURE 15.9   *A prop decapitated head by Carl Fullerton.*

SPASMS BY DICK SMITH

In July 1981, producer John Newton and Bill Fruit of Toronto came to my studio in Larchmont to talk to me about their film, *Death Bite* (later retitled *Spasms*). Earlier I had agreed to help Canadian special effects make-up artist Stephen Dupuis with an arm inflation effect for the film. Now Newton and Fruit had written a more elaborate effect into their modest film about a supersnake and wanted something special.

In their new scene, the bad guy is bitten by the monster and dies slowly and spectacularly. They knew about bladder effects (which I had developed for *Altered States*), but asked if I didn't have something new and exciting that hadn't been used before. I started to lecture them that much time and effort had to be expended to come up with such an effect when an idea floated into my consciousness. Years before I had used trichloroethane (a cleaning fluid) to raise weltlike letters on Linda Blair's foam latex stomach in the *Exorcist*. Recently, friends of mine, Steve Laporte and David Miller, inspired by the *Exorcist* effect, had used trichloroethane together with air bladders to change a lady into a grotesque hag. These associations pointed to an ideal solution because the swelling and distortion trichloroethane produces in foam latex resembles that of a poison.

I excitedly led Newton and Fruit to my basement where I demonstrated the effect on a scrap piece of foam. I knew from experience that the fluid would attack tears in the foam more extremely, so I made some in the foam piece. At first invisible, these tears became gaping wounds as their edges swelled up. It occurred to me that the trichloroethane could probably be colored, enhancing the illusion. We all agreed that the potential was excellent. Later, it was decided that in the first part of the scene, air bladders under a foam skin on the actor's face would cause it to swell up grotesquely. Then the camera would move to his arm, which would also swell. When it cut back to his face, a dummy would be substituted which would take the face to further stages of ghastliness through the use of trichloroethane and mechanics.

There were also a number of other make-up effects including the swelling arm of a sailor who is the first to be bitten, so Carl Fullerton joined Stephan Dupuis and myself on the project. We started by molding the head and arm of the bad guy, Al Waxman (Figure 15.12). On his life mask, an asymetrical mass of lumps was sculpted in plastalene as a means of designing the shapes and arrangement of the bladders. From this, patterns for the various inflatables were drawn on the plaster head as the clay models were removed. The patterns were transferred to flat plexiglass sheets on which the bladders were made of Smooth-On PMC-724.

Meanwhile, thin foam latex appliances were sculpted, molded, and processed to cover the bladders and restore Al Waxman's face to its normal appearance. A set of bladders was glued to his plaster head with the foam latex skin over them. Then on another Waxman head (made of wax), I sculpted a copy of how the first head looked when I inflated the bladders. This sculpture was for the head-and-shoulders dummy that would display the trichloroethane effect.

Next, an Ultracal 30 negative in two halves was made of the dummy sculpture. After removing the wax and plastalene, I modelled a thick layer of plastalene back into the negative, making it thickest in the large facial swellings. A positive core was now cast of Ultracal. After opening the mold and cleaning off the plastalene, a flexible urethane negative was made of the core. In this negative, a fibreglass "skull" about 3/16 inch thick was formed and removed, and the jaw was cut and hinged.

I had decided to increase the horror of the effect head by having the eyeballs bulge and the tongue protrude and swell up. The eyes were rigged to have side-to-side motion and move forward one inch. The tongue was a thick balloonlike structure that was moved forward on a hidden cradle. The head also had ample mouth and neck movement.

The main engineering problem was devising a plumbing system that would pump the trichloroethane through 15 small tubes into the inner surface of the thick foam latex mask that covered the fibreglass skull. Eventually, we used a painter's pressure pot, various mainfold switches, and thick-walled vinyl tubing to do the job.

Assembling the head for a take took hours. The mass of tubes had to be fed through the proper holes in the skull and connected to the manifolds and switches. The working of each small tube had to have a deflector on it so that no stream would squirt right through a mask wound. At each swollen bump on the mask, the foam had to be carefully torn clean through in an irregular, natural-looking way. The insides of these tears were painted bloody with watercolor (not affected by the trichloroethane). Then each tube was hot glued into the back end of the appropriate tear wound. Getting the mask onto the skull was a nightmare; so was the adjustment of the torn foam latex so the tears wouldn't show. Finally, the wig and eyebrows and final touches of make-up had to be applied. A partial test was made with one unpainted mask using an incomplete pumping system. It looked promising.

Finally the time came for a complete test of all systems. The pressure pot (5 gallon) resembles a pressure cooker, and compressed air goes in from an air line. The liquid in the pot is driven out a hose (usually leading to paint spray guns). Our hose led to a manifold that divided the liquid into four smaller tubes, each of which could be switched off. These tubes in turn were divided into four to six smaller tubes attached to the dummy face. We had found that uni-

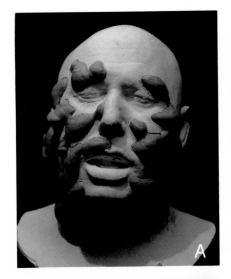

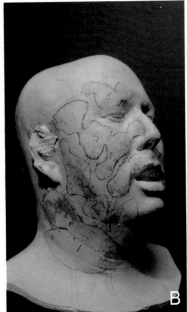

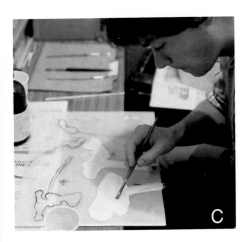

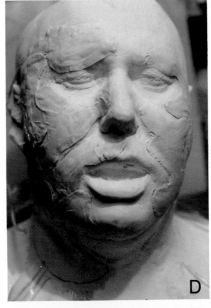

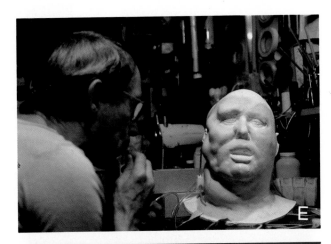

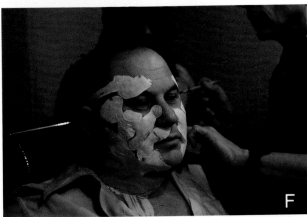

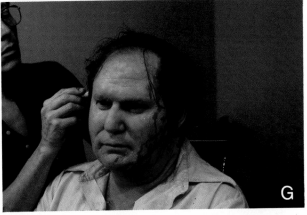

FIGURE 15.12   (A) Plastalene modelled on Al Waxman's life mask, showing the shape of the bladders to be made. (Photos courtesy Dick Smith.) (B) The bladder patterns outlined on the face, showing the necks to which tubing will be attached. (C) Carl Fullerton painting the bladders using Smooth-On PMA-724. (D) Trial bladders attached to the life mask for a test. (E) A thin, foamed latex face was made to cover the face and attached over the bladders. Dick tested the inflation of the bladders on one side of the face by blowing into the attached tubing.

(F) The bladders attached to the actor's face with Prosthetic Adhesive A. (G) The face covered with the thin latex skin and make-up applied produced a natural look. The blood is part of the effect. (H) The facial bladders inflated to their fullest on the set just prior to switching to a dummy to complete the effect. (I) Making the dummy head for the effect (next 11 photos). A plastalene sculpture on the life mask showing the extent of the inflated bladders at the end of the live sequence.

FIGURE 5.12 (cont).   (J) A two-piece Ultracal 30 mold of the head is finished with a Surform tool. (K) The two-piece mold is taken apart, the head and sculpture removed, and plastalene is added on each side. A core can now be made so a foam latex skin can be made to fit on the dummy. (L) Adding coarse weave burlap to strengthen the core of the dummy head. Note that the mold is strapped together firmly around the face portion. (M) The core. A polysulfide rubber mold is made of this to form the skull. (N) The skull or inside of the dummy head, showing the eyes and acrylic teeth inserted. Note the holes drilled for the plastic tubing.

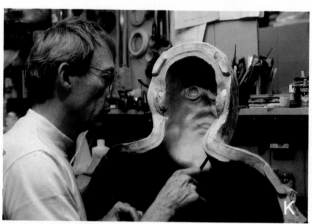

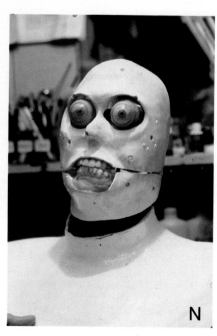

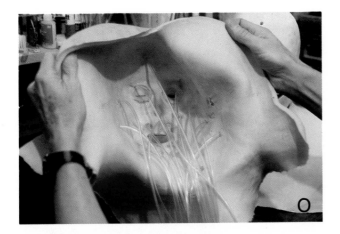

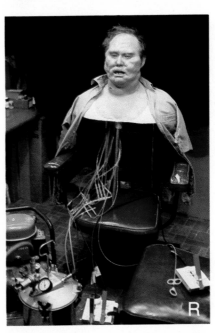

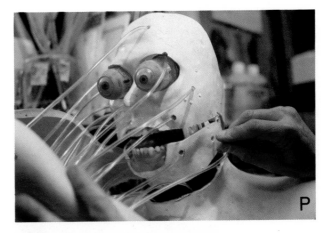

FIGURE 5.12 (cont).   (O) *Tubes for the trichloroethane attached to the inside of the dummy head's latex skin. (P) Fitting the tubes and skin over the dummy skull. This had to be redone a number of times to make the tests, with a new foamed latex mask made and fitted for each sequence. (Q) The first test with trichloroethane pumped into the lower part of the dummy head. (R) The dummy head fully assembled with all the apparatus for a film test. (S) The compressor and paint pot containing the trichloroethane, showing the switches, manifolds, and tubing.*

versal colorant would give the trichloroethane a sickly green color. We knew that the solvent could swell out tubing but hoped the effect would be all over before that became a problem. We had no idea how long it would take for the effect to work, so we set everything to pump as much trichloroethane into the foam latex as possible. To cover our test, we set up a 16mm camera and a still camera.

It was incredible—shocking! It all happened so fast that we could hardly believe it. In 20 to 30 seconds, the bumps swelled, burst open, kept swelling until the whole face was a mass of discolored shaking blobs. The bulging eyes and tongue were activated too late and hardly showed. Some of the wound tears looked too straight and artificial, but we were delighted with the general effect.

As a result of our test, I suggested that the film be overcranked when we shot the effect in Canada. At the same time, we slowed down our pumping. Unfortunately, the combination was too much and problems were had in editing. Also in our on-the-set tension,

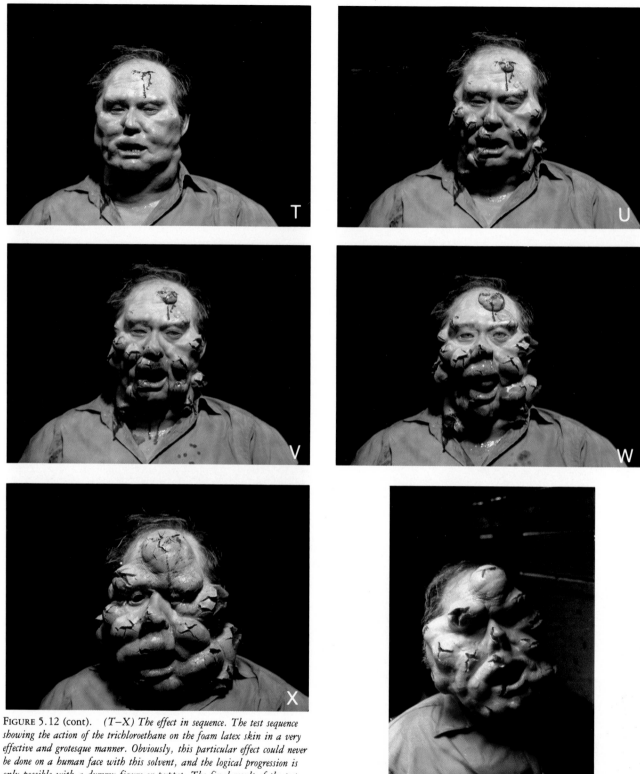

FIGURE 5.12 (cont).   (T–X) The effect in sequence. The test sequence showing the action of the trichloroethane on the foam latex skin in a very effective and grotesque manner. Obviously, this particular effect could never be done on a human face with this solvent, and the logical progression is only possible with a dummy figure or puppet. The final result of the test was spectacular—to say the least! (Y) The final effect as it was seen on screen.

we forgot to shake up the color in the pressure pot just before shooting so no green color came out. Nevertheless, the effect looked so good to the eye that it was not shot a second time.

I suspect this is not the last time that the effect of a hydrocarbon on foam latex will be used. Actually, it all started with a story I heard many years ago from a make-up artist who worked on *The Wizard of Oz*. The MGM make-up department had a diffident apprentice who was supposed to clean the Munchkins' foam latex appliances at the end of each day in a pail of acetone. Some older wags in the lab put carbon tetrachloride in the acetone supply can, and when the apprentice dumped his load of appliances into the carbon tet, they practically exploded out of the pail, to a chorus of feigned dismay and "Boy, are *you* in trouble!" and "Wait until Jack Dawn [department head] hears about this!" When I heard the story, I tried out the effect, and years later I remembered it—but this time used it to advantage!

The special bladders made for *Spasms* were constructed from four-part Smooth-On PMC-724 (also see page 168). First, the outline of the bladders was made in pencil on a sheet of plexiglass, and then two coats of the PMC-724 were painted on, which extended about 1/4 inch beyond the line. Next, a separating coat of plastic cap material was painted on just up to the line but not over the extension. Two coats of Smooth-On Sonite Wax separator were then added. On the long narrow neck made to attach the tubes for inflation, a piece of latex rubber coated with silicone grease was laid down. Then two thin coats of PMC-724 were brushed on, carrying the coating over the previously painted-on extension to seal the bladder edges. The formula for painting the bladders was 50 grams of 724 Part B combined with 15 drops of Part D. Then 5 grams of 724 Part C and 5.4 grams of Part A were added along with the tip-of-a-spatula amount of Monsanto Modaflow to control the flow qualities of the mix.

CHRISTOPHER TUCKER

Christopher Tucker, the leading British make-up artist, resides and works in an elegant manor house in Pangbourn, England, that combines the high ceilings, marble staircases, ornate carved-wood walls and furniture, and spaciousness of the houses of the Elizabethan through Georgian periods, with a complete series of workshops in many rooms, which provide him with a "factory" second to none in the world of make-up special effects. He has rooms to construct mechanicals—radio controlled and manual—a full stillphoto studio, laboratory areas for mixing foam, sculpture rooms, chemical laboratories, a huge conference room, and make-up rooms for applying the finished products. Distinctive in his approach to make-up conceptions, he enjoys creating a different avenue and resultant to achieving a make-up effect.

One of his recent films is *A Company of Wolves* (Cannon Films) for which he created a number of transformations of humans to wolves in various sequence forms. Rather than follow the usual path of humans

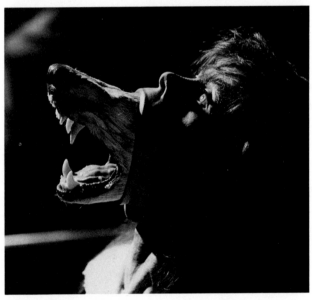

FIGURE 15.13   TOP: *The wolf's head, covered with real wolf's fur, emerging from the puppet head.* BOTTOM: *The wolf's head snarling. The elongation of the foamed latex required special foam construction as well as a lubrication so that it would not split during the action.* (Courtesy of Palace Pictures.)

to werewolves, his character changes were designed to show humans believably changing into wolves rather than wolf-like creatures. Some of the concepts were quite bizarre, such as the head of the wolf emerging from the open mouth of a dummy figure (see Figure 15.13).

This effect was achieved by making a life-sized head of the actor, Micha Borgese, with a foamed latex skin over a hard plastic form. The mouth of the head was lubricated so that a wolf's head could emerge, pneumatically forced through the mouth opening, open its mouth, and snarl. The eyes, neck, and head movements were controlled by cables.

Another metamorphosis was produced as a human went through a series of changes by first peeling off its human skin to show the underlying muscle structure. The features then began to elongate into lupine form as its body assumed an all-fours position to finally become a hairy wolf (see Figures 15.14–15.20).

An underlying appliance with a surface structure of muscle tissue covered by a thin foamed latex prosthesis resembling the actor's own face was applied for the first sequence in which he tears away his outer skin to reveal the muscle tissue beneath it. A skull cap was used to cover the actor's hair and during the sequence, a pair of yellow contact lenses were inserted into his eyes. A skeletal, muscular arm was used in conjunction with the actor's movements, and it was operated by two puppeteers.

From here, Tucker switched to dummy figures or puppets to complete the sequence, and named the first transmogrifications, Bert 1, 2, and 3.

The Berts were constructed of latex, foamed latex, silicone, and polyurethane foams, polyester resins, fibreglass and BJB's Skinflex. The mechanics were made at the same time as the fibreglass skeleton, and pneumatic air rams employing compressed nitrogen operated the movements. Jointly, a system of cables activated the head motions, and the eyes were radio controlled using micro servos. In addition, the eyelids worked, the teeth grew and changed their anatomy, the tongue articulated, the temples pulsated, and tiny tubes were inserted in the inner corners of the eyes so that tears could be simulated. Bert 1 had separate head and neck movements and the neck could also twist laterally and the jaw could open; a very versatilely constructed figure.

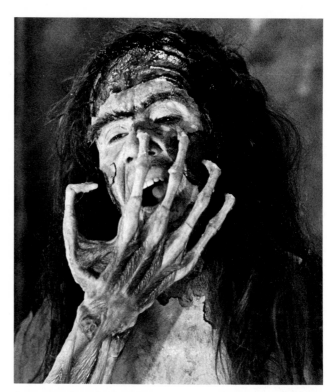

FIGURE 15.15   *The outer skin on Stephen Rea's face being torn away by the articulated mechanical skeletal arm. Note that portions of the face and body have been torn off. (Courtesy Palace Pictures.)*

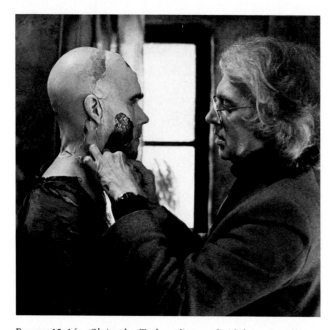

FIGURE 15.14   *Christopher Tucker adjusts and aid the tearing effect on Stephen Rea's face showing the outer skin and underlying muscle tissue structure. (Courtesy Palace Pictures.)*

After Bert 1 had been through his routine of changing as far as it was possible for him to do so, the entire torso was made to tilt forward and the head to come backwards, so that it would approximate the position of a wolf when standing normally. At this point the camera cut to Bert 2, which was designed so that the neck would grow and the shoulders would pulsate and rearrange their anatomy. This was done primarily with air bladders, although the neck extensions were produced by pushing the head and back through the shoulders manually.

Bert 3 had ears and a nose that would grow pneumatically. He also had radio-controlled eyes, a tongue that worked, and full neck movements. Although Berts 1 through 3 were constructed life-size, the next phase where the figure becomes a wolf was made 25% larger to accommodate the inner works. As the audience had no idea of scale, due to camera angles, the size difference did not affect the overall completed effect.

These figures were named Rover 1 and 2 and also had radio-controlled eyes, a working nose that could be raised and snarled together with the lips, the tongue worked and the mouth was capable of frothing. The figure could lunge forward and his flanks had built-in inflatable "lungs." Rover was constructed of latex and foamed latex and required ten people to work him in all his movements. The Berts needed up to fifteen operators as there were so many functions that had to occur simultaneously.

FIGURE 15.16   *The next stage with the Bert 1 head showing the rig control box with its many operators and controls. (Courtesy Palace Pictures.)*

FIGURE 15.17   *Bert 1 undergoing the pneumatical transformation. (Courtesy Palace Pictures.)*

After this plethora of transformations, it will be hard to do another werewolf picture with new effects—but you can be sure someone will try to challenge this!

STAN WINSTON

Two effects by Stan Winston using fibreglass resin underheads or skulls for puppet substitutions are shown in Figure 15.21.

A decapitation effect for the film *The Exterminator* starts with a fibreglass resin body, arms, and skull controlled by various cables (Figure 15.22). A Moto-

FIGURE 15.18   *Bert 1, the skinned man. (Courtesy Palace Pictures.)*

FIGURE 15.20   *Rover 1, the wolflike effect being completed. (Courtesy Palace Pictures.)*

FIGURE 15.19   *Bert 3 with the elongated jaw, teeth and head motion. (Courtesy Palace Pictures.)*

tool was used to drill the various holes needed for mechanical fitments, and the head was rigged to fall to the side during the action.

Foamed latex head, arms, and body appliances were fitted over the fibreglass forms. With hair and eyelashes added, a set of acrylic teeth and a natural make-up, the puppet is ready to perform. Once again, we have the substitution of a realistic dummy for an effect that was not possible with a human performer (Figure 15.23).

BASIC MAKE-UP EFFECTS

A number of basic make-up effects are called for on many occasions, while others often depend upon the ingenuity of the individual artist. Cuts, bruises, burns, tattoos, scars, sunburns, perspiration, tears, and such are simple make-up effects that come under this category. In the make-up artists' examination procedures for union membership, these effects are always called for under the heading of exercises, and most techniques for them can all be accomplished out of the kit that every make-up artist should carry.

Waxes

Of all the effects materials for the make-up artist, wax is one of the oldest and still most useful for temporary use. Old nose putty was a resin-wax mixture that required some manipulation with the fingers to get it soft enough to work with. Being a somewhat heavy material, it often slipped out of place or distorted easily in warm weather. It is seldom, if ever, used in films or television. Related to this is mortician's wax, which

is softer than nose putty but manufactured to be employed by morticians on cold bodies. It softens readily in the heat and has been replaced for professional use by the microwax materials. RCMA makes an excellent plastic wax material that holds its shape better and is far more adhesive than the former outmoded waxes. It comes in No-Color, Light, KW-3, KT-3, KN-5, and Violet shades.

Plastic Wax Material

The No-Color variety can be colored by melting it with a small amount of any Color Process foundation. However, the tinted varieties will suffice for most skins and effects.

One of the basic uses of Plastic Wax Material (PWM) is to make a change in nasal bone structure. First, coat the area to be built up by the PWM with a thin application of Matte Adhesive, and tack with the finger until set. Powder with No-Color Powder. Next, add a small piece of the PWM with a dental spatula to the area, and begin to smooth out and shape the desired nose change. When almost complete, the final shaping and poring can be done with a fine polyurethane stipple sponge or a red rubber one. Blend the edges carefully, and give the surface and just slightly beyond three coats of Matte Plastic Sealer, drying between coats with a hand-held hair dryer. Additional stippling can be done over this sealer coating to finish the nose. Color Process foundation can then be applied with a sponge in a stippling motion or with a brush. Powder with No-Color Powder. Moles, wound areas, or small build-ups can be done the same way.

Very realistic cuts can be made with a dull tool or

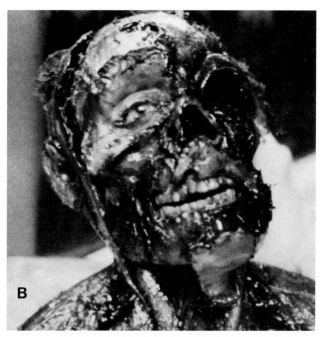

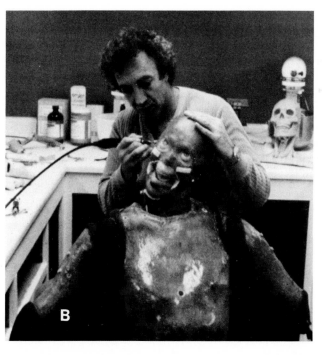

FIGURE 15.21 *(A) From the film,* Dead and Buried, *Stan Winston's basic plastalene sculpture was the basis for making a head that was covered with a foamed latex skin. (B) A burn effect created with gelatin appliances, gelatin blood, and various acrylic paints as well as the foam latex skin, false eye, and acrylic teeth. (Photos courtesy of Stan Winston.)*

FIGURE 15.22 *(A) Stan fitting controls on the skull. (B) On the bench are the latex arms and face that will be fitted over the skull section. (Photo courtesy Stan Winston.)*

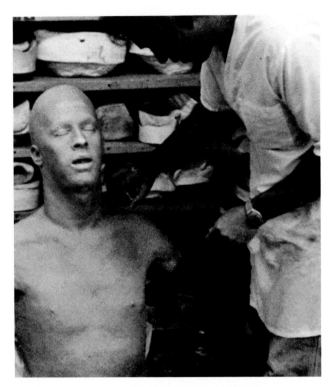

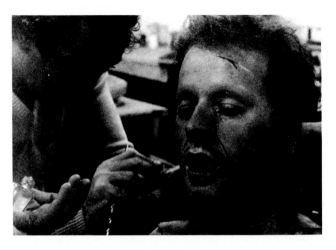

FIGURE 15.23   LEFT: *The foam head and body fitted over the fiberglass inner body.* RIGHT: *Note the decapitation line on the neck where the head of the dummy joins the dummy body. (Photo courtesy Stan Winston.)*

a dulled knife blade right through a Plastic Wax Material build-up made in the preceding manner. If a small blood tube is rigged on the knife with a rubber syringe of blood held in the palm of the hand, a good cut effect can be done. In addition to the PWM materials, a gelatin appliance can be used in the same way (see page 209 for gelatin formulas). For other uses of Plastic Wax Material, see the following sections on bruises, burns, and scars.

Dental Waxes

A number of grades of dental wax are useful. One is a soft pink shade used for simulating gum tissue (see page 40). Dental inlay wax is used to create teeth on a form (see page 40) or to fill a prominent space in the teeth. This is done by warming the wax with a dental spatula in a flame and placing the wax on a dry tooth. As soon as the wax hardens it may be carved with dental tools to the desired shape. A cotton swab can be used to buff the wax to give it a natural shine. This inlay wax is supplied in a number of light shades.

Black carding wax is excellent for the quick creation of missing or broken teeth. Simply cut out a small piece from the sheet of wax and press it over and around a previously dried tooth area. Don't get too much wax on the inside of the teeth as it might make it difficult to talk.

Wax for Molding

Suppliers of sculpture materials can also furnish various waxes for either carving or pouring into a form for certain effects. Breakaway heads can be made with some of these waxes as well as other body parts.

Glycerin

Glycerin can be used to create perspiration and is applied with a stipple sponge or a manual spray bottle. Placed in the corner of the eye (not *in* the eye), it will run and simulate tears. A very heavy grade of this product is furnished by RCMA, called *Tears and Perspiration.*

Burns

There are three categories of burns. First-degree burns redden the skin, second-degree burns blister the skin as well as redden it, and third-degree burns show charred flesh, with bleeding, blisters, and considerable reddening.

Most first-degree burn effects can be done with red foundation and raspberry cheekcolor, using a polyurethane sponge for application. Don't make the edges too even or the color too bright. Powder with No-Color Powder to avoid a shiny surface.

Second-degree burn blisters can be made with RCMA Scar or Blister Making Material (in the tube) for temporary effects or with the gelatin molding material (see page 209) warmed up and added to the desired area with a small spoon in large drops. One of the older ways was to place a blob of petroleum jelly on a piece of fishskin and invert it on to the area and blend off the edges. This technique can also be done with a film wrap (Saran Wrap) in small pieces. Broken blister effects can be made by laying down a coat of pure gum latex or eyelash adhesive, drying it, and lifting the center of the area with a small dental spatula. Always surround second-degree burns with the reddening of the first-degree burn.

Third-degree burns exhibit charred and broken tissue that can be created with Plastic Wax Material or an appliance in latex or plastic, and deep burn areas can be blackened with black foundation. For the broken and cracked tissue, use the raspberry and red colors, and add some RCMA Type C tube Blood as well. A slight run of blood from the cracks can be done with RCMA Type A or Type B Blood. Surround with second-degree blisters and first-degree reddening. RCMA makes a Burn Kit with four colors in one container.

Bruises

Realistic bruises can be simulated with RCMA Color Wheel Violet applied with a brush or sponge. Older bruises should have a surrounding of Color Wheel Bruise Yellow. Bruise areas can be made deeper with a bit of black added to the violet or lighter with some red. RCMA makes a four-color Bruise Kit with these shades. Raised bruises can be made with Plastic Wax Material in the violet shade for temporary buildups or with foamed latex or Plastic Molding Material and covered with the violet foundation.

Blood Effects

There are many forms of blood effects and many materials that appear as blood for production work. Some show the free flow of blood for active bleeding, while others must hold a static effect when the scene takes a while to shoot and the blood must appear in the same place all the time.

Flowing Blood Types

RCMA makes a Blood Type A, which is a liquid that will flow very much like real blood and is easily washed off with water. Similar types have been made with Karo or other corn syrups that have been tinted with vegetable coloration. Some make-up artists use a thin gelatin mixture with color for this effect. The 1-ounce RCMA squeeze bottle is handy as the blood effect can be applied by the drop or in a stream for running blood. The cap effectively seals the plastic bottle, which fits well in the kit.

This variety of blood effect can also be put into clear plastic bags and placed over small explosive charges such as those employed by the special effects personnel to simulate a bullet hit. Take care that they are not overused as the effect will appear ridiculous if the blood bursts out of an area where the bullet hits. Often, it is better to see the exit of the imaginary bullet when using the explosive charge and the blood bags.

Appliances can be made of PMA Molding Material, foamed latex, or similar materials to cover an area of the body in a thin coating under which is imbedded a small polyethylene plastic tube that leads to a syringe full of Blood Type A. At the proper time, a cut can be simulated by passing a dull knife over a precut area or a small precut plug pulled with an invisible nylon thread from a simulated bullet hole and, with pressure

on the syringe, the blood made to appear to flow from the wound. There are many variations of this method that can also be done with the knife blade having the small tube taped to it and a rubber ear syringe held in the palm on the other end of the tube. When the cut is made, the hand squeezes the syringe at the same time and the blood flows from the filled syringe through the small plastic tube and down the knife blade. For safety's sake, the knife blade should always be dulled and rounded smooth so it will never create an injury. Tom Savini makes many such weapons out of rubber and then paints them in a very realistic fashion to simulate the real thing.

Nonflowing Blood Types

The first nonflowing blood type is a thin plastic style that is useful for long scenes as it will dry in place wherever it is applied in a very short time but still appears to be fresh and running (RCMA Blood Type B). Another variety is the soft creme one that comes in a tube and can be used for smeared effects or for quick wound applications. It does not dry out and stays fresh looking (RCMA Blood Type C).

Dried human blood is reddish brown in color, and RCMA Blood Type D is a liquid that dries rapidly and leaves an effect similar to dried blood. It can be used on bandages and wardrobe.

Scabs or clotted blood on wounds can be made by mixing a dark wood flour (such as walnut or mahogany dust from a sander) with RCMA Prosthetic Adhesive A. First, dampen the wood flour with a few drops of Prosthetic Adhesive-A Thinner, then mix in some adhesive until the flour is the consistency of peanut butter. Apply directly to the skin (as it will dry rapidly) in the shapes desired. This mixture will hold to the skin very well, even when wet. It is removed with RCMA Adhesive Remover. These scabs can also be premade on a glass plate and then attached to the skin with Prosthetic Adhesive-A.

Scars

Raised scars can be made with latex or plastic appliances or for temporary use with Plastic Wax Material (Figures 15.24 to 15.26). They can also be formed with the tube variety of RCMA Scar or Blister Making Material by squeezing out some plastic from the tube and forming it with a dental spatula. Thickened latex (see page 189) can also be used to form raised scars. When the surface has set slightly, it can be dried with a hand-held hair dryer to form a skin and then be colored with foundation.

Indented or deeply serrated scars can be made with RCMA Scar Material, which dries semimatte with a slightly pinked tone. Simply apply with a brush on the skin in a line or roughly outlined area, and allow to air dry. The dried scar will pucker the skin and can be slightly colored with red and white foundation to

FIGURE 15.24   *Ray Codi with a long and deep scar. The eye is pulled downward by the shrinking action of the scar material.*

FIGURE 15.25   *Lon Chaney with deep scars under his eyes for the 1944 film,* Dead Man's Eyes *(Universal Pictures).*

FIGURE 15.26   *Raised cicatrix scarring on the character Queequeeg in* Moby Dick. *(Warner Bros./Moulin, 1956.)*

accent it. Cotton batting built up in successive layers with Matte Plastic Sealer and then colored with foundation can also be used for scar tissue.

Nonflexible collodion was formerly used to create incised scars, but it dried with a high shine that was hard to tone down, and the edges often came loose as it dried quite hard on the skin. Most scars for production use are now made with foamed latex or with Plastic Molding Material preformed in molds and attached to the skin in the usual manner.

Tattoos

Temporary tattoos can be made with blue ball-point pens, filling the area with a red lip-lining pencil (Figure 15.27). For a matching tattoo each day, rubber stamps can be made and applied to the skin. The inner parts of the outline can then be filled with a red lipliner.

To get a perspective for a full body tattoo, it is a good idea to sketch out the planned drawings on a store dummy using fine felt-tipped pens before doing the actual work on the body (Figure 15.28). Tattoo artists generally use some form of transfer paper to delineate their designs with holes punched through them and a powder applied to transfer the design.

Reel Creations, a firm formed by two make-up artists in California, Fred Blau and Mike Hancock, has

FIGURE 15.27 *Various tattoo designs.*

produced a kit for professional body art that consists of colors, transfer designs, brushes, and full instructions on their use. They perfected the method doing the make-up for the film, *Tattoo,* and with a little care, the application will last several days. For this long-lasting effect, no oils or lotions should be applied to the body art area, and although one can shower, rubbing with a wash cloth should be avoided. The body can be powdered with talc or No-Color Powder before retiring to absorb any possible body oils. Faded designs can be touched up by reapplication of the colors. Removal can be made with their developer or isopropyl alcohol, followed by soap and water.

Application of the designs with the body art materials is simple (Figure 15.29).

VOLATILE SOLVENTS

A list of some of the volatile solvents is useful for differentiating them. Many are *not* used in the manufacture of any professional make-up products, and

FIGURE 15.28 *Tattoo designs by Ben Lane on a dummy for a film test.*

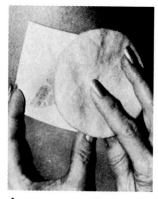

A   B   C   D

FIGURE 15.29   *Application of tattoos. (A) Cleanse skin area where transfer is to be located with special developer/remover to remove all excess body oil or debris. (B) Pre-moisten velour puff with developer. Just enough to dampen. Not dripping wet. Place transfer face down on skin. Press puff on transfer taking care not to move transfer (to prevent smudging). Transfer should be complete in about 20 seconds. Peel off transfer smoothly. (C) Powder your finished transfer design with talc. (Use any talc, body powder, etc.) The transfer design is now set. If at this point the* outline *is not uniform you may outline with the black paint. Or, if lightly smudged, clean the smudge off with cotton swab and developer/remover. (D) Select the colors you wish for your design (be sure to shake colors thoroughly) and begin painting, working each area selected. When total design is finished powder again and then gently wipe off powder with water and your creation is completed. Note: Powder is not necessary. It only acts to set and dry the artwork quickly.*

most should be only used when proper ventilation is available as some produce fumes that are absorbed by certain human organs and will cause serious internal disorders from prolonged breathing.

ALCOHOLS
Ethanol
Methanol
Propanol
Isopropanol
Butanol
Sec-butanol
Amyl alcohol
2-ethyl 1-hexanol
Cyclohexanol

ESTERS
Ethyl lactate
Ethyl acetate
Sec butyl acetate

ETHERS
Ethyl ether
Methyl ether
Ethyl vinyl ether
Isobutyl vinyl ether
Dioxane

HYDROCARBONS
Petroleum ether
Hexane
Stoddard solvent
Cyclohexane
Methycyclohexane
Tetralin (DuPont)

POLYHYDRIC ALCOHOLS
Glycerol or glycerin
Propylene glycol
Polyethylene glycol
Sorbitol
Polyoxyethylene
sorbitol

ETHER ALCOHOLS
GAF Gafcol glycol
ether
Diethylene glycol
Triethylene glycol
Hexamethylene glycol
Polyethylene glycol
400
2.2 thiodiethanol

KETONES
Acetone
Methyl ethyl ketone
(MEK)
Methylcyclohexanone
2 butanone

ALIPHATIC AMINES
Ethylene diamine
Diethylenetriamine

AROMATIC
HYDROCARBONS
Toluene
Xylene
Benzene

CHLORINATED
HYDROCARBONS
Chloroform
Ethylene dichloride
Methylene dichloride
Carbon tetrachloride
Chlorobenzene

ALIPHATIC
HYDROCARBONS
Naphtha
Heptane
Kerosene (refined)
Mineral spirits
Di-isobutylene
Mineral oil

Solvents may also be listed under the headings of nonpolar, or non-H-bonding, medium polar, or high polar:

NONPOLAR	MEDIUM POLAR	HIGH POLAR
Benzene	Linseed oil	Ethanol
Toluene	Dioctyl	Methanol
Xylene	phthalate	Propanol
Turpentine	Cellosolve	Butanol
Mineral spirits	Polyglycols	Ethyl acetate
Naphtha	Polyester resins	Ethylene glycol
Hexane	Epoxy resins	Butyl acetate
Heptane	Alkalyd resins	Glycerol
Mineral Oil	Polyurethane	Acetone
Carbon	resins	Methyl ethyl
tetrachloride		ketone
		(MEK)

The main solvents employed in special materials for make-up purposes are isopropanol (isopropyl alcohol), acetone, MEK, refined kerosene, and trichlorotrifluoroethane. Of these, only MEK has a really pungent residual odor as it is slower drying than acetone (which is the fastest of the solvents to evaporate, with TCTFE not too far behind). These solvents are part of the formulations of most adhesives and sealers and are relatively safe for use (MEK is not recommended for any skin use). Again, always use with proper ventilation. Many of the hydrocarbon products should be used only with extreme care.

THICKENING AGENTS

Dow Chemical makes a number of grades of *Methocel* (methyl cellulose) that are film-forming, thickening agents for water solutions. The standard solution is 2 percent Methocel, first dispersed in about one-third of the amount required of hot water. When the powder is completely wetted, the other two-thirds of the water (cold) can be added. The mixture should then be stirred until smooth.

Carbopol 934 (B.F. Goodrich Chemical Co.) is a water-soluble resin with thixotropic qualities. A 0.5 percent solution mixed in a high shear blender is neutralized (mixing pH is about 3) with 10 percent sodium hydroxide solution. This will produce a stiff gel that may be diluted with water.

Previously mentioned *Cab-O-Sil* (see page 210) is useful in thickening petroleum products and oils in addition to its other uses. A 5 percent dispersion produces a very stiff gel.

Wyoming Bentonite is a colloidal montmorillonite clay that will absorb many times its weight in water, swelling to several times its original volume and forming a thixotropic gel. A mud is prepared by sifting the clay into water while mixing under high shear. These thickening materials and gels can be employed for various effects where inert jellylike substances are called for. Samples for research and developmental use can be obtained from most manufacturers along with technical information and specifications. Also see latex thickening agents on page 189.

Part

V

Hair Goods and the Hair Kit

Fail here and you'll never really become a make-up artist!

ALTHOUGH THERE IS NORMALLY A UNION CATEGORY SEP-aration between hairstylist and make-up artist on most screen productions, essentially the hairstylist does the women's hair and the make-up artist is responsible for men's facial and head hair. The make-up artist can also apply wigs to either men or women but may be assisted with the dressing of the hair (or even the application of a hairpiece or wig) by the hairstylist. As the two departments are very closely allied, there is seldom, if ever, any friction between them, and cooperation and coordination are the general rule.

Also, while hairstylists have their own variety of kit and tools, the make-up artist should always have a *hair kit* for use as well. The carrying case is usually a soft canvas, over-the-shoulder bag with a zipper closing and outside pockets and serves to carry extra make-up supplies as well as hair tools and goods.

No part of character make-up is as important as or used as frequently as the application of hair and hair goods. Many men wear toupees or small hairpieces to augment their natural growth, and, just as the nose is ordinarily the first change area to apply a prosthetic,

so is the moustache used to effect a basic change. Just as the make-up kit should contain all the essentials for the normal application of make-up, plus some contingency materials for special uses, so should the hair kit be supplied with a normal selection of prepared hair, plus possibly a few moustaches on lace in case a change must be made in one of the characterizations. All in all, to excel in the selection and handling of hair takes many years of experience and practice but is a most necessary part of the make-up artist's knowledge.

Few professional make-up artists make their own wigs or lace facial hairpieces as it is a very specialized field, and it is more feasible—and profitable in the long run—to have professionals in the wigmaking business do this for them from the make-up artist's designs. R. Turner Wilcox's *The Mode in Hats and Headdress* (1949) and Richard Corson's *Fashions in Hair* (1965) both contain many drawings of hairstyles and period hairdress that a wigmaker will find easy to employ for suggestions in making a certain or proper style of hairdress or facial hair on lace.

CHAPTER 16

Types of Hairs and Their Uses

HAIR CAN BE DIVIDED INTO MANY CATEGORIES, WITH EACH type having either a specific or a variety of uses. Among the most useful hairs are wool crepe, real hair crepe, yak, angora goat, horsehair, and human hair (Figure 16.1). Human hair is also divided into many classes and grades for various uses (see Appendix B for sources of hair goods and supplies).

WOOL CREPE HAIR

This is the basic and least expensive of the hair types and is the one that is generally most available from suppliers of make-up products. It comes braided over two strands of cord that must be removed to use the hair. One of the best suppliers is Stein's of New York, who imports a superior grade of wool crepe.

To prepare wool crepe hair for use, cut a piece of braided hair about a foot long, and remove the two cords. You will then have a length of creped or wrinkled hair that must have some of the curl taken out. The simplest way is to dip the length of hair in very hot water and squeeze it out on paper towels. While holding it up by one end, dry out the hair with a hand-held hair dryer, reversing your hold from end to end as the water is dried out. Make certain it is fully dried before attempting to separate the hair as it will mat together. It can also be steamed under a damp cloth with a regular iron or run over with a steam iron once or twice. Don't remove all the curl as the hair loses its body; so leave a slight wave in it for use.

Wool crepe hair does not have really long hairs but many hairs held together by overlapping, so a beard, sideburns, or moustache made from this type hair

FIGURE 16.1 *Hair materials.* TOP LEFT: *Wool crepe in the braid and straightened.* TOP RIGHT: *Hair crepe in the braid and prepared.* BOTTOM LEFT: *Angora goat, yak, and horse hair.* BOTTOM RIGHT: *Combings (left) and cut hair (right).*

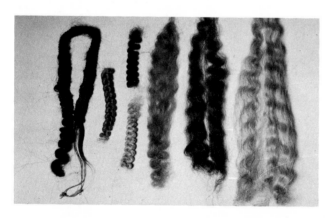

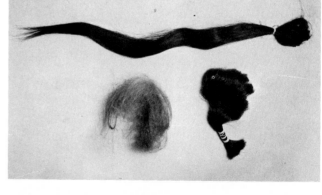

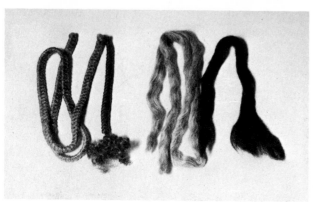

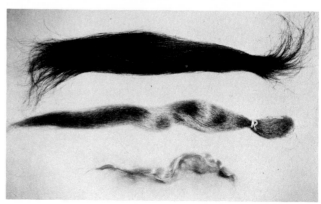

more delicate and will take less combing than yak or real hair. As such, once the skein is dried, it can be cut into lengths of about 4 inches. Each of these lengths is then carefully teased apart laterally to form a sheet of hair about 4 by 4 inches. Next, divide the sheet into approximately 1-inch widths, and pass a wide-tooth metal comb through the ends of the hair (holding the length by the middle). Cut off one end square, and lay it down on a single sheet of facial tissue. Do this with the other three 1-inch lengths, and fold over the tissue and set it aside for later use. In this manner a quantity of wool crepe hair can be prepared for immediate use. A variety of shades of hair should be made ready as such.

Wool crepe hair comes in white, blondes, browns, black, and mixtures of white and black to make various grays or mixtures of brown and gray. The latter is the most useful as it does not have the hard or unnatural look that a one-color wool crepe beard can often have. Special colors such as reds, greens, and so forth are sometimes available or can be dyed from white using regular wool dyes. Mixing shades of wool crepe is easy when the hair has been prepared as described here, simply by placing one mat of hair over another and then pulling the ends to tug apart the hair into two sections. Repeated laying over each other and then tugging apart will mix the hair colors thoroughly to achieve a better match to the performers' beard hair.

During the combing, in preparing the wool crepe hair into mats, one will find that they are left with balls of hair from the combings. These combings can then be end tugged back and forth to remove the longest hairs (normally they will be about 2 inches long) and mats of these can be formed that are about an inch wide and, again, laid across in tissues. Such lengths are good for moustaches or short sideburns. The remainder of the hair combed out will be quite short and should be rolled up into a tight, round length. Holding this tightly between the thumb and forefinger, with the hair scissors, cut across the grain of the roll in fine cuts, so that the hair is trimmed only some 1/64 to 1/32 inch long into a small plastic container. This finely cut hair can be used to make realistic unshaven effects (Figure 12.36).

As wool crepe hair does not have the shine, weight, or body of real human beard hair, it must be sprayed after application to give it a more natural appearance. However, it is seldom used for any principle photography today and is mostly employed for extras or on large stages (see Figure 17.1 for the laying of hair).

One trick of using wool crepe hair for very large casts where beards are seen in profusion in the background is to work with it directly out of the braid. The strings are removed on a length of hair, and without wetting it or steaming it, 6-inch lengths are drawn out of the skein. These are then laterally teased to form mats about 3 inches wide with one end trimmed. This method is used only for speed and a distant effect of beard hair as laying such on the face does not produce a very natural look. Hair used as such would be quite difficult to blend by teasing mats together as one would end up with a large blob of hair due to its curliness.

REAL CREPE HAIR

This is real human and/or animal hair that has been braided together in the same form as wool crepe hair. It is available only from some major hair suppliers. When the cords are removed, the skeins of hair will be 6 to 8 inches in length. To remove some of the curl, the skein of hair can be steamed with an iron and then stretched with the hands. It must be hackled to remove the short ends and then blended with other shades to achieve the desired color. Although there are many shades available for beard use, the hair should always be mixed (unless the beard is pure white or black) in shade to appear natural. (See Chapter 17, Figure 17.2, for application techniques.)

YAK HAIR

Yak hair is from the animal of the same name and comes from Asia. It is usually supplied in long, straight hanks 12 to 18 inches in length and of various grades, from soft to rather stiff, depending upon the part of the animal from which the hair was cut. Although some artists use this hair straight and curl it to the desired dress after application, it is best to run it through the pleating machine to give it body and looseness. It can be mixed with human types of hair to give it more stiffness, and as it is a thick hair, it is excellent for making laid-on beards and other facial hairpieces.

ANGORA GOAT HAIR

This is a specialized type of hair that is quite soft and fine and is sometimes blended with yak hair for Santa Claus beards or period powder wigs. It can also be blended with other hairs to add softness.

HORSEHAIR

British clerical, legal, and judicial wigs are usually made of this very coarse and stiff hair because, once dressed, they require very little care. Horsehair is also useful when animal muzzle hair must be simulated in an appliance because of its stiffness and lack of curl.

HUMAN HAIR

Natural human hair for wigs, hairpieces, falls, and lace beards, moustaches, and sideburns can be divided into two general categories: cut hair and combings (Figure

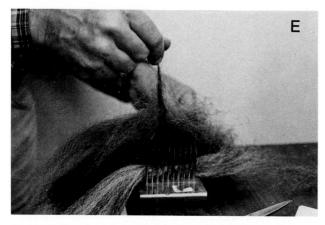

FIGURE 16.2   *Preparing real hair for laying. (A) Leo Lotito shows two colors of tightly crimped European hair. (B) Hand blending the two shades of hair by pulling them apart a few times before hackling. (C) Hair being run through a hackle. First the twist, (D) then pulling it through. (E) Two shades of hair can be held in the hackle at one time by laying in a small twisted skein of hair crosswise to separate one from the other. (F) Lifting up the hair to separate it. (G) The mixed hair.*

(H) Varieties of mixed hair, showing three colors that might be used to make a beard. (I) A small bunch of hair held between the thumb and forefinger is cut on a slant, then twisted to fan out the hair ends for attaching the hair properly to the face.

16.2). Cut hair is normally *rooted* hair—that is, with the natural hair growth all in one direction so that the hair will maintain the ability to be dressed in the same manner as growing hair. If strands of hair are added in an opposite direction, the hair will kink and not lie flat. Leo Lotito, department head of TBS (The Burbank Studios in California), explains that this is because human hair has a smooth direction (roots to ends) and a rough direction (back the other way). All hair used for hairpieces or toupees and fine wigs use rooted hair, as do falls and switches, so the hair can be properly dressed. Prepared hair for laying beards and so forth need not be rooted hair as a pleat or ripple will be added to provide body. Such hair can be combings (the hair combed out of rooted hair during preparation) or mixed hair.

Most real human hair that is used for wigs and other hairpieces today comes from Europe, and fine European hair is preferred by wigmakers over Oriental hair, which is coarser and more difficult to color. Most of this real hair from Europe is first bleached out and then dyed for use in hairpieces or wigs. All human hair is cleaned, matched, graded, and bundled for export by highly trained workers before shipment and, like all hair goods, is sold by weight (except crepe wool, which is available by the yard).

ARTIFICIAL HAIR

Many excellent women's wigs and hairpieces are made with the fibres Elura and Kanekalon, which are artificially created plastic hairs. Early experiments with synthetic hair produced a glossy, unnatural fibre, but today these materials are almost indistinguishable from human hair wigs. Although they lack some of the fine dressability of human hair pieces, they are far cheaper in price when that is a factor on a production. Wiglets, cascades, braids, switches, and falls as well as some men's hairpieces are made with these fibres, and they can be hand tied just like real hair. Much can be done with these new fibres, and many rental period wigs are now made of them as they will withstand rougher treatment than a good hand-tied lace wig of human hair (see Appendix B for suppliers). Synthetic hair is seldom, if ever, utilized for laying facial hair as human or animal hair normally handles better, is easier to curl and dress, and has more workability for this type of use.

Acrylic Hair Shades

Artificial hair comes in a number of colors and blends that are numbered from the darkest to the lightest shades. Single colors are numbered 1, 1B, 2, 4, 5, 6,

8, 10, 12, 16, 18, 20, 22, 23, 24, 27, 30, 32, 33, 101, and 107. Grays are numbered 34, 36, 38, 39, 44, 51, 56, 59, and 60. Frosted shades are 6/22, 8/22, 10/22, 16/22, and 18/22. For easy selection of these fibres, most wigmakers will supply hair swatches with the numbers attached.

One advantage of acrylics such as Elura is that they can be cleaned with a mild wig cleanser in cold water, swishing them gently back and forth. They should then be rinsed in two baths of cold water, drained, and blotted gently with a Turkish towel. They can then be hung over a hair spray can to drip dry. When they are dry, they can be brushed to shape according to René of Paris, a major supplier of Elura and Kanekalon blend wigs and hairpieces.

TOOLS OF THE TRADE

The basic tools are combs and scissors, but since there are many varieties of each, preferences will occur. In the main, a wide-toothed, aluminum tailcomb is the most practical for the make-up artist as it can be cleaned off in strong solvents when gums or resins get on it, and the tail can be used to tease up a flat portion of laid or dressed hair or to press down an edge on a beard. Hair or barber's scissors should be used *only* to cut hair—nothing else! Some prefer a long-bladed type, while others select one that has only 3-inch blades. Whatever feels comfortable and works the best for the individual is the right length of blade.

A hackle is a tool that looks like a bed of nails and is usually about 2 by 6 inches in size with 2-inch spikes. This is clamped to a table top and is used to work, mix, and comb long lengths of human or yak hair. The hackle should be covered after use with a 1-inch board of balsa wood to prevent anyone from inadvertently leaning on it.

A small curling iron electric heater and a series of curling irons, from small moustache size to large curl variety, are required for real hair use. Neither the hackle or the curling irons is ever employed with wool crepe hair. Most make-up artists prefer the smallest curved moustache irons for their beard work.

In addition, a pair each of thinning shears and small pinking shears (for cutting lace) are useful, as are some small hair brushes. Head forms in hard wood, fabric, and balsa wood are all used in the wig-making industry to hold wigs, hairpieces, beards, moustaches, and so on during manufacture, preparation, and dressing. Wigmaker's needles, a small hammer, and some brads and other related tools are part of the wigmaker's tool kit.

In addition, two or three wig block holders that clamp to table tops are useful for holding head and beard forms as are some curling rollers, hairpins, bobby pins, hair clips, and small elastic bands. One odd tool or apparatus for preparing straight yak or human hair into a useable, natural appearing, slightly wavy variety is an oldtime neckcloth pleating machine that can sometimes be hunted down in secondhand or antique shops (Figure 16.3). This machine will quickly crimp or pleat hair passed through its preheated rollers and provide a marvelous workability and body to straight hair. The two brass rollers can be heated with insert metal rods that have been preheated in the curling iron stove, or the machine can be permanently electrified.

FIGURE 16.3   *A pleating machine.* LEFT: *The machine in use.* RIGHT: *Pleated mixture of yak and real hair ready for use.*

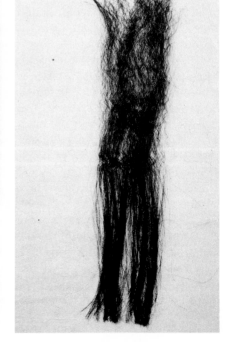

Some make-up artists purchase electric curling irons and insert them into the hollow brass rollers to heat them up for use. In use, the hair is passed through one end of the machine rollers as the handle is slowly turned. The hair will pass through, and the heated rollers will put a nice crimp in the hair.

A can of aerosol hair spray or the pump type is required as are various hair adhesives.

THE HAIR KIT

The on-set hair kit of the make-up artist generally consists of a soft canvas or simulated leather over-the-shoulder or hand-carry convertible zipper bag. A large camera bag or a carry-all style is excellent for this use. The bag should have a number of outside pockets that are zippered (in the case of camera bags) so some of the small items can be carried in plastic boxes or clear plastic bags for ease in finding them. Each make-up artist will work out the best filing system, but like the make-up kit, this hair kit should be kept clean and organized for immediate use.

The large items will fit into the inner compartment of the bag, so it is a good idea to gather all the materials and tools together first and see just how large one wants the case to be. Although some make-up people will carry more than others or the job may require it, it is always best to have a large case for contingencies. Normally, the cases are lightweight, and it will only be what is put into it that will give it the carrying weight.

One of the most instructive things that a make-up artist can do is to compare his or her kit with another artist's to see if and how improvements can be made, to compare products and tools, and to check on the arrangement of the items carried, both in the make-up and the hair kit. A good professional will always share ideas and innovations with others in the field, but do remember to *ask* first as the majority does not take kindly to someone rummaging around in private kits unless permission has been granted.

Basic Items

Heater for curling irons;
Hackle with balsa wood top;
Hackle clamps;
Wig block holder;
Moustache iron and larger iron;
Hair scissors, thinning shears, pinking shears;
Clear Dermicel adhesive tape in 2- and 1-inch widths;
Pair of nylon stockings in flesh color;
Bag of prepared real hair for facial hair;
Box or bag of prepared wool crepe hair plus some in braids;

Small hairbrush with rattail end;
Wide-tooth metal rattail combs;
Box of hair and bobby pins;
Metal and plastic hair clips;
Color spray in brown, blonde, frosty white, and silver;
Hair spray in aerosol or pump-top spray;
Three huck towels, small roll of paper towels, tissues;
Safety razor and shaving cream;
Good hand-held hair dryer (not one with heat cutoff);
Selection of elastic bands;
Needles and thread (heavy linen, clear plastic, and regular cotton);
Plastic make-up capes (two are usually sufficient);
Pump-top spray container for water (small);
Can of Sterno (canned heat) and a package of matches;
Toupee tape (both sides adhesive) like Secure;
8-ounce bottles of RCMA Adhesive Remover and acetone;
Selection of moustaches on lace in a plastic box;
Two wig-cleaning brushes.

Optional Items

Electric razor;
Clip-light and bulb;
Extension cord (heavy duty) with three-way adapter;
Plastic bag with some net for repairing hairpieces, silk muslin for lifts and other uses, pink and yellow tissue for adding to Old Age Stipple;
Small plastic bags and gelatin capsules for blood effects;
Extra make-up supplies that will not fit in the kit;
Business cards, a notebook, an extra pen;
Personal items such as a toothbrush and toothpaste plus some mouthwash for after lunch;
Shooting script, extra timecards for the production;
Polaroid and/or small 35mm camera and film (where permitted);
Wigmaker's needles and cut-head needles (see pages 220 and 257);
Two plastic bald caps;
One-half roll of Bounty Paper Towels.

CHAPTER 17

Application Techniques

MAKE-UP ARTISTS SHOULD STUDY THE GROWTH PATTERN of hair on the human body for basic character work and on animals for any specialized application. Although women's head hair is seldom in the province of the make-up artist as far as dressing and care, men's head, facial, and body hair may have to be duplicated, augmented, styled, or created for a production. These growth patterns govern the application methods that must be used to make the character being portrayed as natural as possible.

FACIAL HAIR

Facial hair can be grown by the actor, but sometimes this is either not feasible because of the style or length of growth required for the production, because changes must be made from shaven to a beard within the day's schedule, because the actor and/or director prefer a temporary growth, or sometimes because the actor cannot grow what is required. Two types of hair applications are possible for facial hair: One type is *laid* on the skin with loose hair and an adhesive, then dressed and trimmed to shape, and the other is a prepared hairpiece on a base, fully trimmed and dressed, ready to be attached. Both have their uses and both are widely employed for production work.

Laid Hair

There are many methods and styles of laying hair on the body or face, and although there are basic techniques that are rather standardized, many make-up artists work out their own special ways and means of doing this. One of the major advances in laying hair has been the advent of the Matte Adhesives as a great improvement over old spirit gums. No matter what one did to avoid the shine of the old resin/solvent gums, the result was always an eventual unwanted degree of reflectivity that was hard to disguise on camera. Matte Adhesives solved many of these problems (see pages 37 and 210 for discussion and information on adhesives).

To lay hair directly on the face for a small beard on a performer, first coat the under-chin area with Matte Adhesive with a #10 round sable brush (use the small brush that may be furnished with the container only for retouching small areas). Then with the finger, tack

the adhesive until it is almost dry. Wipe off the finger with isopropyl alcohol, and coat the area with another thin coat of adhesive; two thin coats are better than one heavy one. Once again, tack the skin with the finger to remove some of the shine and set the surface, then start applying the hair at about the point of the chin. While prepared wool crepe hair can be laid flat to the skin, real hair should be trimmed on an angle and then twisted between the thumb and forefinger to form a cuplike shape so that only the ends of the hair will be attached to the skin (Figure 16.1 (I)).

From the point of the chin, lay the hair in successive layers toward the neck, each layer supporting the next (Figure 17.1). As the hair is applied to the face, it can be pressed into place with the blades of the scissors. Keep a wash bottle of alcohol handy to clean off the scissors often.

Then repeat the two coats of adhesive and tacking procedure on the front of the chin up to the mouth area, and commence laying on hair, a tuft at a time. The blending edge of the beard should be done in a lighter mixture of hair to keep it soft in appearance. Note the growth pattern of the hair in the chin area. Some men will have more or less hair growing under the center of the lip but seldom have much growing under the corners of the lip. This produces two natural dips rather than a full chinline of hair. If the beard is to be full, continue once more, on one side and then the other, with the adhesive/tacking sequence, and lay the hair, graduating upward layer after layer until the sideburn is reached.

If natural head hair is to be used for the character, lift the hairs of the natural sideburns, and adhere one layer of laid hair under them. The natural hair can then be combed over the adhered hair to get a good blend.

Always make certain that the very edges of the beardline are fine, well spread, and not matted to the face. Correct any stray hairs by removing them with college pliers.

If the same beard is to be built on a face for a number of days, sometimes time can be saved by attaching an old lace beard of the correct shade whose edges have been trimmed back so that the original blend is lost, creating a new blended edge with laid hair and combing the two together to form the beard. Always use

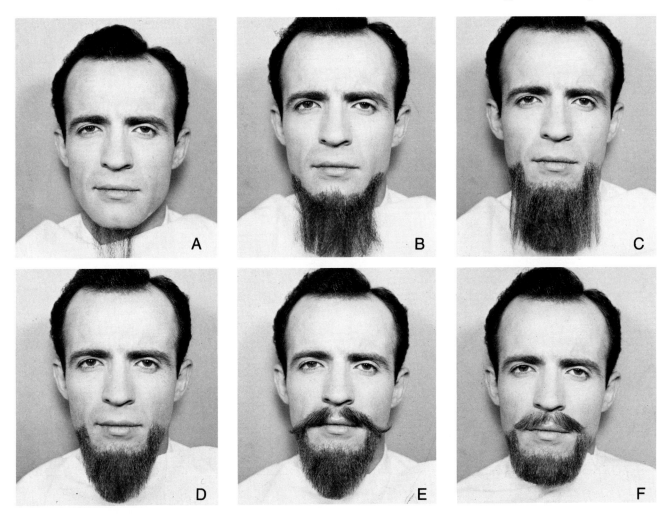

FIGURE 17.1 *Application of a beard. (A) Starting to apply hair under the chin. (B) Buildup continued almost to edge of desired shape by laying hair on the face, layer by layer upward. (C) Lighter colored hair added to form a blending edge. (D) Beard is trimmed roughly to shape. (E) Finished trimming and a moustache added. (F) A different trim of the beard with a different moustache. Both moustaches are on lace.*

the minimum of hair for the maximum effect. The usual fault of beginners with laid hair is that they use too much hair, making the beard too dense and unnatural appearing. The beard is then pressed firmly on the face with a lintless towel that has been slightly dampened. This will further set the gum and remove any excess remaining.

If a moustache is to be laid on, it should be built up with hair layers commencing at the outer end of the upper lip and working inward. Press the moustache to the face with the damp towel, and comb and brush the beard and moustache to remove any excess hairs.

More often than not, however, a moustache is usually applied by attaching a prepared moustache on lace. Always use the same two coats of matte adhesive, tacking with the finger in between to remove all the shine and excess gum, then press on the moustache. Quite often a make-up artist will cut a lace moustache in half at the center and apply each side separately to allow for better lip and mouth movement for the performer. Take care that each side is then adhered at the

same angle, or the effect will be ludicrous. The hair is then trimmed to a basic shape and curled if necessary.

The smallest irons are best for beard work and should not be overheated in the curling iron stove. Some test the iron by crimping a paper towel with it. If it is too hot, the paper (like the hair) will be scorched, and if it is too cold, the crimp won't last. On location where there may not be any electricity for a stove, one can heat the iron over a can of Sterno canned heat. Curl the hair in layers all in a down or under direction in small sections, and carefully comb out the kink. Only real hair can be curled with an iron easily, as wool crepe has a tendency to scorch quite readily. The beard can then be trimmed with scissors, and a number of styles can be achieved with a basic beard application.

Although all wool crepe beards require a spraying of holding hair spray to maintain their shape, real hair beards require less. If the beard must be wet for rain scenes, for example, a spray coat of Krylon will stiffen and protect the hairs.

Both prepared wool crepe and real hair beards can be premade on a rubber or urethane form for quick laying procedures (see page 198). To lay a beard on these rubber or urethane heads, give a coat of Matte Adhesive to the desired beard area in the same pattern and manner as one would for laying a beard on the face, but don't tack the adhesive with the finger (Figure 17.2). As soon as the gum is slightly set, start laying the beard in the usual manner. If a laid moustache is to be part of the beard, attach this also. When it is complete, curl, trim, and dress it as usual. Then, using a Pre-Val Sprayer filled with RCMA Beard Spray or some other type of neoprene adhesive material, spray a fine mist on the beard, taking care not to overspray (as it will mat and whiten the hair). This will dry rapidly. Then the entire built-up beard and moustache may be removed as one piece by either turning the head form upside down and pouring on some acetone to loosen the adhesive, or teasing the beard off the form with a brush liberally soaked in acetone. The beard should separate from the form quite easily and can be placed on a formica table to dry. The beard can

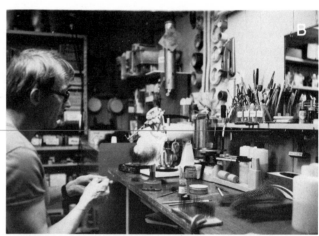

FIGURE 17.2   *Dick Smith lays a small beard. (A) Running the hair through the pleater. Dick has electrified his pleater with two electric curling irons for quick heating. (B) Commencing the beard after adding a layer of adhesive on the head form. (C) Trimming to shape after curling with the iron. (D) The completed beard. (E) Spraying the beard with beard setting spray using a PreVal Sprayer. (F) Using a polyethylene squeeze bottle filled with acetone, he loosens the beard from the form with a brush as he squirts the beard from the bottom upward. (G) The beard removed intact. (H) Testing it on the actor's face cast for fit. (I) A Lincolnesque Dick Smith with his creation held up to his face. (J) A PreVal Sprayer and bottle. Dick prefers the use of RCMA Matte Plasticized Adhesive for laying hair on the face as it gives him better working time and has excellent dry hold.*

then be trimmed on the edges for stray hairs and pinned up on a cork board ready to use. To attach it to the skin, simply coat the skin in the usual manner with Matte Adhesive, tack between the coats, place the beard on the face, and press it in with a damp huck towel.

Care should be taken that the RCMA Beard Spray is used only in a well-ventilated room due to the strong solvent (MEK), and that it is only diluted with this same liquid. Never use this spray on a beard attached to a human face because the fumes are so strong and inflammable.

Some amateurs try to use gum latex on the face as an adherent and then remove the beard intact for a further use, but this is a very unsatisfactory method as the latex becomes very gummy and shows up considerably. What might seem to be a good idea on a large stage certainly is useless for the screen. If the adherent shows, if the beardline is noticeably false, if the hair is matted down or badly laid, or if the density of the hair is too thick and unnatural, the whole thing looks artificial and hokey. There should be no question whatsoever with a laid hair beard; it must look like natural hair.

Prepared Hair Goods

These are beards, moustaches, sideburns, eyebrows, small fill-in pieces for the temples, or toupees for men (Figure 17.3). Only on rare occasions will eyebrows be made for a woman's make-up. They are all usually made by tying real hair (generally human hair) into a netting, or *lace,* with wigmaker's knots in a process called *ventilating.* The net can be made of fine silk, which is used for the best hair goods for the screen or street, or of a coarser and heavier nylon lace used for

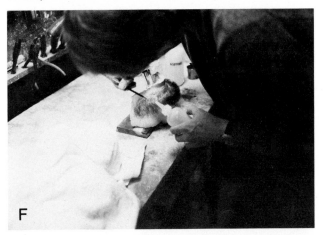

F

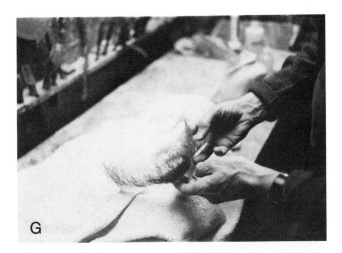

G

H

I

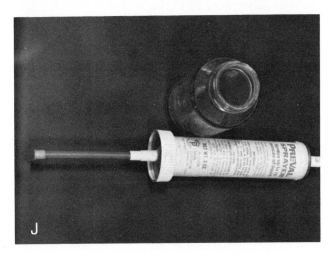

J

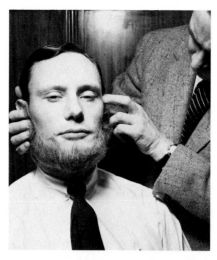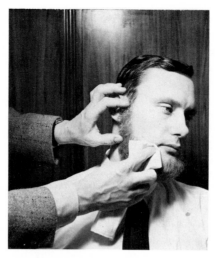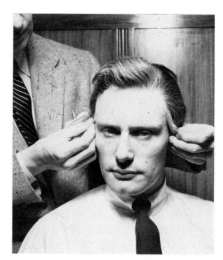

FIGURE 17.3   *Prepared hair goods.* LEFT: *Checking a beard for size.* MIDDLE: *Attaching the beard and pressing the lace to the face.* RIGHT: *Placing a wig in position on the head, holding it properly by the tabs on each side.*

stage hair goods or utilitarian pieces not used in close-ups.

There are several grades of lace, from the finest to the coarsest, and the better it is, the harder it is to find for wigmakers today. Josephine Turner, the grand lady of the wigmakers in Hollywood, says that good hair lace making is a disappearing art and that most workers today don't want to take the time to study the fine points of the trade. However, there are a few good makers of prepared hair goods and wigs both in California and New York who work almost exclusively for the professional industry. There are also many others who make quite adequate pieces and wigs for the opera, stage, and amateur college productions (see Appendix B).

The wigmaker's knot is performed as shown in Figure 17.4 with a special hooked needle, tying the hair into the lace. It is best to nail the lace being worked on to a hard wood block with 5/8-inch fine wire nails to get a tight surface. Wigs can be held in the same manner to a hard wood block for dressing, too. Drive the brad halfway in, and bend it over to hold the lace tightly to the block, stretching the lace as each nail is driven in so that the surface will be smooth and free of wrinkles.

A good toupee has a gauze mesh top with a silk lace front. The hair ties are larger—that is, more hairs per tie—in the gauze, and the ties become spaced more closely together and with fewer hairs in the lace front, down to individually tied hairs for the blend. On camera a fine toupee's front is quite indistinguishable if it is applied correctly. Never put any foundation under a fine toupee's lace edge as it will certainly discolor it when it is adhered to the head.

Prepared hairpieces can be dressed on a block prior to use on the face. Water is used for the dressing and then curling can be done with a hot iron. Beards,

moustaches, sideburns, and eyebrows can be lightly sprayed with a hair lacquer to preserve the set, but a good toupee (like a good wig of human hair) should never be sprayed.

To attach lace goods such as beards, moustaches, and sideburns or eyebrows, hold the piece on the desired area and note where the blend of the lace meets the skin. Remove the piece, coat the area with a thin application of Matte Adhesive, and tack it with the finger until it is set. Then apply a second coat in the same manner, and place the lace piece over the area, pressing gently into place, making certain that it is centered properly. Go over the lace edges with a bit of Matte Adhesive to be certain that they will be adhered to the skin. Don't use an excess of Adhesive, only a bit at a time. Then, with a dampened huck towel, press the edges and the piece into place. Some make-up artists prefer to use a nylon stocking or a piece of nylon cloth for this, but these do not absorb any of the excess gum. A huck towel that is dampened will not adhere to the Matte Adhesive, but will pick up any excess and also aid in setting.

The main mistake many learners make is to use too much adhesive, then not tack it properly. Just waiting for the solvent to evaporate often leaves some shine to the resin, and it does not dry as fast as when it is finger tacked. Live television in the 1950s saw too many lace pieces fall off faces at inopportune times due to adhesive that was not properly set.

As perspiration is the prime enemy of good adhesion, sometimes lace goods should be attached with Matte Plasticized Adhesive, Matte Adhesive #16, or even with Prosthetic Adhesive A when the climate is extremely humid or the performer perspires freely. Only the latter is not a resin adhesive and is truly water resistant. Men's toupees should be applied by first centering the piece on the head area, then holding it lightly, applying Matte Adhesive over the lace, and pressing the lace into the skin with a lintless huck towel that has been slightly dampened until all the RCMA shine is gone. Some make-up artists prefer to use a Matte Lace Adhesive. This product is applied in the same manner over the lace but is not tacked. It will dry rapidly and with no shine. Sometimes a second coat of the same adhesive is necessary to seal the lace edge. In all cases, don't apply any make-up foundation under the lace. With the Matte Lace Adhesive, there is less chance of foundation staining the lace, but it is also more difficult to remove from the lace.

To remove any lace piece from the skin, dampen the lace with isopropyl alcohol to loosen the adhesive, then carefully strip the piece from the skin. The skin can then be cleaned with Adhesive Remover, and the lace piece is best cleaned with acetone. Lay the item to be cleaned face down on a dry huck towel, and with a wig-cleaning brush saturated in acetone, press the adhesive out of the lace onto the towel. A few soakings will remove all the gummy residue. Do not brush the lace vigorously as it can be easily torn by this action. Instead, with a stippling motion with the brush bristles, force the solvent through the lace. Move the lace piece to another area of the towel, and repeat this procedure until the piece is clean. Take special care with toupees that have the finest of edge lace.

To take a hair sample when a wigmaker will make a man's toupee, use the scissors in a sliding motion (on the underneath layer of hair in the back of the head) to the length of the hair while cutting. In this manner, the cut will not be seen as a clipped-out area as it would be if a straight cut were made.

When a made-to-order lace piece is delivered, it

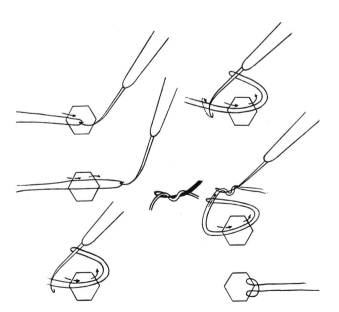

FIGURE 17.4 *Wigmaker's knot. Made with a special hooked needle tying the hair into the lace. The stages are shown in left to right columns of the diagram, top to bottom.*

often has a great amount of extra lace. This should be trimmed off to within about 1/2 inch from the start of the hair ties in the case of a toupee and about 1/4 inch for beards and sideburns. Moustaches and eyebrows can be cut even closer to lessen the likelihood that the lace may be seen. Normally, moustache and eyebrow lace is cut with straight scissors, but beards and sideburns, as well as toupees and other small hairpieces for the front of the hairline, should be cut with small pinking shears to produce a less obvious edge for a better blend. Sometimes after many uses, the lace must be slightly trimmed back to get a less stringy edge. When the lace becomes too worn or fragile in a toupee front, it should be taken back to the wigmaker for refronting with new lace and some replacement of the hair blend in front.

Wig cleaning should only be done with special cleaning fluids and never with shampoo as with natural hair. The piece must be thoroughly dry before any restyling is done.

If a man is quite bald on the top part of his head and the only adhesive holding the toupee is on the front lace, any slight wind is apt to blow the piece up in the back during exterior scenes or any strenuously physical ones. It is a good idea to attach a piece of two-sided clear tape (such as Secure Tape) to the hairpiece gauze in the back part of the toupee and press it firmly to the scalp to avoid any such mishaps. This type of tape is readily removed after each use, and a new piece should be put on every time. It is not a sound idea to use a liquid form of adhesive for this attachment area as the adhesive might stain the gauze or be difficult to remove. When a man has some top hair, instead of this two-sided tape, some wigmakers will provide two small clips, one on each side of the rear of the toupee to affix to the natural hair.

There are also laceless toupees—that is, without a lace front—that are attached either with the toupee tape or by clips to the man's own remaining hair. Many of these are made in artificial fibres such as Elura, with the filaments hand tied to a special base. Such are easier to take care of than real hair ones and are sometimes used when an actor must do a great amount of physical exercise or be in adverse, hot, humid weather conditions as a second hairpiece. As they are far cheaper than real hair toupees, budgetary factors might play a part in their selection as well. René of Paris in North Hollywood makes a number of various stock and to-order styles in the Elura fibre in a great variety of shades, grades, and cuts. Also see page 73 on small frontal hairpieces.

In dressing any prepared lace hair goods, use a wide-tooth comb, taking care that the teeth never reach deep enough into the hair to catch on the lace. Also, after removal, the piece should be examined for any damage to the lace and repairs made with a needle and thread. The hairs should be carefully combed and stray hairs straightened out with college pliers to fall in the flow pattern of the hairs.

Lace moustaches can be stored in small cellophane or plastic bags or, to keep an even better shape, pinned to a series of cork boards for storage in a cabinet. Eyebrows and sideburns can be stored by pinning them to the same type of board or kept in plastic boxes. Beards can also be stored in plastic boxes. The type of covered box sold in department stores for shoes or sweaters makes an excellent dust-free container for any hair goods including toupees and wigs.

Natural Hair

Many times, a performer's natural hair can be dressed, cut, or styled to suit the character being played. Straight hair can be curled or vice versa, dark hair can be lightened, or light hair can be darkened with hair color. Sometimes the hair is bleached out to remove all the color and then retinted to achieve the proper shade. Some actors have even had their head partially or completely shaven to play certain roles—to avoid having to wear a wig or bald cap every day of the shooting. Of course, if there is time, men can grow facial hair if it suits the character, or sometimes the natural facial or body hair must be removed. There are many types of depilatory creams that will do the job quite well on body hair.

Stubble Beards

Stubble beards can be done with a coarse urethane or red rubber sponge that has been touched to a shade of foundation of a dark gray or gray-green color and then lightly stippled on the growth area to simulate hair. This will just produce a lightly unshaven effect.

Finely chopped wool crepe hair can be applied to the face with a damp polyurethane sponge or natural silk sponge after the application of a coat of tacked Matte Adhesive. The damp sponge will pick up the fine hair (various shades can be cut into plastic containers), and transfer them to the face while the dampness of the sponge will set the resin in the gum. Excess hair can be carefully brushed off, and a good unshaven look results.

Blocking Brows

Most brows can be blocked out by combing the hairs out flat and applying a coat of Matte Adhesive, pressing the hairs down with a damp huck towel and smoothing them out after powdering with a small dental spatula. After this first step, a number of methods can be used to complete the blocking:

1. Cover with a light coat of Plastic Molding Wax and then go over this with two or three coats of Matte Plastic Sealer.
2. Cut a piece of cap material slightly larger than the hair area, and attach this over the flattened brows,

blending off the edges with acetone. RCMA Plastic Molding Material can also be used for this, either painting the Plastic Molding Material directly on the skin and drying it or premaking an appliance in the desired shape for attachment over the hairs. PMA Matte Molding Sealer can also be used over the hairs.
3. Make a foamed latex appliance to cover the hair area (which might contain a new brow shape).
4. Attach a piece of silk organza or muslin about 1 by 3 inches in size, cut on the bias so that it will stretch, and cover the flattened hairs with another light coat of Matte Adhesive. Press the silk into the adhesive with a damp towel and then thoroughly dry. A stipple coat of eyelash adhesive or Edge Stipple material can then be applied over the area and dried.

Foundation can be applied over any of these blockers to finish the job. Depending upon the visual effect required, some of these methods produce a stiffer covering than others.

To block out just the ends of the eyebrows to create a racial change for a Caucasoid to an Oriental, comb the first two-thirds of the brow forward toward the nose, and hold back the hairs while applying a coat of Matte Plastic Sealer over the remaining hairs. Press and flatten these out, powder the area, and apply some foundation over the covered, flattened hairs before combing the rest of the brow back in shape. End hairs are far easier to disguise with the simple flattening than the full brow, whose hairs nearest the nose are often more unruly and seemingly coarser. By the way, wool crepe hair right out of the braid, pulled to a thin mat, can sometimes make a very effective heavy or character brow that can be attached on the skin over the brow or over a brow covering as described previously.

BODY HAIR

Body hair is often finer than head or beard hair and sometimes grows in profusion on a man's body, or at times, very little is seen. Most prehistoric men had a profusion of body hair growth, as will of course special character changes in transformations of men to animals. Heavy body hair is sometimes used to delineate a rough male character in a film.

Body hair can be simulated in many ways, but one method is to pull wool crepe hair directly from the braid and form a sheet of hair that is attached to the desired area with Matte Adhesive that has been properly finger tacked. For a large chest section, a hair mat can be made with hair and then one side of it sprayed (using a Pre-Val Sprayer) with a diluted Prosthetic Adhesive A. Use only Prosthetic Adhesive-A Thinner for the dilution because some other solvents are harmful when the vapors are inhaled. The skin area can then be sprayed with the same adhesive and both allowed to set. Carefully place the hair mat over the area, and press it on lightly so as not to mat down the hair. This will hold even if the actor perspires.

Body hair can also be laid on a bit at a time with real hair for close shots (see Figures 15.2a, b and 15.7) or imbedded in an appliance. This latter method is usually done with a needle whose head has been cut off to form a forklike instrument that is used to poke individual hairs or tufts of hair into an appliance. If the growing hair effect is to be done, the hair must be pushed right through the prosthesis (normally it will be 1/8 to 1/4 inch in thickness) and pulled through to the desired outside length. Long hair is used for this so that the hairs can be gathered together and, for the effect, all pulled through the artificial skin appliance.

WIGS, FALLS, AND OTHER HAIRPIECES

The proper fit of wigs for men and women can be assured with a head measurement. Also, for making a correct beard or sideburns, facial measurements can be taken to send to the wigmaker.

Measuring the Head

Following the illustration of the head shown in Figure 17.5, measure as follows:

1. Around the head from the hairline in front to the end of the hair growth in the rear (pole of the neck),
2. From the front of the hairline on the forehead to the pole of the neck,
3. From temple to temple around the back of the head,
4. From ear to ear over the top of the head,
5. From ear to ear across the forehead at the hairline,
6. From ear lobe to ear lobe across the pole of the neck.

Beard measurements are made from:

a. The end of the sideburns to the other side over the point of the chin,
b. The corner of the mouth to the other corner, under the chin as far back as the beard is required,
c. The center of the lip to the depth of the beard.

These measurements are best taken with a cloth tape and given to the nearest 1/8 inch for exactness.

Applying a Lace-fronted Wig

To put a wig on a head, the performer's own hair must be held flat to the head with bobby pins or even a nylon stocking if there is a great amount of hair. Short haircuts seldom require this treatment.

Hold the back of the wig with both hands, and slide

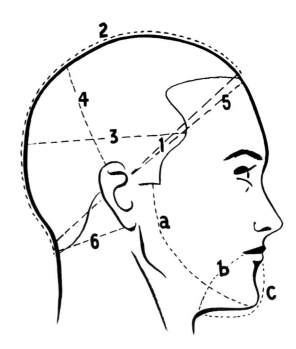

FIGURE 17.5 *Diagram for head measurements.*

the wig front over the forehead. The rear of the wig is then pulled down until it fits snugly on the head. Holding the wig firmly on each side of the head with the open palms of the hands, slide the wig front-back until the desired hairline is reached. Comb back any stray hairs that may be under the lace or from the front of the wig. The lace should be attached to the head just in front of the natural hairline so that the hair growth does not appear to be too far down on the forehead. Take care also that stray natural hairs do not prevent the wig lace from setting flat to the head. Matte Adhesive can then be applied over the lace and pressed into the skin with a damp huck towel. Some make-up artists prefer to roll back the edge of the lace a bit and apply the adhesive to the skin, tack it with the finger, and roll the lace back over the adhesive and press the edge into the head with the towel. Blend the foundation up to, but not over, the lace edge.

The Half-wig or Fall

At times, to preserve a natural look to the front of men's hairlines, the half-wig or fall is used. This is attached to the hair behind the front of the hairline. The first third or half of the hair is combed forward over the forehead, and the fall is attached to the edge of the remaining hair with metal clips that hold firmly to the hair. The front hair is then combed back over this half-wig and blended together so that the hair length is increased. With no lace front, the appearance is very natural.

Smaller versions of this type of fall can be employed to make queues or braids for period hairstyles, or just to lengthen the hair slightly in the back. These are affixed in the same manner, but farther back on the head, by lifting up a section of hair, attaching the clips, and combing the hair back over the piece. Actors often prefer half-wig/falls as they are much more comfortable to wear than wigs and often more realistic in close shots.

Women's Wigs and Hairpieces

Women's wigs and hairpieces are normally handled by the hairstylist, and they can be lace fronted or held on with clips, combs, or various pin arrangements in the manner of the half-wig to utilize the natural hair front. They can be made of human hair or Monsanto's acrylic fibres (such as the Elura and Kanekalon blends), and basic hairpieces can often be styled in a number of fashionable ways.

Many wigs and hairpieces for both men and women are made of *wefted* hair, which is a method of tying hair tufts on one end into three cords to form a sheet of hair. These can then be clustered, coiled, laid in successive overlapping layers on a base, or a number of other ways to make a full or partial wig, fall, braid, ponytail, or switch to cover both modern and period styles. A good studio hairstylist can do marvelous things with a number of switches or wiglets in creating styles (see page 262).

HAIR COLORING

This can be divided into temporary and permanent coloring, but make-up artists normally deal only with the temporary changing of a hair color. Permanent hair coloration is done by the hairstylist for both men and women.

Darkening the Hair

Temporary darkening of light hair, whether it be blonde or gray/white, can usually be done with the following:

1. A darker pencil, held rather flat to the hairs, will temporarily color grayed brows or sideburns on men. It can also be used on the scalp to draw fine lines to simulate hair growth when a man's hair is thin on the top of the head or the front of the hairline. This lining must always be carefully brushed in with a brow brush so that it appears natural and not pencilled.
2. Cake mascara in various colors can be used to color hair, but one must take care that the application is done lightly so that the mascara does not cake or group the hairs as it sometimes does on the eyelashes.
3. The small bald spot that some men get on the crown of the head can be minimized (if the rest of the hair is dark) with various shades of Eyecolor #1 (water applied) in black to light browns. These are best applied with a #12 flat sable brush in a light ap-

plication. Take care that the eyecolor/water mixture is not too thin so as to run freely or too heavy to cake on the scalp and overdo the effect (see Figures 7.27c and d).
4. Nestle LaMaur Company makes many colored hair sprays. The trick is not to use too much spray as it will mat the hair readily. In addition to browns and black, there are some bright colors for fantasy characters.

Lightening the Hair

Temporarily lightening dark hair to blonde can best be accomplished with the hair sprays in blonde, gold, and so forth. Sometimes a spray of white is required before the colored spray, but be careful not to get too much on. It is always best to complete any styling that is required before spraying because brushing or combing the hair after the spray is applied will often reduce the effect. The hair can also be fully bleached out and then tinted (see page 100).

Graying the Hair

Although this is a form of lightening the hair, it is usually in the realm of age character make-up rather than just a hair color change. There are a number of methods used to gray hair, but these are the generally employed ones:

1. Use the white and silver hair sprays when doing a full head of hair. Sometimes a bit of beige will add a good touch of yellowing in the hair for age. The hair should be dressed and then the white and possibly some beige sprayed on to lighten and gray the hair. Highlights can then be added with the silver spray. The face should be covered with a towel while doing this to keep the spray away from the foundation area and from the edge of the hairline. This edge is best done by spraying some white and silver into a container, and using an eyebrow or even an eyelash brush, carefully apply the whitening to the hairline, brushing it in the direction of the hair growth. This spray does not come off on hats as readily as some of the other creme varieties of temporary hair whitening.
2. Hair whiteners in White, Grayed White (HW-1), Pinked White (HW-2), Ochre White (HW-3), and Yellow White (HW-4) are available for graying brows, natural moustaches, or sideburns by applying them with an eyebrow or wig-cleaning brush. Most of the liquid hair whiteners dry too flat for screen use and must be shined a bit with hair spray. This is not a very satisfactory method as the hair spray might make the dried liquid whitener run or the spray cannot be easily controlled. These creme-base hair whiteners are not intended for fully graying or whitening a head of hair but only for spot graying.

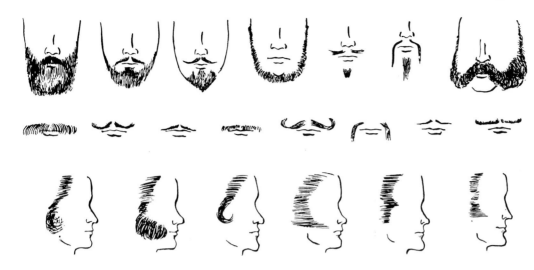

HAIR SPRAYS AND OTHER DRESSINGS

There are many types of hair sprays on the market that are overly perfumed and should be avoided. Some manufacturers do make the unperfumed variety, and be certain to select a hard or heavy-hold type. Instead of the usual aerosol spray, there are some hand-pump sprays that are easier to use on small areas. Solid or liquid brilliantines will give a shine to hair, and there are also aerosol spray varieties of this product. Bandoline or wave sets can be used to dress the hair in certain patterns or styles, with the former giving a far stiffer end product. Curls and points on moustaches can be dressed with plastic wax material in the clear shade. Petroleum jelly can be combed into the hair to give it a very slick look for some 1920s styles or just to keep hair flat to the head.

STYLING HAIR AND HAIR GOODS

Both natural and prepared hair can be cut, dressed, and styled to achieve different looks and periods. Men's styles change less than women's fashions, in the main, for head hair, but men's beard and moustache styles vary in periodic and often age-level times. In the twentieth century we went from copious facial hair in the first decade, through a long period of few beards, no long sideburns, and a smattering of finely cut moustaches to the 1960s when facial hair began to appear more frequently along with longer head hair in many styles from trimmed to totally unkempt.

The trend of more than the usual percentage of facial hair has continued through the 1970s and into the 1980s, but the great majority of men is still clean shaven or sports some form of moustache. A few basic facial hairstyles are shown in Figure 17.6 for descriptive purposes of both modern and period cuts.

Middle Eastern or Arabic beards, moustaches, and hairstyles have been worn in a number of standard cuts and stylings over a great many years. Even today many styles that might be termed as period ones will be seen on men of that area (Figure 17.7).

BASIC WOMEN'S HAIRPIECES

Leon Buchheit describes a number of women's hairpieces that can be made in real human hair or acrylic wig fibres (Figures 17.8 and 17.9). As with those made by Josephine Turner, they are all made to order and require accurate hair samples for a proper match to the hair area to which they will be attached (also see other wigmakers in Appendix B). Many hairpieces are sewn to net caps in various shapes (Figure 17.10).

ATTACHING A HAIRPIECE

Figure 17.11 shows the steps needed to attach a hairpiece correctly.

REMOVING A LACE HAIRPIECE OR WIG

While wigs and hairpieces that are simply clipped or pinned on can be easily removed by detaching them, a lace front wig or toupee should have its lace wet with alcohol to soften the adhesive and then, grasping the back foundation of the wig with a hand on either side, carefully pull it up and forward until it reaches the lace. Then gently peel the lace from the softened adhesive. The skin can be cleaned with RCMA Adhesive Remover and the lace on the wig cleaned as usual with acetone. The wig should then be placed on a block and pinned down to hold its shape and styling.

FIGURE 17.7   (A) Various Arab facial hair styles. (B) Various
Eastern facial hair styles.

FIGURE 17.8   *Some basic women's hairpieces. (Courtesy Leon Buchheit.)*
(A) Hand-wefted cluster of curls. (B) A round base surrounded by one
wire wefted into the hair. (C) A wire cluster hand wefted on a 4-inch
diameter flexible wire foundation. (D) A basic braided pony tail that can
be worn down or opened or chignoned. (E) A basic bun attached with a
barrette that cannot be seen. (F) A clip-on hairpiece on a barrette that can
be styled as a chignon or in soft, loose curls. (G) The French fall is hand
wefted with a hand-ventilated front. Made in many lengths. (H) A
grand fall on a heightened triangular foundation. Adds bulk, body, and
length to the hair.

RENTAL WIGS AND HAIRPIECES

It is always good form and custom to clean carefully and take proper care of a rental wig or hairpiece as if it were the property of the make-up artist. A good and cooperative wigmaker or renter of hair goods is a great asset to a make-up artist who must rely on them to fit the performers sent to them for hair goods. Establishing a reference book (such as Wilcox's *Mode in Hats and Headdress,* 1949), using page and picture references to select and order certain hair goods with a wigmaker, saves much time as orders can be placed by telephone.

One last note on men's toupees: Some actors are very sensitive about not wanting the public to know that they are wearing a toupee, so take care about retouching or re-adhering a loose toupee on set. Get the actor aside or in his dressing room to do any retouching. Always take excellent care of an actor's toupee, cleaning it carefully after each use, recommending cleaning or a new front when required, and the performer will sincerely appreciate it.

FIGURE 17.9 *Finished make-up and hairstyles (with hairpieces). (A) Short, light blond; (B) medium, darker blond; (C) short, medium auburn; (D) curly, long blond; (E) curly, long, light blond.*

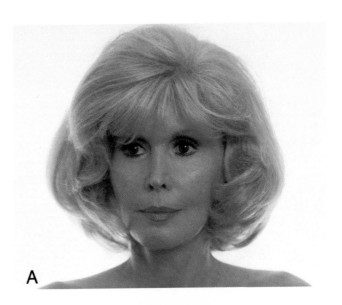

A

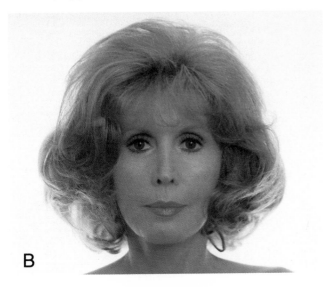

B

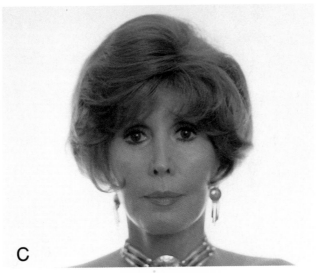

C

D

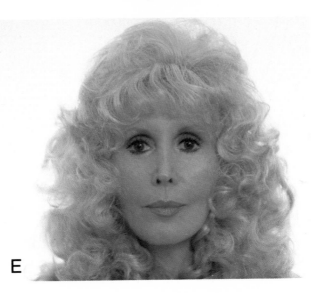

E

FIGURE 17.10   *Net caps. (Courtesy Leon Buchheit.) This roughly triangular type as well as the round style are the most common variety. Sometimes small combs are sewn into one end of the piece to use as an anchor for attaching it to the natural hair.*

FIGURE 17.11   *Attaching a hairpiece. (Courtesy Leon Buchheit.) STEP 1 Comb all but one small section of hair, about one square inch at the front, and one small section at the back (where the hairpiece foundations ends) out of the way. STEP 2 Tease, then flatten these small anchoring sections. STEPS 3 and 4 Insert hairpins from the side, weave hairpin through, flip over, and push through. STEP 5 Repeat, inserting hairpin from the bottom, and reverse to lock the first pin. STEP 6 The crisscrossed pins are the anchoring section. STEP 7 Judge the length of the foundation on the hairpiece, proceed exactly as in Steps 3, 4, and 5, and set up the lower anchoring section. STEP 8 Push comb attached to the hairpiece into the front anchoring section. Tease a small strand of hair from underneath the hairpiece and pin to lower anchoring section. STEP 9 Comb the hair around the hairpiece to blend with the subject's own hair. The closer the hairpiece is anchored to the scalp, the securer it will feel and the more undetectable it will be.*

A New Beginning . . .

AS THE STUDY AND LEARNING PROCESS OF THE ART OF make-up is a constantly changing and desirous direction for the professional, there cannot be a simple conclusion to this book, but rather a challenge to go beyond, where others have not tread, with new ideas, new materials, and new talent. All of us must seek to encourage the beginner with the avid mind and hands so anxious to emulate the accomplishments of those working in the field, and it is this way our art and craft will be perpetuated.

PROFESSIONAL MAKE-UP PRODUCTS

EQUIVALENCY CHARTS

The one basic factor that separates *professional* make-up foundations, shadings, counter-shadings, cheekcolors, and lipcolors from their *commercial,* or even *theatrical,* counterparts is accurate consistency of shades and colors. With commercial and theatrical make-ups, exact day-to-day color matching does not have the importance that mediums of the screen demand, as the former are only seen in person for *that* day, while the latter require day-to-day constants in tonality. Skin testing is a necessity to maintain exact color matches for all professional make-up materials. However, shades of make-up with either the exact names or for similar uses do vary somewhat from manufacturer to manufacturer, depending upon their assessment of values for specific purposes. Actual screen tests are the final method of determination, so that equivalency charts are approximations of the color or shade value as they are expected to appear on screen under normal lighting conditions rather than the same number or name of the shade.

It is not practical to simplify the multitude of foundations made by various manufacturers of professional products in a basic comparison chart so we shall only show some former Max Factor numbers that were quite familiar in the field to the current approximate RCMA equivalents. We shall then list the recommendations of the other two leading professional cosmetics for screen/theater use, and a listing of a theatrical make-up firm as well. (It should be noted that some of the Max Factor numbers are still available.)

RCMA	MAX FACTOR
KW-1	CTV-2W
KW-2	CTV-3W
KW-3	Olive or CTV-4W
KW-4	Deep Olive or CTV-5W
KW-5	725 A
KW-6	725 B
KW-7	725 BC
KW-8	725 CN
KM-1	CTV-6W
KM-2	CTV-7W
KM-3	CTV-8W
KM-5	725 EN
KM-36	665-G
KM-37	665-I

RCMA	MAX FACTOR
KN-3	665-O
KN-4	665-P
KM-8	665-L
F-2	Fair
K-1	K-1
K-2	K-2 or 665-J
K-3	665-K
KN-5	Eddie Leonard (Pancake shade)
KN-6	Negro 1
KN-7	Negro 2 or 665-R
KT-3	Light Egyptian
KT-34	Dark Egyptian
KT-4	Tahitian
RJ-2	KF-7 (Pancake shade)
KW4M3	Golden Tan
KW-38	Natural Tan

The futility of matching printing inks with how foundations appear either in the container or how differently on the skin precludes professional manufacturers from attempting to make commercial *color charts* as exact comparatives but only for advertising purposes in brochures and such.

CUSTOM COLOR COSMETICS BY WILLIAM TUTTLE

An Academy Award, 20 years as head of the make-up department at MGM, and over 400 feature film credits have distinguished William Tuttle in the field of make-up artistry. His line of cosmetics, albeit limited to basic straight make-up materials, has been designed for professional use. His personal preferences in average situations are as follows (Light to Dark):

WOMEN

NORMAL	TAN EFFECT
Medium Beige	Suntone
Deep Olive	Natural Tan
Medium-Dark Beige	Tawny Tan
	Tan-del-Ann

MEN

NORMAL	TAN EFFECT
Dark Beige	Desert Tan
Natural Tan	Xtra Dark Tan
Tan	K-1
	Jan Tan

NEGROES
(MEN AND WOMEN)

All colors from Toasted Honey through Ebony

These recommendations are interchangeable for film or tape, although for live television or talk shows one shade deeper than the norm is employed due to the overall set lighting that is flatter than for normal production lighting.

COLOR FOUNDATIONS

Ivory	Warm Tan
Truly Beige	Tan-del-Ann
Fair & Warmer	Tawny Tan
Light Peach	Fern Tan
Pink 'n' Pretty	Bronze Tone
Peach	Jan Tan
Rose Medium	TNT
Light Beige	Light-Medium Peach
Light-Medium Beige	Deb-Tan
Medium Beige	Suntone
Medium Dark Beige	Medium
Bronze	Beige Tan II
Tan	Shibui
Xtra Dark Tan	Deep Olive
Hot Chocolate	Chinese I
Sumatra	Chinese II
N-1	Western Indian
Natural Tan	Tawn-Shee
Dark Beige	BT6
Toasted Honey	BT7
Cafe Olé	11019 (Dark Tawny Tan)
Rahma	
Bronze	K-1
Natural Tan	Hi-Yeller
Chocolate Cream	Tan
Tan Tone	Ebony
CTV-8W	Desert Tan
BT5	

CORRECTIVE CONTOURING

Special Hi-Lite (For erasing circles under eyes)
Hi-Lite (For contouring)
Shadow I (For contouring)
Shadow II (Darker than Shadow I)
Blusher
Tan Rouge (Used with dark foundations)
Sunburn Stipple (For an outdoor look)
Red-out (For erasing darkened areas)
CTV Rouge
Mauve Blusher
Persimmon Piquant
007

EYE ESSENTIALS
20-30-40 Eye Shadow
CoCo Bare
O-Dee
Frosted Eye Shadow Duos
    Chamois/Burnt Spice
    Pearl Blue/Caribbean Blue

PROFESSIONAL PENCILS

Black	Light Brown
Charcoal Grey	Auburn
Midnight Brown	Taupe
Dark Brown	Beige
Medium Brown	

LIP ESSENTIALS
Mango Mist
Lip Glosses
    Ming Coral
    Bacchante Brown
    Castiliana Red

SKIN CARE ESSENTIALS
Cleansing Lotion (Dry and Normal Skin)
Freshener (Dry Skin)
Freshener (Normal Skin)
Oily Skin Cleanser
Oily Skin Astringent
Moisturizer (Dry and Normal Skin)
Skin Conditioner (Very Dry Skin)
Facial Scrub (All Skin Types)
Sun Screen Moisturizing Lotion

Also, Extra Fine Translucent Powder and Prosthetic Make-up can be had on special order to match any listed color.

CREME CHEEK ROUGE
DRY CHEEK ROUGE

Red	Red
Dusty Rose	Raspberry
Raspberry	Coral
Dark Tech	Dark Tech
Coral	
Blush Coral	

CHEEK BLUSHERS	CREME HIGHLIGHT
Dusty Pink	Extra Lite
Nectar Peach	Medium
Golden Amber	Deep

CREME BROWN SHADOW	FACE POWDER
#40 Character (Purple-Brown)	Translucent
42 Medium (Brown)	Coco Tan (For dark skins)
43 Dark (Warm Dark Brown)	Special White (Clowns)
44 Extra Dark (Rich Dark Brown)	

LIPSTICK

#1 Lip Gloss	9 Plum
3 Coral	10 Bordeaux
5 Garnet	12 Natural Brown
6 Plum Pink	14 True Red
7 Natural	15 Siren Red
8 Cranberry	

PRESSED EYESHADOW

Toast	Navy Blue
Silver	Rust
Lavender-Mist	Gold Leaf
Plum	Iridescent Taupe
Burgundy	Dark Brown

CAKE EYELINER

Black	Blue
Brown-Black	Iridescent Turquoise
Brown	Green
Plum	Iridescent Green
Silver	Iridescent White

PEARL SHEEN LINERS

White	Bronze
Green	Copper
Turquoise	Mango
Royal Blue	Dusty Rose
Sapphire	Cabernet
Shimmering Lilac	Rose
Ultra Violet	Brown
Amethyst	Walnut
Gold	Charcoal

BEN NYE MAKE-UP
Ben Nye was the director of make-up for 20th Century Fox for many years, and his make-up line is a reflection of his high standards of artistic quality. Now operated in Los Angeles by his son, Dana, this firm has an extensive list of products for use in screen make-up as well as being the leading supplier to college drama and visual arts departments around the country.

For the screen, Dana Nye recommends N-3, Deep Olive, N-6, and T-1 for Caucasoid women and M-1, M-2, M-3, Y-3, and M-5 for men. The most popular Negroid shades are 27, 28, and Dark Coco. In the N series, N-1 is the fairest for light complexions.

CREME-CAKE FOUNDATIONS
White
Black
Ultra Fair
Old Age
Bronzetone
Dark Coco

Tan Series: Natural Tan colors

T-1 Golden Tan	T-2 Bronze Tan

Twenty Series for Asians, Latins, and Negroids

22 Golden Beige	White
23 Fawn	Forest Green
24 Honey	Green
25 Amber Lite	Yellow
26 Amber	Orange
27 Coco Tan	Sunburn Stipple
28 Cinnamon	
29 Blush Sable	
30 Ebony	

L Series: Lighter value shades
L-1 Juvenile Female
L-2 Light Beige
L-3 Rose Beige
L-4 Medium Beige
L-5 Tan-Rose

Y Series: Olive skin tones
Y-3 Medium Olive
Y-5 Deep Olive Tan

American Indian Series
I-1 Golden Copper
I-3 Bronze
I-5 Dark Bronze

Mexican Series: Olive brown tones
MX-1 Olive Brown
MX-3 Ruddy Brown

Natural Series
1, 2, 3, 4, 5, 6
Deep Olive

M Series: Medium warm brown tones
M-1 Juvenile Male
M-2 Light Suntone
M-3 Medium Tan
M-4 Deep Suntone
M-5 Desert Tan

CREME LINING COLORS

Dark Sunburn	Gray
Fire Red	Beard Stipple
Blood Red	Black
Maroon	White
Misty Violet	Forest Green
Purple	Green
Blue	Yellow
Sky Blue	Orange
Blue-Gray	Sunburn Stipple

COLOR LINING PENCILS

Red	Iridescent Violet
Maroon	Iridescent Blue
Rust	Iridescent Green
Violet	Gold
Blue	Silver
Green	White

EYEBROW PENCILS

Light Brown	Dark Brown
Auburn	Charcoal Gray
Medium Brown	Midnite Brown
Black	

Mellow Yellow (To mute redness)
Five O'Sharp (Beard cover in Olive and Ruddy)
Clown White
Nose and Scar Wax
Mascara (creme) (In Black and Brown)
Silver Hair Gray
Custom Flat Brushes (Synthetic Bristle) #2, 3, 5, 7, 10, and 12
Sable Brushes (Round) #1 and 2
Dry Rouge, Lipstick (Retractable), and Powder Brushes
Latex Foam Sponges, Nylon Stipple Sponges, and Velour Powder Puffs
Crepe Hair in White, Blond, Light Brown, Auburn, Medium Brown, Dark Brown, Light Gray, Medium Gray, Dark Gray, and Black
Liquid Latex (Casting)
Spirit Gum
Stage Blood
Make-up Remover
Brush Cleaner

Ben Nye also furnishes student make-up kits for individuals in the following:

White Female	Olive Male
White Male	Black Medium Brown
Olive Female	Black Dark Brown

Each kit contains three foundations plus an assortment of cream rouges, highlights, shadows and lining colors, silver hair gray, natural lipcolor, nose and scar wax, make-up remover, translucent face powder, two custom flat brushes, two pencils, dry rouge and brush, stipple sponge, powder puff, latex sponge, and spirit gum. This is the best-buy kit for student make-up in its field and far better than others offered by some competing theatrical manufacturers.

STEIN'S MAKE-UP
Since 1883, the M. Stein Cosmetic Company has been making theatrical make-up for both amateurs and professionals. Many items have been on their lists for years, and they still will supply both the old stick greasepaint, soft tube greasepaint, colored face powders, liquid make-up, stick liners, and other classic items.

GREASEPAINT
(Cardboard Tube)

1 Pink	14 Gypsy
2 Pale Juvenile	15 Othello-Moor
3 Juvenile	16 Chinese
4 Juv. Flesh Auguste	14 Amer. Indian
	18 Carmine (red)
5 Dpr. Flesh Clown	19 Negro
5L. Ivory Yellow	20 East Indian
6 Robust-Juvenile	21 Vermilion
7 Light Sunburn	22 White
8 Dark Sunburn	23 Yellow
9 Cream Sallow	24 Brown
10 Middle Age	25 Black
11 Sallow Old Man	26 Japanese
12 Robust Old Man	27 Cinema Yellow
13 Olive	28 Cinema Orange
13L. Red Brown	

LIQUID MAKE-UP
(Plastic Bottles, Water Soluble)

2 Pink Natural	15 White
3 Flesh	16 Negro Brown
5½ Orange	17 Black
6 Tan	17½ Green
6½ Blue	18 Carmine (Red)
7 Olive	18½ Vermilion
7F Dark Sunburn	19 Light Negro
8 Light Sunburn	20 Dark Negro
8½ American Indian	21 Peach Bloom
	22 Rachel
9 Indian Brown	23 Brunette
9½ East Indian	24 Sun Tan
41A Special Indian	27 Lt. Egyptian
10 Hindu Brown	28 Dk. Egyptian
11 Mulatto Brown	29 Light Creole
12 Mikado Yellow	30 Dark Creole
12½ Bright Yellow	31 Hawaiian
13 Purple Sallow	32 Frankenstein
14 Gypsy Olive	Gray

Bright Colors

6½ Bright Blue	17½ Green
17 Black	15 White
12½ Bright Yellow	18 Red

LIQUID MAKE-UP METALLIC

25 Gold	26 Silver

CAKE MAKE-UP PAINTING PALETTE
Five colors; in water soluble White, Black, Blue, Yellow, and Red

CAKE MAKE-UP

Natural Blush	44 Middle Age
Natural A	45 Robust Juvenile
Natural B	46 Green
Cream Blush	47 Blue
Cream A	48 Red
Cream B	49 Lavender
Tan Blush	50 Sallow Old Age
Tan Blush No. 2	51 Negro Brown
Tan Blush No. 3	52 Dk. Negro
Tan A	53 Sunburn
Tan B	54 Pink
2 Russet	55 Olive
7 Sun Tan	56 Spanish Mulatto
9 Red Brown	(New)
21 Peach Bloom	7N
22 Rachel	725 A
23 Lady Fair	725 B
24 Brunette	T.V. 1
25 Rose Brick	T.V. 2
26 Toffee	T.V. 3
27 Burnt Toffee	T.V. 4
28 Clove	T.V. 5
29 Mesa Brown	T.V. 6
30 Toast	T.V. 7
31 Cinnamon	T.V. 8
32 Othello	T.V. 9
33 Black Minstrel	T.V. 10
34 Gray Frankenstein	T.V. 11
	T.V. 12 (TV Black)
35 Light Creole	C-1 Light
36 Dark Creole	C-2 Very Fair
37 Hawaiian	C-3 Fair
38 Light Egyptian	C-4 Medium Fair
39 Dark Egyptian	C-5 Lt. Brunette
39½ Japanese	C-6 Brunette
40 Chinese	C-7 Dk. Brunette
40½ Mikado Yellow	C-8 Golden Tan
	C-9 Copper
41 Indian	C-10 Bronze Tone
41A Special Indian	C-11 Walnut
42 East Indian	C-12 Ebony
43 White Clown	

SOFT GREASEPAINT
(Collapsible Tube)

1 Yellowish Pink	11 Mulatto Brown
1½ Light Pink	12 Mikado Yellow
2 Pink-Natural	13 Purple Sallow
3 Flesh	14 Gypsy Olive
3½ Deeper	15 White
4 Cream	16 Negro Brown
5 Ivory Yellow	17 Black
5½ Orange	18 Carmine (Red)
6 Tan	19 Light Negro

7 Olive	20 Dark Negro
7F Dark Sunburn	21 Peach Bloom
8 Light Sunburn	22 Rachel
9 Indian Brown	23 Brunette
10 Hindu Brown	24 Sun Tan

SOFT SHADOW LINERS

1 Lt. Flesh	17 Black
2 Medium Flesh	19 Bright Green
3 Lt. Gray	19¼ Lt. Green
4 Medium Gray	19½ Dk. Green
5 Dark Gray	20 Blue-Green
6 Lt. Brown	21 Purple
7 Dark Brown	23 Lavender
8 Lt. Blue	24 Pearl White
9 Med. Bright Blue	25 Gold
10 Dark Blue	26 Silver
11 Special Blue	27 Silver Blue
15 White	28 Rose Sallow

VELVET STICK
Matching
Cake Make-up in Colors

SPECIAL SOFT LINERS
(Do not use in eye area)

12 Crimson	16 Bright Yellow
13 Dk. Crimson	18 Carmine Red
14 Vermilion	22 Red Brown

LINING COLORS STICK

1 Golden Yellow	15 White
2 Flesh*	16 Yellow*
3 Light Gray	17 Black
4 Medium Gray	18 Carmine* (Red)
5 Dark Gray	19 Green*
6 Light Brown	20 Blue-Green
7 Dark Brown	21 Purple*
8 Light Blue	22 Light Green
9 Medium Brown	(New)
10 Dark Blue*	23 Brunette
11 Sky Blue	24 Tan
12 Crimson*	25 Red Brown*
13 Dark Crimson*	26 Lavender*
14 Vermilion*	

*Do not use in area of eyes.

METALLIC LINING STICK

Gold	Silver

CREAM (MOIST) ROUGE

1 Twilight	10 Youth Flame
2 Light Red	11 T1 Dk. Pink
3 Medium Red	12 T2 Rusty
4 Dark Red	13 Orchid
5 Youth Blush	14 Light Rose
6 Real Red	15 Rose
7 Royal Red	16 Brunette
8 Pink Tone	17 Carmine
9 Brown (Male)	

DRY ROUGE

5 Youth Blush (T1)	18 Evening Glow
12 Dawn (T2)	20 Tropical
14 Twilight	22 Youth Flame
16 Raspberry	

BLUSH-ON STICK

1 Pink Pearl	4 Bronze
2 Summer Gold	5 Peach Pearl
3 Dawn	

LIPSTICK

1 Twilight	13 Green (New)
2 Evening Glow	14 Blue (New)
3 Raspberry	15 Purple (New)
4 Dark Red	Video Pink 1 Lt.
5 Youth Blush	Video Pink 2 Med.
6 Real Red	Video Pink 3 Dark
7 Royal Red	Video Blush 1
8 Pink Tone	Video Blush 2
9 Brown (Male)	Video Light Red
10 Lip Gloss	Video Medium Red
11 Youth Flame	Video Dark Red
12 Witches Black	

LIPGLOSS

1 Cherry Red	3 No Color
(Regular)	
2 Frosted	

EYESHADOW STICK
(Swivel Case)

1 Natural	7 Taupe
2 Tan (Video)	8 Lilac
3 Blue (Video)	9 Silver
4 Pale Blue	10 Gold
5 Irr. Blue	11 Green
6 Turquoise (Video)	12 Pearl

CAKE EYESHADOW
(With Applicator, Wet or Dry)

1 Aqua	5 Blue
2 Turquoise	6 Jade
3 Lavender	7 Brown
4 Green	8 White

CAKE EYELINER
(With Brush)

1 Black	6 Lt. Blue
2 Brown-Black	7 White
3 Brown	8 Hi-Lite
4 Smoke	9 Green
5 Med. Blue	

FLUID EYELINER

1 Black	6 Green
2 Brown-Black	7 White
3 Brown	8 Hi-Lite
4 Smokey Gray	9 Navy
5 Ultra-Blue	10 Plum

PRESSED PEARLITES
(Pressed Shadows in High Shimmering
Colors for Lids, Face, and Body)

1 Copper	6 Temptation
2 Silver	(Blue-White)
3 Grape	7 Blue
4 Raspberry (Red-	8 Dark Gold
Blue)	9 Pale Gold
5 Champagne	10 Bronze

PEARLITES POWDER
(Loose Powder Shadows
in High Shimmering Colors
for Lids, Face, and Body)

1 Plum	6 Temptation
2 Silver	(Blue-White)
3 Amethyst	7 Sapphire (Blue)
(Purple)	8 Autumn Leaves
4 Raspberry	(Gold)
5 Turquoise	9 Sea Green
	10 Bronze
	11 Pale Gold

NAIL ENAMEL

Black	Green
Purple	Yellow
	Blue

NAIL ENAMEL GLITTER

Red	Silver
Blue	Purple
Gold	Multicolor

BODY GLITTER (CREAM)
(Only the Glitter Shows)
(Do Not Use in Eye Area)

1 Gold	4 Green
2 Silver	5 Blue
3 Multicolor	6 Red

LUMINOUS MAKE-UP

Phosphorescent	Fluorescent
Glows in The	Glows in U.V.
Dark	Light

MAKE-UP CREAM
(Collapsible tube, water soluble, 1.25 oz.)

Minstral Black	Green
Light Creole	White Clown
Glow Face	Blue
Dark Creole	

LIP LINING PENCIL
(7 inch)

Tawny (Maroon)	Red

EYEBROW PENCIL
(7 inch)

Black	Auburn
Blue	Silver Gray
Brown	

MASCARA (CAKE)

Black	Brown
Blue-Green	White

FACE POWDERS
(Plastic Jar)

1A Neutral	13 Othello Dk.
(Colorless)	Brown
1 White Clown	14 Chinese
2 Light Pink	14½ Japanese
4 Flesh	15½ East Indian
5 Brunette	24 Light Brown
5½ Dark Brunette	24½ Red Brown
7½ Suntan	25 Black
8 Tan	26 Gray
9 Sallow Old Age	Frankenstein
11 Healthy Old	27 Blue
Age	28 Green
12 Olive	

RUBBER MASK GREASEPAINT

C-1 Light Skin	C-7 Medium Skin
C-4 Fair Skin	C-11 Dark Skin

COLOR CROWN TEMPORARY HAIR SPRAY
Fluorescent Colors

Ripe Red	Shining Silver
Yummy Yellow	Glowing Green
Blazing Blue	Gorgeous Grape
Jet Black	Wild White
Passion Pink	Obvious Orchid
Outrageous Orange	Glimmering Gold

Glitter Colors

Blue	Burgundy Red
Green	Gold
Silver	Multi-Color

CREPE HAIR WOOL

1 White	7 Light Gray
2 Light Brown	8 Medium Gray
3 Medium Brown	9 Dark Gray
4 Dark Brown	10 Blonde
5 Auburn	10½ Dark Blonde
6 Black	11 Red

In addition, Stein's also furnishes toupee tape, collodion (flexible and nonflexible), nose putty, black tooth wax, Derma (mortician's) wax (three shades), latex, spirit gum, clown white (regular and water soluble), blemi-sticks (three shades), stage blood, hair wax sticks (three shades), temporary liquid hair color touch-up, eyebrow and lipliner pencils (several shades), sponges, puffs, paper stomps, brushes, cold cream, liquid make-up remover, spirit gum remover, and a variety of student kits and masquerade, clown, and face-painting kits.

RCMA COLOR PROCESS MAKE-UP

The Research Council of Make-up Artists (RCMA) was formed as a membership organization in autumn 1955 by Vincent J-R Kehoe with a group of leading New York make-up artists. In 1962, it was incorporated as a membership corporation, and in 1965 it changed to a sales corporation under the laws of the State of New York. Since its inception in the sales field, it has serviced only the professional industry with products designed especially for their use. Research and Supply offices were set up in Lowell, Massachusetts, in 1966, and the company was moved in 1985 to Mar Vista, California, to better serve the industry while new laboratories were built for advanced research in plastic and latex appliances, modern adhesives, and special products for make-up artists. Although RCMA Professional Make-up cannot be purchased in any drug or department stores, it is available through authorized distributors in major production areas as well as sales outlets in many parts of the United States. (See Appendix B.)

Color Process Foundations

These are some shades matched to some other manufacturer's discontinued numbers that are still being requested by certain make-up personnel:

Ivory	K-1	K-3 (655-K)
Tantone	K-2	RJ-2 (KF-7, Pan-Cake match)

There are also some four-part containers with special colors:

Bruise Kit Violet, Black, Raspberry, Bruise Yellow
Burn Kit Red, Raspberry, Black
Clown Kit Red, Yellow, Blue, Black
Old Age Spots Kit Four shades to simulate old age discolorations and spots.

Lipcolors

These lipcolors contain no staining dyes and no perfumes.

NUMBER		APPROXIMATE COLOR
CP-1		Coral Pink
CP-3		Browned Pink
CP-4		Bright Pink
CP-5		Orange Red
CP-6		Blue Red
CP-7		Blue Pink
CP-8		Light Pink
CP-9		Orange
CP-10		Grape
CP-11		Brown Red
CP-12		Special Red
CP-13		Mix CP-1 and 3
CP-14		Rose
CP-15		Dark Red-Brown
CP-1		Pearl
CP-3	Pearl	
CP-4	Pearl	
CP-5	Pearl	
CP-6	Pearl	
CP-7	Pearl	
CP-8	Pearl	
CP-9	Pearl	
CP-10	Pearl	
CP-11	Pearl	
CP-12	Pearl	
CP-13	Pearl	
CP-15	Pearl	
CP-6	Copper	
CP-12	Silver	
CP-1	Gold	
CP-4	Gold	
CP-6	Gold	
CP-7	Gold	
CP-11	Gold	
CP-12	Gold	
CP-14	Gold	

Gena Special Red
Gena Special Red-Gold
Gena Special Red-Silver
Gena Special Red-Copper
DP Special
Pink Frosting
Special Pink #1
Special Pink #2
Special Pink #3
Pink Cocoa
Pink Gloss
Gena Pink
Gena Coral
Gena Pearl
Tomato Soup
Fantasy Red
Candy Apple
Pure Pearl
Bronze
Copper
Silver
Gold Sparkle

LS-1		Fuchsia
LS-2		Garnet
LS-3		Cranberry
LS-4		Orange Brown
LS-5		Peach Pearl
LS-6		Rotten Apple
LS-7		Natural Rose
LS-8		Light Pink Pearl
LS-9		Orange Pearl
LS-10		Salmon Pearl
LS-11		Dusty Rose
LS-12		Dusty Coral
LS-13		Peached Brown
LS-14		Browned Peach
LS-15		Rust
LS-16		Plum
LS-17		Purple Brown
LS-18		Bright Red
LS-19		Dark Blue-Pink
LS-20		Special Rust
LS-4	Pearl	
LS-7	Pearl	
LS-13	Gold	
LS-16	Gold	
LS-2	Silver	
LS-3	Silver	
LS-6	Silver	
LS-18	Silver	
LS-639		Special Coral
LS-739		Light Special Coral

Lipgloss

CP-1	CP-10	CP-12
CP-4	CP-11	LS-12

Gena Special Red
No-Color
Clear
RCMA Lip Balm (Semigloss)

Within these shades and colors, one can find lip make-up for every period and fashion time. New shades are added from time to time according to fashion trends.

Eyecolors

#1 WATER APPLIED

Black	Silver Green
Midnite Brown	Olive
Gray Brown (Dark)	Aqua
Royal Brown	Turquoise
Light Brown	Silver Turquoise
Charcoal	Silver Lilac
Light Gray	Silver
Midnite Blue	Gold
Light Blue	Cream
Silver Blue	White
Green	Yellow
Light Green	Plum

#2 DRY APPLIED

Royal Brown	Turquoise
Light Brown	Silver Turquoise
Royal Blue	Silver Lilac
Silver Blue	Silver
Wedgewood	Cream
Silver Green	White
Gold Green	Yellow

DRY BLUSHES

Plum	Brown Pearl
Red	Gold Brandy
Mocha	Blush Pink
Pink Pearl	Flame
Peach Pearl	Candy Pink
Bronze Pearl	Garnet
Rust Pearl	Claret
Raspberry Pearl	Lilac

#3 PEARLIZED AND GOLD SHADES

Pearl Black	Gold Huckleberry
Pearl Charcoal	Gold Brown
Pearl Smoke	Gold Burgundy
Pearl Brown	Gold Emerald
Pearl Royal Brown	Pearl Wine
Pearl Light Brown	Pearl Mocha
Pearl Cinnamon	Pearl Plum
Pearl Rust	Pearl Pink
Pearl Cafe	Pearl Rose
Pearl Yellow	Pearl Apricot
Pearl Cream	Pearl Salmon
Pearl Olive	Pearl Violet
Pearl Green	Pearl Lilac
Pearl Turquoise	Pearl Blue
Pearl Teal	Pearl Light Blue
Pearl Pure White	Pearl Slate Blue

These also come in kits of 12 colors of EC #1 or EC #3 and kits of 6 in EC #2 or Dry Blush, made to order for make-up artists in their choice of colors in a flat plastic box.

Special Materials

RCMA has the most extensive and varied selection of special materials, and although mentioned throughout the text, this is a complete list of the items that are available in a number of sizes. Where designated, PMA is the abbreviation for a special line of Professional Make-up Artist materials.

PMA Plastic Wax Material

Light
Women (KW-3 shade)
Men (KT-3 shade)
Negro (KN-5 shade)
Violet No-Color

Also, there is *Dental Wax* in Red Gum shade and Black.

Latex

Pure Gum
Casting Latex
Casting Filler
Foamed Latex (four-part formula)
Foamed Latex (three-part formula)

Mold Release

PMA Silicon Mold Release
MR-8 Urethane Mold Release
PJ-1 Mold Release
FL-1 Mold Release (four-part foamed latex)

Old Age Stipple

KW-2 (Light)	KM-2 (Deep)
KW-4 (Medium)	KN-5 (Dark Brown)

Adhesives

Prosthetic Adhesive A (Solvent based)
Prosthetic Adhesive B (Water based)

Adhesives continued

Special Adhesive #1
Eyelash Adhesive
PMA Matte Lace Adhesive
Matte Adhesive
Matte Adhesive #16
Matte Plasticized Adhesive
Special Matte Adhesive (Two-part)

Sealers

Matte Plastic Sealer
PMA Matte Molding Sealer

Molding Materials

PMA Press Molding Material
PMA Molding Material (Light, Deep, and Dark shades)

Thinners

Prosthetic Adhesive A Thinner
Foundation Thinner
PB Foundation Thinner

Scar Materials

Scar Material
Scar or Blister Making Material

Plastic Cap Material

This comes in clear and pale tinted shades.

PMA Blood Series

Type A (Water soluble)
Type B (Plastic base)
Type C (Cream variety in tube)
Type D (Dried blood effect)

RCMA also makes *Beard Setting Spray, Artificial Tears and Perspiration, PB Foundation Remover, Alginate Separator,* and two-part *Flexible Urethane Foam* and *Solid Urethane* materials.

Appliance Foundation Series

RCMA originally made a PB (Prosthetic Base) series which was similar to the old Rubber Mask Greasepaint of Max Factor and others. However, this has been replaced with an advanced type of material which has higher coverage, no oil separation, and is extremely opaque while spreading very easily with a brush or sponge. (See page 188 for full discussion of this series.)

The *Appliance Paint* (AP) series includes paint-on shades matched to the RCMA Color Process series for latices or urethanes in certified colors. There is also a clear *Acrylid* for coating latices or urethanes. RCMA also furnishes Universal Colors in 1-ounce squeeze bottles that dispense just one drop at a time to tint any type of casting or slush latex or latices and urethanes during foaming operations.

Make-up Cases

RCMA furnishes three styles of make-up cases (also see Appendix E). The accordion style is lightweight, compact, and easy to transport but holds a considerable amount of make-up materials. Also there are two solid walnut drawer styles that can carry more make-up than the accordion style.

RCMA also sells a variety of raw materials for molding and casting, dental acrylics, chemicals, and so forth under Make-up Lab Specialties Company. They also will furnish empties of all their containers for make-up artists who wish to mix special colors or add to their kits.

There are a number of other firms that supply theatrical make-up with similar products to some of those mentioned herein. However, those included are generally representative of the major items employed by professional make-up artists or the procedures and make-ups employed in this book.

SUPPLIERS AND PROFESSIONAL ADDRESSES

It is not possible to list every supplier of materials that is or can be employed for professional make-up. As well, new sources are always being found as technology advances and old sources either discontinue items or businesses. However, listed here are a number of representative manufacturers and suppliers that will be useful as a current source of materials.

CHEMICALS
General
City Chemical Corp., 132 West 22nd St., New York, NY 10001

Kem Chemical Co., 545 S. Fulton Ave., Mt. Vernon, NY 10550

Ruger Chemical Co., Box 806, Hillside, NY 07205

Special
Carbopols Goodrich Co., 3135 Euclid Ave., Cleveland, OH 44115

Methocels Dow Chemical Co., Midland, MI 48192

Plasticizers Monsanto Chemical Co., St Louis, MO 63166

Surfactants BASF Wyandotte Corp., Wyandotte, MI 48192
Rohm and Haas Co., Philadelphia, PA 19105

Separators (Silicons) Dow Corning Corp., Midland, MI 48640

COSMETIC SURGEONS
John Williams, M.D., and James Williams, M.D., Century City Medical Plaza, 2080 Century City East, Los Angeles, CA 90067

Francis G. Wolfort, M.D., Deaconess Medical Building, 110 Francis St., Boston, MA 02215

DENTAL SUPPLIES
The *Yellow Pages* of the telephone book will provide local sources for both medical and dental supplies. However, dental and prosthetic grades of alginates can be found at the following:

Mid-America Dental, 160 Center Dr., Gilberts, IL 60136

Teledyne Getz Co., 1550 Greenleaf Ave., Elk Grove Village, IL 60007

HAIR AND HAIR SUPPLIES
Leon Buchheit, Inc., 217 Greenwich Ave., Goshen, NY 10924   Fine women's hairpieces in real and acrylic hair.

Frends Beauty Supply, 5202 Laurel Canyon Blvd., N. Hollywood, CA 91607   Wool crepe and yak hair and hairdressing supplies.

Seigfried Geike (Ziggy), 14318 Victory Blvd., Van Nuyes, CA 91401   Real and prepared hair goods of all kinds.

Ideal Wig Co., 38 Pearl St., New York, NY 10004   Rental and sales of theatrical-grade wigs.

Polly Products Co., 1 Plummers Corner, Whitinsville, MA 01588   Styrofoam heads.

Ira Senz Co., 13 East 47th St., New York, NY 10017   Wigs, hairpieces, beards, and other lace pieces.

Stein Cosmetics Co., 430 Broome St., New York, NY 10013   Wool crepe hair of excellent quality.

Josephine Turner, 12642 Kling St., N. Hollywood, CA 91607   Real hair lace-front wigs and falls.

LIGHTING MATERIAL
(Special Sunlight-Rated Bulbs)
Luxor Lighting Products, 350 Fifth Ave., New York, NY 10001

PROFESSIONAL AND THEATRICAL MAKE-UP SUPPLIES
Alcone Co., 575 8th Ave., New York, NY 10018

Ben Nye Inc., 11571 Santa Monica Blvd., Los Angeles, CA 90025

Cinema Secrets, Inc., 2909 West Olive St., Burbank, CA 91505

Custom Color Cosmetics, Box 56, Pacific Palisades, CA 90272

Frends Beauty Supply, 5202 Laurel Canyon Blvd., N. Hollywood, CA 91607

Joyce Fox Make-up, 97 Jamestown Pk., Nashville, TN 37205

Kryolan Corp., 747 Polk St., San Francisco, CA 94109

MIS Retail Corp., 736 7th Ave., New York, NY 10019

RCMA (Research Council of Make-up Artists, Inc.), P.O. Box 66282, Los Angeles, CA 90066

Stages Unlimited, Inc., 635 Dee Road, Park Ridge, IL 60063

Stein Cosmetics Co., 430 Broome St., New York, NY 10013

In United Kingdom: Screen Face Dist. Ltd., Marc Dahan, 52 Wardour St., London W13HL

In Canada: Mavis Theatrical Supplies, 697 Glasgow Rd., Kitchener, Ont. N2M 2N7

In Australia: Josy Knowland, 46 Young St., Annandale, NSW 2038

In Far East: Modern Film & TV Equipment Co., 23 Chatham Rd. S. Ocean View Ct., Tsimshatsui, Kowloon, Hong Kong

MAKE-UP EFFECTS STUDIOS
(Prostheses, Special Make-up Effects)
Rick Baker Studio, 12547 Sherman Way, N. Hollywood, CA 91607

TBS, 4000 Warner Blvd., Burbank, CA 91505

The Burman Studios, 4706 West Magnolia Blvd., Burbank, CA 91505

Terry Smith Studio, 330 South Myers St., Burbank, CA 91506

Werner Keppler Studio, 23740 Albers St., Woodland Hills, CA 91364

Dick Smith Company, 209 Murray Ave., Larchmont, NY 10538

Stan Winston Studio, 19201 Parthenia, Northridge, CA 91324

MEDICAL SUPPLIES
(Plaster Bandages, Tubing, Scissors)
Lomedco, 55 Church St., Lowell, MA 01852

Also see local telephone book for other suppliers.

PLASTICS
(Casting Plastics, RTV, Silicones, Acrylics, Urethanes)
Adhesive Products Corp., 1660 Boone Ave., Bronx, NY 10460   Foamart, Adrub RTV, Kwikmold, Casting resins, adhesives, and so on.

BJB Enterprises Inc., Box 2136, Huntington Beach, CA 92647   Various solid and foamed urethanes.

Devcon Corp., Endicott St., Danvers, MA 01923   Devcon WR, metallic varieties, Flexane 60, and others.

Dow Corning Engineering Products Division, Midland, MI 48640   Silicone elastomers, RTV rubbers.

General Electric Co. Silicone Products Division, Midland, MI 48640   Silicone adhesives and sealants, RTV silicone rubber.

Hastings Corp., 1704 Colorado Ave., Santa Monica, CA 90404   Various urethanes.

Industrial Arts Supply Co., 5724 West 6th St., Minneapolis, MN 55416   IASCO products: Plastics, molding supplies, and so on.

Perma-Flex Mold Co., 1919 East Livingston Ave., Columbus, OH 43209   Blak Tufy and Stretchy, Gra-Tufy, UNH, Regular CMC, P-60s.

Smooth-On Corp., 1000 Valley Rd., Gillette, NY 07933   PMC-724, Sonite Release Agents.

RUBBER
(Latex Products Including Slush, Foamed, and Pure Gum)

A-R Products Inc., 8024 Westman St., Whittier, CA 90607

Burman Foam Latex (Sandra Burman), 20930 Almazan Road, Woodland Hills, CA 91364

General Latex Co., 11266 Jersey Blvd., Cucamonga, CA 91730

General Latex Co., High Street, Billerica, MA 01853

R & D Latex Co., 5901 Telegraph Hill Rd., Commerce, CA 90040

R.T. Vanderbilt Co. (latex chemicals), 30 Winfield St., Norwalk, CT 06855

SILICAS AND CLAYS

Attapulgus Clays, Englehard Minerals, Menlo Park, NJ 08817   AT-40, Pharmasorb.

Cabot Corp., Tuscola, IL 61953   Cab-O-Sil grades.

Degussa, Rt. 46, Teterboro, NJ 07608

Grace Chemical Co., Lexington, MA 02173

Wyoming Bentonite, National Lead Co., 111 Broadway, New York, NY 10006

SCULPTORS' SUPPLIES

Chavant, Inc., 42 West St., Redbank, NJ 07701   Plastalene, many varieties.

Dick Ells, 908 Venice Blvd., Los Angeles, CA 90015   General supplier of all types of materials.

Knickerbocker Plaster Co., 588 Myrtle Ave., Brooklyn, NY 11205   Plaster and Ultracal.

Sculpture Associates, 40 East 19th St., New York, NY 10003

Sculpture House, 38 East 30th St., New York, NY 10016

Waldo Bros., 202 Southampton St., Boston, MA 02118   Plaster and Ultracal.

Westwood Ceramic Supply Co., 14400 Lomitas Ave., City of Industry, CA 91744

SPECIAL ITEMS
Artificial Eyes

Jonas Brothers, Inc., 1037 Broadway, Denver, CO 80223

Schoepfer, Inc., 138 West 31st St., New York, NY 10001

Tech Optics, 2903 Ocean Park Blvd., Santa Monica, CA 90405

Van Dyke Co., Woonsocket, SD 57385

Mechanical Materials

Edmund Scientific, 101 East Gloucester Pike, Barrington, NJ 08007

Hobby Lobby International, Rt. 3, Franklin Pike Circle, Brentwood, TN 37027

Jerryco, Inc., 601 Linden Place, Evanston, IL 60202

The Joint Works, P.O. Box 9280, Marina Del Rey, CA 90295

Small Parts, Inc., 6901 NE Third Ave., Miami, FL 33138

Techni-Tool, 5 Apollo Road, Plymouth Meeting, PA 19462

SURGEONS
(AESTHETIC AND COSMETIC)

Although there are many excellent aesthetic and cosmetic surgeons throughout the United States, we have only selected three, who are eminently representative of their art as East and West coast examples.

Francis G. Wolfort, M.D., PC, 170 Commonwealth Avenue, Boston, MA 02216

John and James Williams, M.D., FACS, 2080 Century Park East, Los Angeles, CA 90067

UNIONS AND ORGANIZATIONS

IATSE, Make-up Artists and Hairstylists
Local 706, 11518 Chandler Blvd., North Hollywood, CA 91601
Local 798, 1790 Broadway, New York, NY 10019

NABET, Make-up Sections
Local 15, 1776 Broadway, New York, NY 10019
Local 531, 1800 North Argyle, Hollywood, CA 90028

Aestheticians International Association, Inc., 4818 Cole Avenue, Dallas, TX 75205

SCHOOLS AND SEMINARS
IN MAKE-UP

Years ago, when the major film studios in California were thriving and live television was centered in the New York networks, most had a form of apprentice program to bring eligible people into the unions. However, this entrance to the field is now very limited, and attending a make-up school or studying with individual make-up artists is more prevalent today. College-level make-up classes are mostly for student performers who wish to learn the basics for self-use and for degree credit. There are no degrees in professional make-up artistry. Hairdressing/beauty schools do not teach studio make-up artistry but prepare students to take a state license in hairdressing and/or skin care to work in a beauty salon—not a film or television studio. Studying with an established union member provides a better insight into the professional field, but choose one who has teaching ability and has never failed a union exam!

Small class groups are best (eight to ten) rather than unwieldy large ones, and adequate facilities for all phases of the sessions are necessary. As well, take care before enrolling that one is guaranteed hands-on demonstrations and practical work rather than a steady stream of prepared videotapes of normal make-up procedures. Tapes should always supplement, but not supplant, teaching methods. Of course, videotapes of important special character make-up effects from films are valuable for illustration of specific make-ups as an adjunctive tool.

Finally, students should not expect miracles and should devote a great amount of time practicing all make-up principles before asking to take a union examination. Know your profession first and thoroughly prepare for it. For information on union examinations, see Appendix F.

Vincent J-R Kehoe holds two-week intensive professional level seminars in basic and advanced make-up, and information on these is available through RCMA. Marvin Westmore Academy of Cosmetic Arts, 12552 Cumpston Ave., North Hollywood, CA 90010, specializes in training for cosmeticomedical procedures.

LIGHTING AND FILTERS

COLOR FILM AND TELEVISION

Normally, professional-type films are balanced for studios that use lighting that is rated at 3200° Kelvin, just as sets in television use the same quality of lighting. For outdoor daylight use, a #85 orangy filter is placed in front of the lens to balance the light to its proper quality.

For minor changes in quality and color, *color-compensating* (cc) filters can be utilized. These special filters may be used to introduce slight intentional changes or corrections in the overall color balance. The numbers assigned are an indication of the depth of the introduced color, with the lowest numbers being the lightest shades. They are furnished in six color series—Yellow, Cyan, Magenta, Green, Red, and Blue—to cover both the subtractive and the additive systems.

YELLOW (Absorbs Blue)	MAGENTA (Absorbs Green)
cc 05Y	cc 05M
cc 10Y	cc 10M
cc 20Y	cc 20M
cc 30Y	cc 30M
cc 40Y	cc 40M
cc 50Y	cc 50M

GREEN (Absorbs Red and Blue)	BLUE (Absorbs Red and Green)
cc 05G	cc 05B
cc 10G	cc 10B
cc 20G	cc 20B
cc 30G	cc 30B
cc 40G	cc 40B
cc 50G	cc 50B

CYAN (Absorbs Red)	RED (Absorbs Blue and Green)
cc 05C	
cc 10C	cc 05R
cc 20C	cc 10R
cc 30C	cc 20R
cc 40C	cc 30R
cc 50C	cc 40R
	cc 50R

Other basic filters that are usually for black-and-white photography affect color directly overall in the scene as to their shade; that is, a red filter placed over the lens of a camera will turn everything on the screen during viewing red. This effect is sometimes used for filming naval productions to simulate the red night lighting sometimes employed to accustom the eyes of naval personnel for seeing at night. In the main, as the scene is entirely red, the audience is not aware of any change made in the face by the color filter.

BLACK-AND-WHITE FILM

A number of colored filters were used in black-and-white filming (little used in television) for creating special visual effects. For example, a yellow filter was sometimes used to deepen blue sky and make puffy white clouds stand out more strongly for scenic shots. A red filter placed over the lens would darken the blue sky considerably more to enable shooting in the daytime but having the sky appear dark as it would at night. A green filter would deepen skin tone (especially on men), darken the skies somewhat, and lighten the foliage. Gray neutral density filters were

also employed to cut down the brightness of exterior light as well.

The basic filters were:

8 (K-2)	Yellow: For proper sky, cloud, and foliage rendering.
8N5	Yellow: No. 8 yellow plus 0.5 neutral density.
11 (X-1)	Yellowish Green: For making outdoor portraits, darkening skies, and lightening foliage.
15 (G)	Deep Yellow: For overcorrection in landscape photography.
23A	Light Red: For contrast effects.
25 (A)	Red: For contrast effects on both indoor and outdoor scenes. Cuts distance haze and is often used in conjunction with underexposure to effect night scenes in daytime on panchromatic materials.
29 (F)	Deep Red: For strong contrast effects.

Also see color definitions and charts in Chapter 2.

FILTER COLOR	NUMBER	EFFECT			
		YELLOW	RED	GREEN	BLUE
Yellow	8	L	L	L	D
Yellowish Green	11	SL	D	ML	D
Light Red	23A	ML	VL	MD	VD
Tricolor Red	25	VL	W	VD	VD
Deep Red	29	VL	W	B	B
Tricolor Blue	47	VD	B	L	W
Tricolor Green	58	ML	VD	W	VD

Key:
SL, slightly lighter VL, very light   MD, much darker   W, white
L, lighter   SD, slightly darker VD, very dark   B, black
ML, much lighter   D, darker

RESEARCH FILE SYSTEM

To do character make-up, every professional make-up artist should set up a research picture file for reference purposes. One efficient way is to file cutout pictures, photos, postcards, and so forth in expanding files with multiple pockets, labeling them on the tabs for ease in category and type. Five such 21-pocket files should suffice for general use, while special subjects or interests can be added in other such files, breaking them down to individual characters or productions.

FILE 1

Musicians and Composers
Authors, Poets, and Writers
Artists
Scientists, Doctors, Nurses, and Inventors
Philosophers, Saints, Gods, and Other Religious Personalities
Fantasy, Story Book, and Such Characters
Comic Book Characters
Space and Extraterrestrial Types
Sports Personalities
Convicts and Criminals
Women to Men and Men to Women
Age in Men
Age in Women
Youth to Old Age
Multiple Make-ups
Wigs, Beards, Mustaches, and Sideburns
Women's Wigs
Bald Heads, Partial Wigs, and Foreheads
Double Chins and Necks
Ears and Eyes
Noses and Nasolabial Folds

FILE 2

Cadavers, Witches, Skulls, Magicians, Devils, Dracula, and so on
Mummies, Genies, Phantoms
Frankenstein Monsters
Dr. Jekyll and Mr. Hyde and Such Changes
Animal People and Other Horror Effects
Stone Age People
Cuts, Bruises, Tattoos, and Battle Make-ups
Masks, Dolls, and Puppets
Plants and Insects
Clowns
Plays, Classics, and Literature
Ballet
Opera
Shakespeare: Histories, Comedies, and Tragedies
Renaissance
Other Production Pictures
Oldtime Theatrical and Film Personalities
Modern Actors and Actresses

FILE 3 U.S. HISTORY AND POLITICS

Discoverers and Explorers
Vice-Presidents, Cabinets, Chief Justices
Washington and Lincoln
Presidents: Adams, Jefferson, Monroe, Madison, J.Q. Adams, Jackson, Van Buren, Harrison, Tyler, Polk, Taylor, Fillmore, Pierce, Buchanan, Johnson, Grant, Hayes, Garfield, Arthur, Cleveland, Harrison, McKinley, T. Roosevelt, Taft, Wilson, Harding, Coolidge, Hoover, F. Roosevelt, Truman, Eisenhower, Kennedy, Johnson, Nixon, Ford, Carter, Reagan
Political Personalities: 1600–1750, 1750–1810, 1810–1880, 1880–1920, 1920 to date
Military Men: 1600–1700, 1700–1800, 1800–1900, 1900 to date
British, French, and other Military Personalities during Colonial and Revolutionary Wars.
Foreign Rulers of the Period

FILE 4 NATIONALS

Canada
Eskimos
American Indians
Mexico
South American Indians
Central and South Americans
Negroes
Black Africa
Tunis, Morocco, Algeria
Egypt
Historical England
United Kingdom—Modern
France, Italy, Greece
Spain and Portugal
Germany, Austria, Hungary
Yugoslavia, Czechoslavakia, Bulgaria, Rumania
Iceland, Finland, Sweden, Norway, Denmark
Poland
USSR
Belgium, Netherlands, Switzerland

FILE 5

Mongolian Make-ups
Japan
Korea
Ancient China
Modern China
Tibet, Nepal, Afghanistan
Vietnam, Thailand, Indonesia, Melanesia, and so on
Story Book Arabia, Ancient Persia, and so on
Saudi Arabia, Iran, Iraq, Syria, Lebanon, and Other Arab States
Israel and Turkey
India
Australia and New Zealand
Philippines, Hawaii, and Other Islands
Personal Production Photos
Miscellaneous

PROFESSIONAL MAKE-UP KITS

MAKE-UP CASES

Two basic styles of cases are generally utilized for make-up kits by professionals. While many East Coast artists prefer the accordion-style case, the West Coast people generally carry the wooden, drawered variety.

RCMA stocks both of these cases in the most popular and versatile sizes. The accordion-style case is black fibre with chromed fixtures and is reinforced with steel frames (Figures E.1 and E.2). It is large enough to hold a full professional assortment of make-up materials. The wooden case with drawers is often hampered in capacity in the style with many small drawers (and often contains more wood than space!). However, RCMA stocks two kinds of these in solid walnut that have only three wide, full-depth drawers plus a generous hinged-top space on top. One has an overall size of 16 inches long, 12¼ inches high, and 8½ inches deep. It has chromed fixtures and a comfortable handle for carrying. This type of case has a lock (the accordion-style fibre case does not) in addition to three strong clasps. Its capacity is larger than that of the accordion case, but its carrying weight is more due to its solid construction.

The second wooden case is a combination design of the old on-set kit and the regular wooden one in use for many years (Figures E.3 and E.4). However, this new concept case made exclusively for RCMA combines all the good features of both and eliminates the main problems of weight and loss of space. Essentially, by removing much of the metal hardware, which is unnecessary, the case weighs empty only 8 pounds and fully loaded, with tissues attached, only 21 pounds, an easy weight for anyone to carry.

A spring clip on the side holds a box of tissues for easy use, and all the drawers are metal lined for ease in cleaning. The top section does lock, and all the drawers are on spring catches so they will not fall out when opened. One can open the drawers without opening the top or cover of the case. This case is not only elegant but also practical for professional make-up artists and is especially designed to fit the newest professional containers without waste of space.

Two types of make-up kits of materials and tools are listed here. The first is a basic RCMA kit recommended by Vincent J-R Kehoe for his Professional Make-up Seminars. It contains all the primary materials and tools and allows room for individual selection of additional items for expansion. The items will fit in either of the case styles, with the multiple color ones such as foundations, lipcolors, pencils, shadings and countershadings, cheekcolors, beardcover, lipgloss, and eyecolors, as well as brushes and tools, all boxed as units for ease in both storage and use.

THE BASIC KIT

CP Foundations: KW-13, KW-3, KW-14, KW-4, KW-37, KW-38, KM-2, KM-3, KM-37, KM-38, KT-3, KN-1, KN-3, KN-5, Superwhite, and Gena Beige.

Countershading: CS-1, CS-2, and Beardcover BC-3, Hair Whitener HW-4

Shading: S-1, S-13, S-14, S-4, S-6, S-8

Cheekcolor: Genacolor Rose and Genacolor Pink

Eyecolor Kit #1: 12 water-applied eyecolors in a plastic box

Pencil Kit #1: 10 haircolor pencils in a plastic box

Brush Kit #2: Two each: 1R, 3F, 4F, 7F, 10R, Eyebrow, and Eyelash
One each: 12F, Wig-cleaning brush

Lipcolors: CP-1, CP-3, CP-4, CP-5, CP-6, CP-14, LS-5, LS-12, DP, LS-16G, Gold Sparkle, Copper

Lipgloss: No-Color Gloss

Lipliner Pencils: Maroon, Dark Red, and Lake Red

Pencil Sharpener

Mascara: Black and Brown

Sponges and Puffs: 5 packs of polyurethane, 1 stipple, and 2 puffs

Powder: No-Color Transparent and PB Powders

Tools: Stainless steel spatula, dental college pliers, and tweezers

Special Material: Plastic Wax Molding Material, (Light) Scar Material, Matte Adhesive, Matte Plasticized Adhesive, Eyelash Adhesive, Matte Plastic Sealer, Tears and Perspiration, Foundation

Thinner, PMA Molding Material (Light), Color Process Blood (Types A and C), Prosthetic Adhesive (A and B), Scar and Blister Making Material

Appliance Foundations: KW-2, KM-2, KN-5.

THE FULL-CONCEPT KIT

This list is based on a full kit for both straight and character make-up use to be carried along with the Hair Kit described in Chapter 16. It expands the basic kit previously described with additional materials and tools and is a full working kit for a professional make-up artist.

CP Foundations: Shinto I to IV, Ivory, F-2, F-3, F-4, KW-23, KW-24, KW4M2, KW4M3, KT-1, KT-2, KT-4, K-3, KW-36, KN-2

Shading: S-2, S-3, S-7, S-9, S-10

Cheekcolor: Flame, Dark, Pink, Lilac, GC Plum

Special Shades: 1667, 6205, Green Clown Kit: Red, Yellow, Blue, Black Bruise Kit: Raspberry, Violet, Yellow, Black and the Old Age Spot Kit

Eyecolor Kit #3: 24 different shades in two plastic boxes

Pencils: Dark Silver Gray, Navy

Brush: Wig-cleaning brush

Lipcolors: CP-7, CP-9, CP-10, CP-11, CP-12, LS-17, LS-18, LS-19, LS-20, CP-4G, CP-11G, CP-12G

Tools: Nail file, aluminum tail comb, hair scissors, curved scissors

Plastic Wax Molding Material: Clear, KW-3, KT-3, KN-5, Violet

PMA Lace Adhesive, Prosthetic Adhesive Thinner, Blood Type B, Acetone

Appliance Foundations: Kits #1, #2, and others, or with single containers of these shades.

Other Items: Liquifresh, RCMA Lip Balm; Visine Eyedrops; Small Package of Aspirins; One Plastic Container Each of Q-Tips, Hair Clips, Bobby Pins, Straight Pins, and Rubber Bands; and a Single-edge Razor Blade

Dental Wax: Package each of Black and Red

False Lashes: One Each Black and Brown in Plastic Box

KIT TRICKS

1. Replace the regular low-density polyethylene stock bottles with linear high-density types. Especially suitable for lotions, acetone, alcohols, and so on.
2. Place about a dozen mascara remover pads in a regular 1-ounce foundation container for ease in carrying.
3. Remove sponges from glassine shipping envelopes, and place in plastic sandwich bags.
4. Use shipping cardboard boxes for holding foundations, lipcolors, and small-sized countershading, shading, cheekcolor, etc., containers.
5. Label tops of bottles for easy selection.

Although the full-concept kit will serve for most general applications and work, there are times when the kit must be adjusted for special occasions or specific assignments. For example, those make-up artists who are only doing commercial straight make-up might wish to remove some of the special materials and replace them with extra lipcolors and eyecolors. In addition, when it is known that only one person is to be made up (as in the case of politicians who hire a make-up artist to improve their appearance for camera or public appearances), one might transfer the known use materials to a small attaché case where a make-up cape, towel, and so forth may also be carried rather than carrying the normal hair kit.

There are other available types and varieties of make-up cases. However, these two are well adapted for weight, capacity, and long-term suitability.

FIGURE E.1   *Accordian case closed.*

FIGURE E.3   RCMA *Professional Case.*

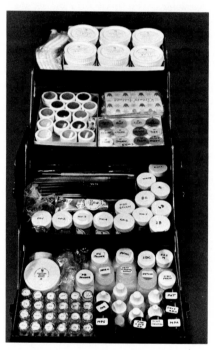

FIGURE E.2   *Accordion case filled.*

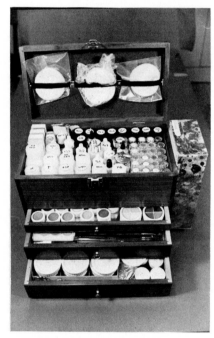

FIGURE E.4   *The RCMA Case filled.*

MAKE-UP ARTIST EXAMINATIONS

Qualification in a union representing make-up artists consists of being able to prove a certain number of days of employment in the field as well as to take an examination in practical make-up artistry before a union committee to demonstrate the applicant's ability. As the main production areas are New York City and Hollywood, both represented by different Locals of two different unions, IATSE and NABET, and all with different examination procedures, one must first select the work area in which he or she wishes to gain employment and the opportunities afforded by each. Intelligent business letters to each should be the first method of approach, asking for an application and/or an appointment with the business manager for an interview if feasible.

SCREENING COMMITTEES

Unions normally have screening committees that review applications before interviewing the prospects. It is to any Local's advantage to consider carefully such applications so that the best talent is made available for their ranks and to increase the artistic potential of the Local. However, much depends upon the progressive attitude of the Business Manager (whose opinion is always given on these matters) as to how unknown or new applicants are handled. It has always proven, in the long run, a serious disservice to itself for any unionized group to seriously restrict membership at any time, even in slow periods of work, as talent, like time, does not rest or wait as they will *make* opportunity. If any applicant feels that unfair treatment or undue favoritism was exhibited in the screening procedure, one does have access to the National Labor Relations Board (who have often ruled in favor of the individuals when a narrow attitude of employment is afforded by the local in question).

EXAMINING COMMITTEE

There are many union members who refuse to be on the Examining Committee because it often takes much volunteer time for the work entailed. However, it is the most important committee function of any Local, and great care should be taken with its membership. All examiners, without exception, should either have taken such an examination or have performed all the work entailed in its procedures. It is enormously unfair for any applicant to take an examination before Examining Committee members who have not done and possibly could not do every phase of the examination during a work day themselves.

The Examining Committee should consist of three make-up artist members in good standing (who are not union officers, paid, or craft representatives), with two alternates selected to fill in when a regular member cannot attend an examination for one reason or another. This Examining Committee need not be rotated on a fixed periodic limit if it operates properly, although the membership of the Local should have the privilege of asking for a new committee to be formed if they feel that the present one is not operating properly.

One of the most difficult problems of an examiner is to be completely fair in judging an applicant's work. Examiners cannot inflict personality problems (either their own or the applicant's) into the procedure of talent judgment and must, in all instances, judge the completed work of an applicant with proper perspective and fairness. The main question should always be: "Can the applicant's work fully represent the art and craft of the Local in a job on the very next day if he or she is allowed to pass?" Only "Yes" should suffice.

REQUIREMENTS FOR APPLICANTS

After filling out the application for the union and requesting an examination, the applicant must appear before the Screening Committee. The applicants should be ready to show pictures of their work, proof of their employment requirements, and maybe show what their make-up kit looks like. Completeness of these essentials should be judged along with the personality and appearance of the applicant. Just having the tools of the trade does not make a professional make-up artist, but it does show a basic step toward this end.

Once the applicants have been screened and are ready to take an examination, then a date should be set up at which at least two and not more than six applicants are to be examined at the same time. The frequency of such examinations should be at least every six months when applicants are available.

On the day of the examination, applicants must present themselves ready to work, with two models, one male and one female, furnished at their expense and selection. A tip for applicants: A very attractive female with few, if any, facial problems and a male with some character lines in the face, no facial hair, and a normal haircut are best. It is also best to select Caucasian models as more changes can be made on their faces for the examination procedures. Applicants must have a full make-up kit and the hair goods necessary for the examination as well as any appliances that may be required. They must have paid all the fees necessary to the Local and otherwise be in good standing with the union's requirements. It is the strict responsibility of the union to provide a proper space, lighting, chair, and mirror for each applicant. It is extremely unfair to expect applicants to work in inadequate conditions when they are asked to show their work to enter the union. Union members have the right to complain if the producer does not furnish proper make-up rooms, so why shouldn't the applicants have the same privilege?

PAYMENTS BY APPLICANTS FOR EXAMINATION

Although some Locals do not pay Examiners, it is wrong for any applicant or Local to expect that a make-up artist examiner should give up a day or more of time to serve on the Examining Committee or to expect the Local to pay the Examiners to serve. Such expense should be borne by the applicants. A fee of at least $100 should be required, with at least half this amount going to the examiners and the remainder to the Local for administrative work and securing a proper place for the examination. The fee should be an individual one and not be returnable in case of failure in the examination. Once again, applicants will think twice before requesting an examination if they must bear its cost and will have the tendency to prepare themselves fully. It also places the Local under the responsibility to provide adequate examination premises and conditions. An applicant who fails the examination should have to wait at least 90 to 180 days before being allowed to take another.

SCORING THE EXAMINATION

Although most applicants for any form of examination will be subject to apprehension and nervousness, it is only in the *work completed* that an examiner may judge the talent of the applicant. Whatever system of point grading is employed, the percentage rating of a passing grade should not be less than 80 percent. If the applicant fails to complete any project or part of the examination in the required time, no points should be allowed and, if done, only for one project or part of the examination. These circumstances must be agreed upon by all examiners present.

No applicant should be informed as to the progress or marks of the examination until it is completed. Telling an applicant that he or she is passing or failing at any time during the examination is unfair.

CATEGORIES WITHIN A LOCAL

Some Locals may set up certain categorical groups such as department heads, charge make-up artists, assistant make-up artists, and so on, employing seniority of membership or status, work ability, or area of work (film, television, or stage), and may require an examination to rise to a higher category within the Local's work structure.

Such categories should not be harmful or restrictive to the talent advancement of any member, and they should be carefully administered and designed to provide fairness and opportunity potential. In today's similarity between the make-up requirements of all mediums, any examination should be as strict as another and standards not lowered in any manner or for any medium.

PERIOD OF EXAMINATION

The suggested examination (which is designed for one full day) may be extended to two or even three days of time, with all elective projects required as well as additional hair and prosthetic work, if it is believed that such a competitive level of demonstration of ability is necessary. This is based, of course, on the needs and requirements of an individual Local, but the decision should be the norm and not used solely to make one particular examination more difficult than another. Due to considerably more character work being performed in the California area of production, the extended examination time is the case, while in New York City, both unions require only a one-day examination at this time. No examination should, however, consist of less than the stated requirements and projects.

EXAMINERS' MARKING PROCEDURE

All three examiners must mark independently and not compare individual scores until the examination is fully completed. A scoring card should be made for each applicant and the final score for each be an average of the three examiners' marks. Applicants

SUGGESTED TYPICAL EXAMINATION		
Time: 8:30 A.M. to 5:30 P.M. with a 1-hour lunch break.		
APPLICANTS GRADED ON:	POINTS	TIME (Minutes)
Attitude and Conduct	2	
Appearance	1	
Make-up Kit	2	10
Oral Exam: At least three questions from each examiner	5	20
Written Exam: A series of 10 questions (selected from a standard list of about 50)	10	30
Set-up Period for Materials and Examiners Marking Written Exam		20
Practical Exam		
Straight Beauty Make-up on a Woman	10	45
Straight Make-up on a Man	5	25
Fashion Make-up on a Woman (including false lashes)	5	45
Lunch Break		60
Age a Man 20 Years (include hair graying but no appliances)	10	30
Lay a Beard and Mustache on a Man	10	30
Apply Eyebags or Other Aging Prosthesis and Make-up	5	30
Apply a Prosthetic Nose and Make-up	5	30
Note: The last three exercises can be combined in one make-up.		
Apply Old Age Stipple on Half a Face	5	45
Exercises: Required		
Apply cuts, bruises, tattoo, scars, burns, tears, and sweat	5	30
Attach a men's toupee, lace beard, or sideburns	5	30
Exercises: Elective	15	60
Attach a Bald Cap and Make-up		
Attach an Appliance to Show a Bullet Hit and Blood Flowing		
Apply Oriental Eyelids and Make-up		

should have the right only to total scores, and such should be posted in the Local's office within three days. As most examiners know within an hour or so what the final scores are, there should be little reason to make an applicant wait two or three weeks for the marks. Examiners' score cards must be filed with the Local after being signed by the person who scored the card.

NOTIFICATION OF APPLICANTS

Those who pass the examination should then follow the union's procedure for membership, while those who fail should either ask to be placed on the list for the next examination or seriously re-assess their desire to be a professional make-up artist by reading once again Chapter 1.

To make the examination more comprehensive, the three electives can be made required, with 5 points given for each exercise and the time extended an additional hour. If this is done, it should be standard and constant and not done for just a certain or single group of applicants.

As can be totalled, the score available is 100 points, and at least 80 points are required for a passing grade. Written, oral, and kit account should go for 20 points, straight make-up for 20 points, and 60 points for character make-up.

*   *   *   *

Mr. Kehoe's qualifications to discuss these procedures are based on his 25 years of membership in the New York City IATSE and having been on the Examining Committees for over ten years, in addition to heading and writing their Apprentice Training Program. As well, he formulated the first examination procedures for the New York NABET Local as well as holding a card in that union under various categories for 20 years. His original professional book on make-up, published over 25 years ago, was the bible of study for the majority of union members all over the world.

BRITISH AND EUROPEAN MAKE-UP MATERIALS

It will be found that some of the products mentioned in this book are not available in the United Kingdom. However, there are, in most cases, corresponding materials available that will fulfill a similar purpose. Most of the theatrical cosmetic preparations are available both in the United States and in the United Kingdom, although not all manufacturers are represented in the United Kingdom. Wherever possible, ranges that are imported from the United States or indeed from Europe will be listed by those companies stocking their products in the United Kingdom. There are certain differences in techniques where special effects processes are concerned. Individual make-up artists do have individual ways of doing many of these processes, which have evolved because of the availability of certain products and compounds in both countries. Most often it will be found that problems arise when certain proprietary materials are used for which the companies do not have an agent or supplier in the United Kingdom. Whenever possible, equivalents are listed or alternative techniques suggested. For simplicity, the materials in this field are dealt with according to manufacturers and suppliers in general. Addresses of manufacturers or suppliers are for convenience in an alphabetical list at the end of this appendix. There is also a chart listing equivalent products. However, Screen Face Dist. Ltd. in London will take orders for any of the special materials listed in this book which are not otherwise available in the European area or the United Kingdom.

MAX FACTOR

Max Factor has now regrettably ceased all production of theatrical make-up in the United Kingdom and Europe, as well as in the United States. Throughout this section, as with the rest of this book, the former Max Factor numbers are used in some cases as a guide to color matching for other manufacturers' ranges where available. At the time of going to press, however, Charles H. Fox Ltd., 22 Tavistock Street, London WC2, does have a limited remaining stock of Factor's products.

LEICHNER

There are two companies by the name of Leichner in Europe, but this one refers to the London based company, not to that in Germany.

Leichner produces make-up for stage, still photography, film, and television. Several ranges of products are available. Leichner also produces a Grease Paint make-up called Spot Lite Klear, supplied in a tube, and is applied with a slightly moist sponge in the normal way. Leichner also makes a Creme Cake foundation, which comes in different colors for body make-up and is applied as most cake foundations with a wet sponge. The company still manufactures the old-fashioned Grease Paint Sticks wrapped in foil: Form C, the thicker of the two, and Form G, the thin liner, both of which are used in theater. Leichner also makes a special range for Casualty Simulation when training medical personnel. This contains such items as Blood Capsules, White Frothing Capsules, and Scar Material. In addition, the range of products also contains the more normal items, including Blending Powder, Clown White, Negro Black or Negro Brown, Removing Cream, and Spirit Gums.

BLENDING POWDER

Rose	Neutral
	Brownish

CREME CAKE

Biscuit	Light tan
Beige	Olive brown
Tan	Red brown
Gold	Golden sand
	Brown sand

EAU DE LYS BODY MAKE-UP

Medium pink	Yellow tan
Ivory	Peach special
Deep brown	Blithe spirit
White	Incredible hulk
	Light olive

GREASE PAINT FORM C

Light pink	Chrome yellow
Medium pink	Dark ivory
Dark pink	Peach special
Coffee brown	Light tan
Reddish brown	Golden tan
Ivory	Deep brown
Shallow pink	Light peach
Brick red	Medium peach
Black	White
Yellowish ochre	Golden tan
Brownish ochre	Brownish tan
Light olive	Yellowish tan

LINER FORM G

Black	Carmine 1, 2, 3
Dark brown	Carmine camillion
White	Light blue
Crimson lake	Medium blue
Light brown	Dark blue
Red brown	Light green
Light grey	Medium green
Dark grey	Dark green
Chrome yellow	Light mauve
	Medium mauve
	Dark mauve

SPOT LITE KLEAR

Medium pink	Brownish tan
Ivory	Yellow tan
Shallow pink	Peach special
Brick red	Gold
Light tan	Silver

MEHRON

A limited range of Mehron is imported into the United Kingdom by Charles H. Fox Ltd. who has a small range of twelve matching Pan Sticks and Pan Cakes of foundation colors in stock.

BEN NYE

Ben Nye products are imported into the United Kingdom by certain outlets such as Theatre Zoo, 21 Earlham Street, London WC2, and Charles H. Fox Ltd. Both have limited quantities and ranges such as:

Cream Cheek Rouge
Cream Highlight
Cream Lining Colours
Mellow Yellow
Dry Cheek Rouge
Pearl Sheen Liners
Pressed Eye Shadow

RCMA PRODUCTS

RCMA products are now available in the United Kingdom at Screen Face Ltd., which carries a full line of regular products in addition to all RCMA Special Materials.

CHRISTIAN DIOR

Christian Dior film and television range "Visiora" is not yet stocked in the United Kingdom. However, its launch is planned sometime in 1985. Their products are available in Germany, France, Australia, and Japan. The most popular ranges are the Liquid and Compact Foundations, which are applied in the normal way with a slightly moist sponge. In addition to these, they make a liquid body make-up, face coloring creams, as well as cleansing milk, pre make-up base, powders, etc.

KRYOLAN

The range of Kryolan is a very large one, and was originally devised to supply the needs of theaters within Germany and near European countries. Although their products do not always fulfill the requirements of a professional range of make-up, *as their colors can vary somewhat from batch to batch,* they do, however, maintain a very useful function. Kryolan now produces a matching range of colors to the old Max Factor CTV W Series and other commonly used colors, but not the Factor 725 range. Following is a list of some equivalents together with those of RCMA. (Note that these are approximate color matches for comparative use only.)

KRYOLAN	FACTOR	RCMA
1W	CTV 1W	CS-1
2W	CTV 2W	KW-1
3W	CTV 3W	KW-2
4W	CTV 4W	KW-3
5W	CTV 5W	KW-4
6W	CTV 6W	KM-1
7W	CTV 7W	KM-2
8W	CTV 8W	KM-3
9W	CTV 9W	—
10W	CTV 10W	—
11W	CTV 11W	—
12W	CTV 12W	—
K1	K1	K-1
K2	K2	K-2
070	White	Superwhite
071	Black	Black
Grey	1742	—
Shader AL	AL	S-1
EF 085	085	—
EF 080	080	—
Ivory	Ivory	Ivory
Negro 1	Negro 1	KN-6
Negro 2	Negro 2	KN-7
Negro 2	Negro 2	—
Oriental	Oriental	See Shinto series 1 to 4, also KT series 1 to 4
Chinese	Chinese	—
Light Indian	Light Indian	—
Dark Indian	Dark Indian	—
Light Egyptian	Light Egyptian	KT-3
Dark Egyptian	Dark Egyptian	KT-34
2880 Natural beige	2880 Natural beige	—

The full Kryolan range includes several hundred colors available in various formulations such as grease paints, paint sticks, pan sticks, pan cakes, dry powders, body make-ups, rubber mask grease paints, and more. In addition, they produce a large range of special effects materials such as Glatzan, which is a cap plastic material; Hydro Oil Remover, which is a very efficient remover especially for silicone adhesives and certain other hard-to-get-rid-of adhesives; latex, blood, wig-making accessories, solvents, spirit gums, glues, adhesives, tooth enamels, Ultra Violet make-ups, etc.

Their products are available from many retail outlets in London as well as in the provinces. It is not possible to list the entire Kryolan range, since different outlets stock different items; however, some examples of the range are:

Aquacolour: 250 shades
Bald cap make-up: 100 shades (rubber mask grease paint)
Body make-up powder: 33 shades
Cake make-up: 20 shades
Cake Rouge: 20 shades
Collodion
Eyebrow pencil: 15 colors
Face powders: 20 shades
Glatzan: bald cap material
Glitter
Grease sticks: 130 shades
Hydro oil remover
Liquid latex
Magic blood
Make-up palettes
Sealor
Solvent
Spirit gums
Supracolour: 250 shades
Tooth enamels: 6 shades
Ultra violet make-up: 6 shades

GENERAL LIST OF SUPPLIERS

Screen Face Dist. Ltd.

Screen Face Dist. Ltd., 52 Wardour Street, London W1 3HL, carries a full line of RCMA products for professional use, including all RCMA Special Materials.

Alec Tiranti Ltd.

Alec Tiranti Ltd., 21 Goodge Place, London W1, and 70 High Street, Theale, Berkshire, produces a large range of sculptors' tools, materials, and equipment. Many of the items in their range are especially useful to the make-up artist, particularly to those make-up artists working in the special effects field. A large range of sculptors' tools—chisels, modelling tools, modelling clays, fibreglass resins, fillers, pigments, armature wire, plasticine, spatulas, callipers, scrim, burlap, shellacs, release agents, epoxy resins, protective clothing, sculptors' easels and stands, and so on—are stocked.

Fulham Pottery

Fulham Pottery, 184 New Kings Road, London SW6, provides a useful range of modelling clays, tools, and equipment. Although these are primarily designed for use in pottery, they are very useful to make-up artists. Wet clay is normally supplied in various quantities up to 25 kilo bags. The most commonly used clay for modelling is DBP Modelling Clay. The advantage of using clay instead of plasticine is that clay, being greaseless, does not leave any deposits in the plaster molds.

Fulham Pottery also produces many modelling and sculptors' tools in addition to Crystacal Plasters.

DENTAL MATERIAL SUPPLIERS

There are many dental suppliers throughout the United Kingdom. Most of these provide a very comprehensive range of products, tools, and equipment, mostly oriented toward dentistry, but many of them are useful to the make-up artist. Dental instruments in particular can be very useful. Most of them will supply alginate; acrylic compounds; various waxes for molding, modelling, and impression taking; plaster bandage; cellophane separator sheets; dental drills; various burrs; and polishing materials for acrylic teeth, eyes, etc. A few names and addresses are listed at the end of this section.

HAIR AND HAIR SUPPLIERS

There are many wigmakers in the United Kingdom, most of whom in addition to preparing all kinds of hair products such as wigs, moustaches, sideburns, and eyebrows will also supply hair, crepe hair, crepe wool, human hair, or yak hair.

A.H. Isles Ltd., 146 Lower Road, London SE16, will supply all wigmaking materials including wigblocks, knotting hooks, mixing hackles, nylon net, caul net, gauzes, iron heaters, chemicals, human hair, crepe hair, crepe wool, yak hair, angora, etc.

CHEMICALS

Many of the commonly used chemicals can be obtained from the High Street chemists or similar stores throughout the country. John Bell & Croydon Ltd., 50 Wigmore Street, London W1, stocks many items that are useful for make-up artists, such as petroleum jelly, spatulas, surgical gloves, acetone, alcohol, green soap, stearic acid, oleic acid, ammonium hydroxide, distilled water, mineral oil, gelatin powder, kaolin, glycerine, talc, etc.

As far as the more specialized chemicals are concerned, it will probably be necessary to approach industrial or commercial chemical companies. These can be found listed in the Yellow Pages under their appropriate headings. Many of them will supply only through their subsidiary companies, particularly where small quantities are concerned, so it will be necessary to first approach the manufacturer, and then ascertain who their distributor may be. The problem make-up artists encounter sooner or later is the difficulty in getting small enough quantities of materials, since many commercial houses are interested only in supplying in bulk.

MEDICAL SUPPLIES
There are several companies in London and in all the major cities supplying the medical profession with surgical tubing, silicones, gloves, instruments, plaster bandages, scissors, etc. John Bell & Croydon Ltd. has a very extensive range of medical products, as does McCarthys Surgical Ltd., Selinas Lane, Dagenham, Essex.

POLYMERS, PLASTICS, AND RUBBERS
There are many commercial companies throughout the United Kingdom either importing proprietary products or supplying their own; however, it should be noted that neither Permaflex nor BJB products are represented in the United Kingdom or Europe.

W.P. Notcutt Ltd., 44 Church Road, Teddington, Middlesex, is the agent for all Smooth-On products. It is also the agent for an excellent silicone rubber made by Wacker GmbH in Germany, in addition to keeping various polyurethane products annd silicone fluid release agents, etc.

Dow Corning Ltd., Bridge Road House, Reading, Berkshire, makes a huge range of silicone rubbers, fluids, release agents, and all silicone materials. They also make a Surgical Silicone Foam Compound, which can be very useful.

Elastomer Products Ltd., Wharf Way, Glen Parva, Leics., supplies both flexible and non-flexible polyurethane foams, which are generally available in large quantities only.

Alec Tiranti Ltd. is the agent for General Electric Products in the United Kingdom and has a comprehensive range of many silicone rubbers, fluids, etc., as well as stocking many other elastomers such as Vinamold.

Charles H. Fox Ltd. supplies many cap plastic materials, liquid latex, and foam latex formulations, especially from Kryolan, many of which are obtainable in kit forms.

Tom McLaughlin (Chemical Sales), 26 Avenue Road, London E7, supplies an excellent foam latex kit of 25 litres of latex together with instructions. Other items include High Rise Foaming Agent, a cure "Accelerator," Acrylic Latex Thickener, etc.

Casting Latex and Pure Gum Latex are pure Natural Latex Compounds, which are supplied in the United Kingdom by Macadam & Co. Ltd., Monument House, Monument Street, London EC3, or William Symington & Son Ltd., Bath House, Holborn Viaduct, London EC1. Both of these companies will supply in gallon (5 litre) quantities.

PLASTERS
In the United Kingdom all plaster products are made by British Gypsum Ltd., so in case of any difficulty, it would be advisable to contact them regarding whom they supply in your area. Most make-up artists use two different types of plaster: Superfine, Fine Casting Plaster or Dental Plaster; and for making foam latex molds, Crystacal Plaster.

Foam latex molds made of Crystacal Plaster seem to stand up reasonably well to repeated use in ovens, providing the oven temperature is not allowed to rise too high. This material is not exactly comparable with the Ultracal range in the United States.

Plaster bandage is available generally from medical supply companies. Crystacal Plaster and Dental or Fine Casting Plasters are available from various retail outlets, including Fulham Pottery, Alec Tiranti Ltd., and several good builder's merchants; in fact, most builder's merchants will order to your own requirements from British Gypsum.

As mentioned, there are no direct equivalents to American Ultracal Plasters. The nearest equivalent plasters obtainable in the United Kingdom are probably the dental stone products, which can be obtained from dental supply companies. These are normally packed in small quantities, since they are only used for dental purposes.

GENERAL ART MATERIALS AND ARTISTS' SUPPLIERS
There are at least three major manufacturers of artists' materials in the United Kingdom, notably Winsor Newton, Rowney, and Reeves. They all produce a range of pigments, oil colors, water colors, brushes, etc. Most of their products are normally obtainable in good art shops throughout the United Kingdom.

In addition to these companies' products, you will find the materials of most major European and American companies, such as Liquitex and Pebeo, also easily obtainable in art stores throughout the United Kingdom.

Brodie and Middleton Ltd., 68 Drury Lane, London W.C.2, has an excellent range of brushes, pigments, and shellac, etc.

DUPLICATING MATERIALS
As far as duplication of molds is concerned, in the United Kingdom, most make-up artists use silicone rubbers or Vinamold. Polyurethane compounds such as those manufactured by Permaflex Company, for example, are not exported to the United Kingdom. Certain Smooth-On products are, however, imported by Notcutt Ltd. who is their agent in the United Kingdom.

A full range of alginates is available from all dental supply companies. Screen Face Dist. Ltd, 52 Wardour St., London W.1, carries Mid-America's Prosthetic Grade Alginate.

Vinatex Ltd. produces Vinamold, which is stocked by Alec Tiranti Ltd., and which is a Hot Melt Compound frequently used for duplicating either molds, plaster casts, or in certain cases plasticine and other models. This is a vinyl-based flexible mold making material which is fairly tough, and can in fact be re-used by simply melting it down when there is no further use for the mold. The material is designed for reproducing molds for casting polyester resins, plasters, low melting waxes, etc. It has good tensile strength, and is available in three grades varying in flexibility and resilience: Natural, Red, and Yellow. Most molds can be made with Natural or Red, but in some cases the Yellow is more suitable; that is, a flat open mold can be made in Yellow, and then it would not require a plaster case as Yellow is self-supporting. Melting temperatures vary from about 150°C to 170°C.

The basic principle behind Vinamold is that the flexible PVC material is melted down in a melting pot, and poured over the original cast or model. It is then allowed to cool, and the resultant flexible mold is backed with a plaster case. The material is then removed from the model. Vinamolds in general are relatively harmless to handle, provided the manufacturer's instructions are followed. Tiranti publishes a very comprehensive booklet on Vinamold and it is recommended that it should be consulted before using this material. In addition, Tiranti also supplies both the Vinamold and all the melters and necessary equipment for using it.

MODELLING MATERIALS
Most British make-up artists are now changing over to modelling in wet clay for its obvious advantages. Plasticine is similar to Plastalene in the United States.

EUROPEAN MAKE-UP MANUFACTURERS AND SUPPLIERS
The best known European make-up manufacturers are either in Germany or France, from where the ranges are distributed to other parts of Western Europe and Scandinavia. Suppliers are usually located in the capitals, where film and television studios are generally situated. Some of the commonly used chemicals such as alcohol and acetone are not always so freely obtainable as in the United Kingdom or United States, and consequently it might be necessary to get a prescription. Following is a list of suppliers, by country, with products supplied.

Austria
Fritz Brennig, Magdalenenstr.22, A-1070 Vienna. Make-up supplies.

Belgium
Maison Jean Avonstondt et Fils, 13 rue Melsens, B-1000 Brussels. Make-up supplies.

Denmark
Anker & Lovbo, Amagertorv 15, II, 1160 Copenhagen K. Make-up supplies.
Bertello, Trianglen 4, 2100 Copenhagen O. Make-up supplies.
Frida Davidsens Eftf., Nr.Farimagsgade 17, 1364 Copenhagen K. Make-up supplies.

Chr. W. Mende & Co., Kronprinsensgade 9, 1114 Copenhagen K. Make-up supplies.

Finland
Koukkanen Peruukkiliike Ky, It.Teatterikuja 5B, Helsinki 10. Make-up supplies. Wigs and hair supplies.

France
Adam, 11 Blvd. Edgarquinte, 75014 Paris. Fibreglass, urethanes, resins, foam latex, plasters.

Berty, 49 rue Claude Bernard, 75005 Paris. Same materials as above plus paints, acrylics, tools, etc.

Bogard, François, 131 rue l'Université, 75007 Paris. Make-up supplies.

Deruelle, 34 av. Paul Doumer, 92500 Rueil Malmaison. Make-up supplies, their own range.

Dorin SA, 36 rue des Renouillers, 92700 Colombes. Make-up supplies.

Isker, 24 rue des Petes Ecures, 75101 Paris. Dental supplies, alginate, tools, equipment.

Merle (Parfumerie Lucienne), 9 av. Matignon, 75008 Paris. Make-up supplies and chemicals.

Make-up Studio, 45 rue Saint Honore, 75001 Paris. Make-up supplies.

Parfumerie des Vedettes, 85 rue de Fabourg Saint-Denis, 75010 Paris. Make-up supplies and chemicals.

Paris-Berlin, 30 rue Chaptal, 75009 Paris. Make-up supplies.

Rosier Atelier Pascal, 10 rue du Rendez-vous, 75012 Paris. Make-up supplies.

Hair and Hair Supplies
The following companies supply wigs, moustaches, beards, etc.

Bertrand SA, 31 Fg. Montmartre, 75009 Paris.

Carita, 11 Fabourg Saint-Honore, 75008 Paris.

Cuverville Figaro, 117 rue de la Convention, 75015 Paris.

Denis Poluene, 10 Cite Trevisse, 75007 Paris.

Neant-Drieaux Jeanne, 43 av. Aristide Briand, 78270 Paris.

Wig Studios, 45 Rue de Lille, 75007 Paris.

Germany
Bela GmbH, Herzogstr.97,8 Munich 40. Make-up supplies.

Herhausen Drogerien, Tempelhofer Damm 122, 1 Berlin 42. Make-up supplies.

Kryolan-Brandel GmbH, Papierstr.10, 1 Berlin 51. Make-up supplies for theater, film, and television as mentioned in detail in the previous section. Special effects materials such as latex, foam latex, modelling tools, clay, plasters, urethanes, etc. are supplied.

Pauls & Co. Stretzeuggasse 6, 5 Koln. Make-up supplies.

Sternenmode GmbH, Bergisch Gladbacher Str.1027, 5 Koln 80. Make-up supplies.

Hair and Hair Supplies
The following companies supply wigs, beards, moustaches, hair, etc.

Fischbach & Miller, Sudd. Haarveredelung, Poststr.1, 7958 Laupheim.

Haar-Praxis, Klaassen, Alter Markt 5, 46 Dortmund.

Herzig, Gustav, Postfach 1440, Carl-Benz-Str.9-11, 6830 Schwetzingen.

Lehmkul-Import Agenturen, St. Benedict Str.30, 2 Hamburg 13.

Plessow, Karl GmbH & Co. Kg., Jebelstr.1, 1 Berlin 12.

Italy
Paruccheria Filistrucchi, Via S. Verdi 9, Florence. Make-up supplies.

Rocchetti-Carboni Srl., Via Monte della Farina 19, Roma. Make-up supplies (their own range), wigs, beards, moustaches, hair, etc.

The Netherlands
The following companies supply make-up.

Cladder van P.E., Utrechtsestr.47, Amsterdam.

Coelho, Nieuw Prinsengracht 7, Amsterdam.

Partyhaus, Rosengracht 68, Amsterdam.

Volen van Leon en Zolen, Woustraat 8, Amsterdam.

Norway
A/S Frisor, Youngsgatan 11 B, N-Oslo. Make-up supplies.

Sweden
Teaterspecialisten, Gosta Jonsson, Storgatan 12, Stockholm. Make-up supplies.

Switzerland
Salon P. Tansini., Limmatquai 4/Tordasse, Zurich 1. Make-up supplies.

ADDRESSES OF MANUFACTURERS AND SUPPLIERS

Aesthetic Productions, 111 Tudor Avenue, Worcester Park, Surrey.

Amalgamated Dental Co. Ltd., Amalco House, 26-40 Broadwick Street, London W.1.

Ash, Claudius Sons & Co. Ltd., Casco House, Moon Lane, Barnet, Herts.

Austenal Dental Products Ltd., 622 Western Avenue, Park Royal, London W3.

Banbury Postiche Ltd., Little Bourton House, Southam Road, Banbury, Oxon.

Baxter Dental Co. Ltd., 6/7 Crystal Centre, Elmgrove Road, Harrow, Middlesex.

BDH Chemicals Ltd., Freshwater Road, Dagenham, Essex.

Bell, John & Croydon Ltd., 50 Wigmore Street, London W.1.

British Gypsum Ltd., Westfield, Singlewell Road, Gravesend.

Boots Chemists Plc., Station Road, Nottingham.

Brodie & Middleton Ltd., 68 Drury Lane, London WC2.

Cottrell & Co. Dental Ltd., 15-17 Charlotte Street, London W.1.

Dior, Christian, Siege Social, 30 Avenue Hoche, 75008 Paris.

Dow Corning Ltd., Bridge Road House, Reading, Berkshire.

Dunlop Rubber Co. Ltd., Dunlop House, Ryder Street, London SW1.

Elastomer Products Ltd., Wharf Way, Glen Parva, Leics.

Eylure Ltd., Grange Industrial Estate, Cwmbran, Gwent.

Factor, Max Ltd., 75 Davies Street, London W.1.

Fox, Charles H., Ltd., 22 Tavistock Street, London WC2.

Fulham Pottery, 184 New Kings Road, London SW6.

Ici PLC, Imperial Chemical House, Millbank, London SW1.

Isles, A.H., Ltd., 146 Lower Road, London SE16.

Kryolan-Brandel GmbH, Papierstr.10, 1 Berlin 51.

Macadam & Co. Ltd., Monument House, Monument Street, London EC3.

McCarthys Surgical Ltd., Selinas Lane, Dagenham, Essex.

McLaughlin, Tom (Chemical Sales), 26 Avenue Road, London E.7.

Notcutt, W.P. Ltd., 44 Church Road, Teddington, Middlesex.

Reeves & Sons Ltd., Lincoln Road, Enfield.

Rowney, George & Co. Ltd., Bracknell, Berkshire.

Screen Face Dist. Ltd., 52 Wardour St., London W.1.

Simon Wigs Ltd., 2 New Burlington Street, London W.1.

Smith, R.H., & Sons (Wigmakers) Ltd., Attifer Works, Gainsborough, Lincs.

Symington, William, & Son, Bath House, Holborn Viaduct, London EC1.

Theatre Zoo, 21 Earlham Street, London WC2.

Tiranti, Alec Ltd., 22 Goodge Place, London W.1. and 70 High Street, Theale, Berkshire.

White, S.S., Dental Co. Ltd., Bedford House, 32-34 Clarendon Road, Harrow, Middlesex.

Wig Creations, 12 Old Burlington Street, London W.1.

Wigmakers, The, Unit 24, 44 Earlham Street, London WC2.

Wig Specialties Ltd., 173 Seymour Place, London W.1.

Winsor & Newton, 51 Rathbone Place, London W.1.

Wright Dental Group Ltd., Dunsinane Avenue, Kingsway West, Dundee.

EQUIVALENT MATERIALS OR SUPPLIERS

UNITED STATES	UNITED KINGDOM
Prosthetic Grade Cream	Screen Face
Hemp	Scrim
Koroseal	Vinamold
Smooth-on PMC-724	Notcutt Ltd.
Petroleum jelly	Chemist
Wet cellophane	Dental companies
Silicon mold release	Notcutt Ltd. and Screen Face
Dow Corning Compound DC-7	Tiranti
Cap Plastic	Glatzan Eylure and RCMA (Screen Face)
Stearic acid	Chemist
Ivory soap	Chemist
Orange shellac	Art shops
Polyurethane laquer	Hardware shop
Mineral oil	Chemist
Burlap	Tiranti
Casting plaster	Fine casting
Ultracal	Crystacal
Plastalene	Plastacine
Casting latex	William Symington and Screen Face
Acryl 60 Cement Hardener	Polypond
Fibre	Tiranti
RCMA Prosthetic Adhesive A	Screen Face
RCMA Prosthetic Adhesive B	Screen Face
Liquidtex Acrylic Colors	Art shops
Urethane Stipple Sponge	Fox, Theatre Zoo and Screen Face
Zinc oxide	Chemical company
Tincture of green soap	Chemist
BJB Polyurethane Foam (similar type)	Screen Face (RCMA polyurethanes)
Molding material	Eylure Cap Plastic and Screen Face
Scar or blister making material	Tuplast–Kryolan and Screen Face
Silicon grease	Dow Corning–Tiranti and Screen Face
Jet Denture Repair Acrylic	Simplex Rapid Acryl
Flexacryl Soft Rebase Acrylic	Flexibase
Matte Plasticized Adhesive	Mastix–Kryolan and RCMA (Screen Face)
Matte Plastic Sealer	Eylure Adhesive and RCMA (Screen Face)

Selected Bibliography

The professional make-up artist's library can vary from a sparse number of reference books to an extensive collection covering a wide range of subjects. However, the following lists a selection of basic books and periodicals that will serve as a sound start for make-up purposes. In addition to the periodicals mentioned here, many manufacturers will furnish, on request, excellent technical bulletins on their products to aid both in research and in understanding the chemical functions and properties of the items. Addresses for some of these suppliers are found in Appendix B.

ANATOMY
American Academy of Facial Plastic and Reconstructive Surgery, "How to Select a Cosmetic Facial Surgeon." Chicago: 1979.

Sheppard, John. *Anatomy: A Complete Guide for Artists.* New York: Watson Guptill, 1975.

BEAUTY AND AESTHETICS
Gerson, Joel. *Standard Textbook for Professional Estheticians.* New York: Milady, 1979.

Lord, Shirley. *You Are Beautiful.* London: Sidgwick & Jackson, 1978.

Lubowe, Irwin. *Modern Guide to Skin Care and Beauty.* New York: E.P. Dutton, 1973.

HAIRSTYLING AND WIGDRESSING
Botham, Mary, and Sharrard, L. *Manual of Wigmaking.* New York: Funk & Wagnalls, 1964.

Corson, Richard. *Fashions in Hair.* London: Peter Owen, 1965.

Wilcox, R. Turner. *The Mode in Hats and Headdress.* New York: Scribner's, 1949.

HISTORICAL STUDIES
These studies can be broken down into two categories: books on the history of make-up and books on historical periods of time. Many of the latter can be found in the remainder displays in chain bookstores at a reduced price. Unless one wants to build a vast library, these should be purchased as they are required for researching a particular production. Many can also be taken out on loan from public libraries.

Westmore, Frank, and Davidson, Muriel. *The Westmores of Hollywood.* New York: J.B. Lippincott, 1976.

Lorant, Stefan. *The Presidency.* New York: Macmillan, 1952.

Some of the book series published by Time-Life contain period illustrations in many areas that are also useful for research of historical figures and personalities.

MAKE-UP
Although many books on make-up have been published over the years, a considerable number are designed for amateur or some college-level self-application studies for aspiring performers. Most should be considered as secondary sources as many are based on outmoded principles and methods for professional work.

Three basic books that have been recently published bring different approaches to the field and cover basic to advanced character make-up principles.

Baygan, Lee. *Techniques of Three-Dimensional Make-up.* New York: Watson Guptill, 1982.

Savini, Tom. *Grande Illusions: Special Effects Make-up.* Pittsburgh: Imagine, 1983.

Westmore, Michael. *The Art of Theatrical Make-up for Stage and Screen.* New York: McGraw-Hill, 1973.

NATIONAL AND RACIAL
The National Geographic Society has published a magazine for many years with excellent photographs of ethnic studies as well as a number of books on the subject. Also see Time-Life Books on this subject and the Smithsonian Institute's Bureau of Ethnology *Reports.*

TECHNICAL AND SPECIAL SUBJECTS
Clark, F. *Special Effects in Motion Pictures.* New York: SMPTE, 1966.

Clarke, Carl Dame. *Molding and Casting.* St. Louis: C.V. Mosby, 1945.

Clarke, Charles G. *Professional Cinematography.* Hollywood: ASC, n.d.

Dental Technician, Prosthetic. Washington: Government Printing Office, 1965. (A U.S. Navy Training Course book.)

The Elements of Color for Professional Motion Pictures. New York: SMPTE, 1957.

Grabb, William C., and Smith, James W. *Plastic Surgery.* Boston: Little, Brown, 1979.

Other books of interest are:

Taylor, A., and Taylor, S. *Making a Monster.* New York: Crown, 1980.

Wilcox, R.T. *Five Centuries of American Costume.* New York: Scribner's, 1936.

Wilkie, Bernard. *Creating Special Effects for TV and Films.* London: Focal, 1977.

Also, Milady Publishing of New York has many books on various phases of hairdressing, cosmetology, aesthetics, and related books in the beauty salon field.

PROFESSIONAL AND TRADE PUBLICATIONS
The major periodicals in the professional field include the *Journal of the Society of Motion Picture and Television Engineers* and *The American Cinematographer* of the American Society of Cinematographers, while the manufacturing area is covered by *Modern Plastics, Drug and Cosmetic Industry,* and *HAPPI* (Household and Personal Products Industry) magazines.

In addition, there are a number of magazines devoted to special film effects that cover make-up, electronics, mechanicals, computer science applications, and so forth:

Cinefantastique, P.O. Box 270, Oak Park, IL 60603.

Cinemagic and *Fangoria,* Starlog Press, 475 Park Ave., New York, NY 10016.

In the field of professional theatre, the leading magazine is *Theatre Crafts,* 135 Fifth Ave., New York, NY 10010.

There are also many books on cosmetic manufacture and processing that are available, and one of the best is Sagarin, Edward, ed. *Cosmetics: Science and Technology.* New York: Wiley-Interscience, 1972, 2nd edition (3 volumes).

It is also recommended to purchase from time to time various fashion magazines to keep up with the current fads and directions in style, while news magazines furnish a source for photographs of historical as well as current personalities. Visits to second-hand bookstores are often rewarded with useful books to build a make-up artist's working library with out-of-print books and magazines at reduced prices. Steel-cut engravings in old history books form an excellent source of illustrations for a research file system, as described in Appendix D, as the low cost of these books allows one to take out the cuts and discard the rest of the book.

Index

Acetone, 37
Acryl 60, 60, 166 185
Additive primaries, 12
Adhesives, 27, 37, 210ff.
    eyelash, 34, 39, 108
    Matte, 37, 42, 72, 250ff.
    Matte Lace, 37, 40, 42, 72
    Matte Plasticized, 38, 211ff.
    Matte #16, 38, 254ff.
    Prosthetic A, 38, 42 121, 211ff.
    Prosthetic B, 38, 42, 211ff.
    Remover, 38, 255
    Special #1, 38, 215
    Special #2, 38
    spirit gum, 37
    tape, 37
    Thinner, 38
Adler, Felix, 157
Aesthetician, 4, 78, 79
Africans, 127
Age Make-up, 97ff.
    progressive, 98, 99, 110, 116, 117
    reversal, 110, 118
    spots, 101, 108
Albinoes, 127
Alcohol, 38
Alginate, 165ff.
    Separator, 185, 186
Allergies, 29
Alpine types, 122
Ambient light, 13
American Academy of Facial Plastic and
        Reconstructive Surgery, 81
American Optical Co., 30
Americans
    Central, 125
    Indians, 121–125
    North, 122–125
    South, 125
Animal men, 146–154
Appliances, 102, 110, 165ff.
Armenians, 125
Artificial Blood, 38, 41, 235
    Tears and Perspiration, 38
Ashkenazim, 125
Asiatics
    Central, 126
    Eastern, 126
    Northern, 126
    Southern, 125
    Southeastern, 126
    Southwestern, 125
Australoid stock, 126, 127

Baker, Rick, 110, 142, 144, 147–152,
        191, 217–222

Elaine, 148
Bald caps and heads, 102–104, 107–108,
        124–125, 212–216
    Mold, 181ff.
    Partial, 103
Ballet make-up, 76
Bandoline, 41
Barrymore, John, 142
Barnett, Vince, 102, 108
Bau, George, 39, 107, 196
Bausch and Lomb Co., 80
Baygan, Lee, 207
Beard
    Covers, 32, 70–72
    Setting Spray, 38
Beards
    Application, 250ff.
    Stipple, 256
    Stubble, 256
Beauty Make-up, 51ff.
    Clinics and spas, 78
Ben Nye Co., 29, 31–34, 38, 55, 71,
        101, 125, 268, 269
BJB Enterprises, 201
Black-and-White, 10, 43, 47, 52, 57, 59,
        60, 70, 100
    Television, 12, 14, 16, 56
    Film, 13, 16–19, 27–28, 275
Blair, Janet, 14
Blair, Linda, 224
Blau, Fred, 236–238
Blends, 52
Blepharoplasty, 82ff.
Blister Making Material, 29
Blood, 38, 41, 235
Blush, See Rouge, dry cake or blush
Body
    hair, 256–257
    make-up, 74
Boham, Gary, 108, 111–112, 198–200
Bolger, Ray, 154
Borgese, Micha, 229
Botin, Rob, 217
Brachycephalic head shape, 49
Brilliantine, 41
British and European make-up materials,
        281–285
Bruises, 235
Brushes, 35–36
    badger hair, 35
    blending, 36
    bristle, 35–36
    camel hair, 35–36
    eyebrow, 36, 63–64
    eyecolor, 36
    eyelash, 36, 65

eyelining, 36
fitch hair, 35
goat hair, 35
hair whitening, 36
lipcolor, 36, 43
ox hair, 35
plastic, 35–36
powder, 35
sable, black, 35
sable, red, 35–36
sculpture, 173, 178
shading and countershading, 36
Buchheit, Leon, 259ff.
Burlap, 172ff.
Burman, Tom, xii, 162, 164–165, 184,
        196, 209, 214–218
Burns, 232–235
Burnt cork, 41
Burton, Richard, 146–147
Byzantines, 134

Caprice lenses, 80
Cap, bald, 102–104, 107–108, 124–125,
        212–216
    material, 203
Carbon tetrachloride, 41
Carotene, 84
Casting plaster, 165ff.
    molds, 178ff.
Caucasoid, 29, 119ff.
    make-up, 30–31, 51ff.
Cell therapy, 79
Central Americans, 125
Chambers, John, 147–149, 166, 182,
        203, 217
Chaney, Lon, Jr., 39, 41, 143, 236
Chaney, Lon, Sr., 93, 94, 141, 217
Character make-up, 73, 76, 91ff.
Charts or records, 24–25, 91, 93
Cheekbones, 121
Cheekcolor, 27, 29, 30, 32, 33, 41, 49,
        52, 54, 59–61, 72, 75
Chemical peel, 84
Chemosurgery, 85
Chiaroscuro, 100
Children's make-up, 73
Chins, 102, 105
Chroma-key, 16
Chrominance, 13
Churchill, Sarah, 212
Clairol Co., 66
Clark, Bobby, 153
Clay
    Attapulgus, 37
    sculpture, 172ff.
Cleansers, 27, 34, 72

Cleansing tissues, 107, 277
Clown white, 41
Clowns, 157
Codi, Ray, 236
Collagen, 82ff.
College pliers, 36, 37
Collodion
    flexible, 39, 42
    non-flexible, 42
Color
    accent, 50, 61
    charts, 50
    contouring, 50
    coordination, 49, 50, 61
    correcting, 28, 70
    framing, 50
    highlights, 49, 50
    matching, 54
    negative film, 17–19
    perception, 29
    pigments, 29, 33
    principles, 49
    quality, 5, 22
    rendering index (CRI), 23
    response, 50
    reversal film, 18, 19
    sensation, 50
    shift, 16, 24
    television, 12ff.
    temperature, 15, 22, 23
Color Process, 29ff.
Coloration
    facial, 51
    latex, 188
Combs, 36, 248, 255
Contact lenses, 80
Contouring
    color, 52
    natural, 32, 52, 56ff.
    reflective, 52
Cosmetic
    dentistry, 86–87
    lenses, 80
    surgery, 58, 78ff.
Cosmetique, 41
Cosmetician, 4, 78, 79
Cosmetologist, 4, 78, 79
Countershading, 27, 29, 32, 41, 52, 56,
    59, 70-72
Cotton balls, 61
Crepe hair, 244ff.
    real, 244, 245
    wool, 244ff.
CRI, 23
Cronyn, Hume, 109
Crosby, Bing, 102
Crowns or caps, 86
Curling irons, 248ff.
Cuts, 235

Dark Ages types, 133–134
Dawn, Jack, 143, 146, 154, 192
DeCarlo, Yvonne, 110, 116
Dental
    bonding, 87
    bridges, 87
    caps or crowns, 86
    carding wax, 40

college pliers, 36, 37
    inlay, 40
    periodontics, 87
    plaster, 165ff.
    spatulas, 36, 37
    stone, 166ff.
Dentistry, cosmetic, 86–87
Depilatory cremes, 81
Dermabrasion, 78ff.
Dermatologists, 81ff.
Devils, 144, 146
Diffusion, 20
Dr. Jekyll and Mr. Hyde, 142, 144
Doherty, Dr. Jean, 86–87
Dolichocephalic head shape, 49
Dolls and toys, 152, 156
Dracula make-up, 141, 143
Dry rouge or blush, 32, 33, 41
Duplicating materials, 200, 203
Dupuis, Steven, 224

Ears, 70, 72
    aging, 99
Eastman Kodak Co., 5, 10, 15, 17–19,
    21, 42
Eighteenth century types, 134–138
Electrolysis, 81
Electronic flash, 14
Elastin, 84
Elephant Man make-up, 143, 144
Elliot, Peter, 150, 151
Epicanthic fold, 120ff.
Eskimoes, 122
Europeans, 122, 125
Eye or Eyes, 48, 49, 102, 103
    brow blocking, 123, 256
    brows, 52, 63, 64
    colors, 27, 33, 41, 42, 49, 50, 52, 54,
        61, 62
    drops, 277
    lashes, 34, 52, 65, 66, 75
    lines, 52ff
    shadow, 33, 41
    shades, 120ff.
    sloe, 121
Eyelash
    adhesive, 39, 65, 66
    curlers, 37
    make-up, 52
Egyptians, ancient, 129–131
Expressionism, 74, 75

Face cast, 169ff.
Facial
    anatomy, 47–49
    balance, 47, 48
    coloration, 51
    contours, 52
    definitions, 52
    distortion, 91, 94
    lifts, 118
    proportions, 47–49
    shapes, 47–49
FDA, 78
Fadjo, Dr. Lawrence, 86
Falls, 256ff.
False lashes, 52, 65, 66
Fantasy types, 151–152

Father Time make-up, 155
Filing system, 91, 276
Film
    color negative, 17–19
    early make-up, 158
    orthochromatic, 27
    panchromatic, 17–19, 27, 28
Filters, 11, 15, 275
Flat plate molds, 178, 180
Flexible impression materials, 167
Fluorescent lighting, 22–24
Foamed latex, 35, 39
Fonda, Henry, 139
Footcandles, 13, 14
Foreheads, 102, 106
Foundations or bases, 27–29, 32, 34, 38,
    41, 52, 54ff.
    cake, 27, 42
    creme, 41
    PB, RMG, or AF, 32, 34, 39, 188, 272,
        277
    stick, 41
    Thinner, 32, 74, 120
    tube, 41
Frankenstein Monster, 39, 141, 142
Frontal bone, 58, 59
Fullerton, Carl, 197, 222, 223, 225

Gauze (silk), 118
Gelatin, 11, 235
    capsules, 38
    materials, 209, 210
Gianis, Dr. George, 80
Glycerin, 234
Grant, U.S., 139
Gray scale, 10, 12, 17
Greasepaint, 41, 56
Greeks, ancient, 131, 132, 140
Guinness, Alec, 102
Gypsum, 165ff.

Hackle, 246, 248
Hair, 105–107, 244ff.
    angora goat, 244, 245
    artificial, 247ff.
    bleaching, 100, 107
    body, 256–257
    color spray, 74, 107, 258
    colors, 48, 105
    crepe, real, 244, 245
    crepe, wool, 244ff.
    cut, 244
    dressing, 41, 259
    facial, 250ff.
    goods, 27, 76, 244ff.
    horse, 244, 245
    human, 244ff.
    kit, 27, 249
    line, 47
    men's, 71–73
    natural, 256
    prepared, 246, 253ff.
    racial types, 119, 120
    removal, 81
    scissors, 36
    shades and forms, 120ff.
    shine, 41, 259
    spray, 39, 41, 74
    styles, 49, 259ff.

tinting or coloring, 66, 121–122, 258
transplants, 81
types, 48, 120, 122
whitening or graying, 39, 105, 108, 258
yak, 244, 245
Halation, 51, 52, 101
Haley, Jack, 154
Hancock, Mike, 236–238
Hand make-up, 72, 74, 107
Hawkins, Jack, 104
Hayward, Louis, 142
Head, 102
    shapes, 119, 120
    forms, 248
Hebrews, ancient, 131, 133
High definition, 20
High fashion make-up, 54ff.
Highlights, 41, 52, 101
Historical characters, 129
Hoffman, Malvina, 128
Horror make-up, 141
Hurt, John, 143, 144
Hurt, William, 191
Hydrocal, 166ff.

IATSE, 3, 43
Image structure, 20
Impression materials, 165ff.
Impressionism, 74, 75
Incident light, 13, 16
Indians
    American, 121–125
    East, 125, 126
Inflatable bladders, 191, 192
Ironside, Michael, 214, 215, 218

Jawlines, 102, 105, 121ff.
Jerome, Paul, 157

Kaolin, 37
Karloff, Boris, 39, 142, 143
Karo syrup, 235
Kaufman, Andy, 151, 156
Kelvin scale, 15, 22, 23
Kelly, Emmett, 157
Keppler, Werner, 166, 186, 192, 198, 202
Kits, make-up, 27, 40, 43, 277–278
Kleenex tissue, 107
Kramer, Sylvia, 33

Lace, 248ff.
Lacquer, clear, 172, 180, 184, 185
Laden, Robert, 92, 93
Lahr, Bert, 147
Lanolin, 29
LaPorte, Steven, 224
Laplanders, 122
Lashes
    curlers, 37
    eye, 34, 52, 65, 66, 75
    false, 34, 75
Latex, 27, 39, 187ff.
    appliances, 187ff.
    caps, 212
    casting, 39, 187ff.
    coloration, 188, 189

eyelash adhesive, 39
    fillers, 189
    foam, 35, 39, 112ff.
    pure gum, 39, 187ff.
    softeners, 189
    thickeners, 189
LeClaire, Jackie, 157, 158
Lee, Robert E., 139
Life mask, 169ff.
Light, 10–16, 22–24, 75
    primaries, 11, 12
Lighting ratio, 14
Lifts, 110, 114, 118, 249
Lines, 52
    character, 97ff.
    eye or lash, 52ff.
    lip, 53ff.
Liners, 33, 41
Lincoln, Abraham, 139
Lip
    colors, 27, 30, 33, 41, 42, 49, 53, 54, 61, 66–69, 75
    glosses, 33, 42, 61
    liners, 33
    pomades, 42
    shapes, 49, 66–68, 121ff.
    Stae, or Stay, 42
Lipstick, 41, 43
Lotito, Leo, 195, 246, 247
Lugosi, Bela, 143

MacLane, Barton, 156
Make-up
    age with, 97ff.
    application sequence
        children, 73
        men, 71ff.
        women, 52ff.
    artist examinations, 279, 280
    ballet, 76
    beauty, 51ff.
    body, 74
    character,
        screen, 89ff.
        stage, 76
    hand, 72, 74
    high fashion, 54ff.
    kits, 27, 40, 43, 277–278
    lab, 163
    license, 54
    men's, 70ff.
    production, 52ff.
    racial change, 120ff.
    records or charts, 24, 25, 93
    room, 22ff.
    seminars and schools, 274
    showgirl, 76
    special effects, 94, 217ff.
    stage or theatrical, 26, 55ff.
    styles, 53
    women, 51ff.
Manson, Maurice, 94, 96, 155
March, Frederic, 142, 144
Mascara, 27, 34, 41, 52, 64, 65, 75, 105
    Remover Pads, 34
Massey, Raymond, 139
Matte
    Adhesive, 37, 42, 250ff.

Adhesive #16, 38, 254
    Lace Adhesive, 37, 255
    Plastic Sealer, 40, 42, 114, 211, 256
    Plasticized Adhesive, 37, 42, 210–213, 254
Matthau, Walter, 104
Max Factor Co., 27, 28, 31, 59, 267
McCorkindale, Simon, 153
McDowall, Roddy, 92, 93, 153
Medical approaches, 79ff.
    cell therapy, 79, 80
    holistic, 79, 80
    problems, 79
Medieval types, 134
Mediterranean types, 122
Melanin, 84
Melanoid, 84
Men's make-up, 70ff.
Merrick, Jon, 143
Mesopotamians, ancient, 131, 132
Metallic powders, 155
Mexicans, 121, 125
Mid-America Dental Co., 167ff.
Middle age, 97ff.
Milland, Raymond, 102
Miller, David, 224
Mineral oil, 29, 51
Mixers
    Sunbeam, 192ff.
    Hobart, 192, 197
Modelling tools, 37, 172
Mold releases, 168, 169
Molds
    bald cap, 181ff.
    casting, 179ff.
    duplicate, 167, 168
    eyes, 181, 182, 204
    flat plate, 178, 180
    noses, 179, 180, 185, 186, 201, 206
    paint-in, 181
    slush, 179, 183, 190
    teeth, 186, 208
    two-piece, 183ff.
        foamed latex, 185, 186, 193, 194, 195, 206
        foamed urethane, 185, 198–203, 206
        press molding, 206
Molding
    PMA Material, 39, 204
    PMA Press Material, 39, 205–207
    Wax, 40, 42, 93, 232, 234
Mongoloid or Oriental, 29, 30, 119ff.
Monochromatic (B&W) Film & Television, 12–19
Monitors, 7, 9
Moors, 127, 134
Mortician's Wax, 42
Morehead, Agnes, 114, 115
Morton, Cavendish, 91, 94
Moulage, 165, 167
Moustache wax, 42
Mozart Family, 138
Mummy make-up, 141, 143
Muni, Paul, 100

NABET, 3
Nasolabial folds, 102, 104
National types, 122ff.

Naughton, David, 218–222
Neck
    aging, 105, 111–117
    age reversal on, 118
Negroid, 29, 119ff.
    make-up, 30, 31, 55ff.
Nestle-Lamaur Co., 39, 258
Niehans, Dr. Paul, 79
Nineteenth century types, 135–140
Niven, David, 102
Nordic types, 122
Norin, Gustaf, 166, 205
Nose putty, 42
Noses or nasal changes, 121

O'Bradovich, Robert, 93, 109, 146–147
Obsolete terms, 41, 42
Oceania types, 127
Old Age make-up, 97ff.
    Stipple, 39, 100, 101, 107ff.
Older age, progressive, 99, 116, 117
Ophthalmologists, 80
Oriental or Mongoloid, 119ff.
    make-up, 55ff.
Orthochromatic film, 27
Orthodontics, 86
Otolaryngologists, 80ff.
Over Powders, 34
Oxyhemoglobin, 84

Paint primaries, 10, 11
Painting-in molding, 179ff.
Pan-Cake make-up, 27, 42, 51, 56, 59,
    60, 74
Panchromatic film, 17–19, 27, 28
    make-up, 28
Pan-Stik make-up, 27, 59
Pasteur, Louis, 139
Peck, Gregory, 139
Pencils, 27, 33, 52, 62–64, 71–73
    sharpener, 33
Perfumes, non-use, 29, 40
Period characters, 129
Perlman, Ron, 130, 132
Perma-Flex Co., 167, 168, 182
Perspiration, simulation of, 38
Peters, Bernadette, 151, 156
Petroleum jelly, 169, 206
pH value, 33, 189
Phantom of the Opera make-up, 141
Philippe, Robert, 110
Phenol, 78, 85
Phosphor plate, 12
Pierce, Jack, 39, 141–143, 146, 217
Pirates, 156, 157
Plastalene, 165ff.
Plastics, 27, 167ff., 187ff.
    bald cap material, 39, 203
    special, 209
    tooth, 207–209
    vinyl, 39
Plastic Wax Molding Material, 40, 42, 93,
    232, 234
Plaster
    casting, 165ff.
    dental, 165ff.
Pleating machine, 248, 249, 252
Polaroid photos, 21, 25

Popular types, 141ff.
Powder, 27, 29, 34, 41, 52, 61
    Neutral, 34
    PB, 34
    puffs, 35, 43
    Translucent, 34, 41
    Transparent "No-Color," 34, 41, 61
Prehistoric types, 129–131
Press Molding Material, 39
Printing flow charts, 18, 19
Production make-up, 5ff.
Prognathism, 121
Progressive older age make-up, 99, 116,
    117
Prominent ears, 70
Prosthetics, 27, 110, 165ff.
    Adhesives, 38, 42, 121, 211ff.
    foundations, 39

Q-Tips, 35, 62, 67

Racial types, 119ff.
Randall, Tony, 94
Razor blades, 33, 277
Ramus, Armando, 216
RCMA (The Research Council of Make-up
    Artists Co.), xi, 29, 35ff., 270–272
Rea, Stephen, 230
Reardon, Craig, 189
Records or make-up charts, 24, 25, 93
Reflectance, 16, 20
Reflective contouring, 52
Reflectors, 14
Reiner, Robert, 112
Removal of make-up, 34
Renaissance period make-up, 133, 134
René of Paris Co., 235
Research file system, 276
Resolution, 20, 21
Reversal techniques, 110, 114, 118
Reynolds, Burt, 102
Rhinoplasty, 82ff.
Rhytidoplasty, 83ff.
Rigid impression materials, 165ff.
Robards, Jason, 110, 116
Romans, ancient, 131, 132, 133
Rouge, 41
    creme, 41, 59
    dry cake or blush, 32, 33, 41, 59, 61
    lip, 41
    under or wet, 41, 59, 60
Ruder, Dr. Robert, 81ff.

Sagarin, Edward, 33
Santa Claus, 94, 155
Saracens, ancient, 134
Savini, Tom, 174–177, 208, 217, 222
Scalp Masque, 42
Scars, 234–236
Scar Material, 39, 42, 91, 93, 204,
    234–236
Scissors, 36
Sculpture, clay, 172ff.
Schiffer, Robert, 155, 182
Schwartzenegger, Arnold, 179
Sealers, 27, 42, 169
    collodion, 27, 142
    Matte Lace, 40

Matte Molding, 40
Matte Plastic, 40, 42
Sealskin, 27
Seed oils, 51
Separators, 168, 169
Sephardim, 125
Sequence of application, 52
Seventeenth century types, 133–136
Shading, 27–29, 32, 41, 52, 56, 60, 70,
    101
Shadows, 44, 52, 101
Shellac, 169, 172
Shore Durometer, 187, 202
Sideburns, 140, 244ff.
Silicone Mold Release, 169, 204ff.
Silk muslin, 93, 110, 118
    organza, 256
Skin
    astringents, 35
    care, 27, 35, 77ff.
    cross section, 84
    dry, 34, 35
    fresheners, 34, 41
    normal, 34
    moisturizers, 34, 35, 41
    oily, 34, 35
    peels, 78ff.
    tightener, 58
    tones, 29, 49, 51
    treatments, 79
    types, 48
Skull make-up, 144–146
Slezak, Walter, 107, 135
Sloe eyes, 121
Slush molding, 179, 183, 190
Smith, Dick, 43–44, 104, 112, 115, 166,
    172, 185, 188, 189, 191, 195–198,
    205, 210, 217, 219, 222, 224–229,
    252, 253
Smith, Kenneth, 210
Smith, Terry, 166, 195, 218
Smooth-On Co., 168
SMPTE (Society of Motion Picture and
    Television Engineers), 11, 27, 28
Soaps
    Ivory, 171
    Neutrogena, 34
Solvents, 237, 238
Sorolla, Joaquin, 99
Spatulas, 172, 178
Special effects, 6, 7, 25, 161ff.
Special historical characters, 138ff.
Special materials, 27, 37–42, 271
Sponges, 35, 42, 43
Spirit gum, 42
South American types, 125
Stage make-up, 26ff.
Statuary effects, 152, 155
Stave, Jack, 142
Stein's Co., 38, 101, 244, 269, 270
Stipple
    beard, 256
    Old Age, 39, 97ff.
    sponges, 35
    sunburn, 72
Strange, Glenn, 143
Stubble beards, 256
Studio Brush Cleaner, 68

Studio Make-up Remover, 67
Subtractive primaries, 12
Sunburn, 78, 101
Suntan effects, 101
Sydney, Basil, 213
Syrians, 125

Tattoos, 236–238
Taylor, Vaughn, 112
TCTFE (Trichlorotrifluoroethane), 41
Tears and Perspiration, 124
Technicolor, 78
Teldyne Getz Co. 167ff.
Theatrical or stage make-up, 26ff.
Thickeners, 189, 238, 239
Tracy, Spencer, 141
Trichloroacetic acid, 85
Tri-Ess Sciences Co., 189
Tools, 27, 35–37
  sculpture, 172, 173
Tooth
  enamel, 42
  molds, 186, 208
  plastics, 207–209
Toupees, 255ff.

Toys, make-up, 152, 156
Tucker, Christopher, 130, 131, 143, 192,
  229–232
Turks, 125
Turner, Josephine, 254, 259
Twentieth century types, 138
Tyson, Cicely, 110, 113

Ultracal, 166ff.
Undertone value, 29, 54ff.
Urethane
  elastomer, 202ff.
  foam, 198ff.
U.S. Gypsum Co., 166ff.

Vaseline, 169, 206
Vegetable oil, 51
Velcro, 38
Ventilating, 253, 255
Vistanex, 170
Vitalite (Luxor Co.), 23–25

Wardrobe and make-up, 16, 50
Waxes
  Dental, 40, 42, 234

Molding, 40, 42, 93, 232, 234
  Moustache, 42
Waxman, Al, 224–228
Westmore Bros., 47, 137
  Bud, 110, 114–116
White henna, 66
Wig Cement, 42
  maker's knot, 253–255
  maker's needles, 253–255
Wigs, 257ff.
  cleaning, 255
Wilcox, R. Turner, 131
William Tuttle Co., 29, 31–34, 40, 55,
  62, 71, 93, 101, 125, 267, 268
Winchell, Paul, 93, 95
Winston, Stan, 110, 112, 113, 116, 150–
  154, 179, 217, 218, 231–234
Winters, Jonathan, 155
Witches, 110, 144–147
Wolfman make-up, 143, 146
Women's make-up, 51ff.
Wool crepe hair, 244ff.

Yak hair, 244, 245